Plasma Processing of Nanomaterials

T0225589

Nanomaterials and Their Applications
Series Editor: M. Meyyappan

Plasma Processing of Nanomaterials

Edited by
R. Mohan Sankaran

CRC Press
Taylor & Francis Group
Boca Raton   London   New York

CRC Press is an imprint of the
Taylor & Francis Group, an **informa** business

CRC Press
Taylor & Francis Group
6000 Broken Sound Parkway NW, Suite 300
Boca Raton, FL 33487-2742

First issued in paperback 2017

© 2012 by Taylor & Francis Group, LLC
CRC Press is an imprint of Taylor & Francis Group, an Informa business

No claim to original U.S. Government works
Version Date: 20111101

ISBN 13: 978-1-138-07743-0 (pbk)
ISBN 13: 978-1-4398-6676-4 (hbk)

Library of Congress Cataloging-in-Publication Data

Plasma processing of nanomaterials / [edited by] Mohan Sankaran.
    p. cm. -- (Nanomaterials and their applications)
  Includes bibliographical references and index.
  ISBN 978-1-4398-6676-4 (hardback)
    1. Nanostructured materials. 2. Plasma engineering. I. Sankaran, Mohan. II. Title.

TA418.9.N35P53 2011
620'.5--dc23   2011038445

Visit the Taylor & Francis Web site at
http://www.taylorandfrancis.com

and the CRC Press Web site at
http://www.crcpress.com

Contents

Preface

Plasma processing is a well-established technology that is vital to materials manufacturing in numerous industries including electronics, textiles, automobile, aerospace, and biomedical. The revolution in microelectronics over the last 30 years has been largely enabled by the ability of plasma-based tools to etch, deposit, and sculpt thin films of metal and semiconductor materials into extraordinarily precise digital circuits. Despite these achievements, fundamental research and industrial implementation of plasma technology now sit at a critical juncture. Because of limitations associated with photolithography, the conventional approach to patterning materials from the top down is approaching a physical limit. This has serious consequences in the electronics industry where the pattern resolution (i.e., feature size) determines the speed of a computer processor or density of a memory device. However, there are even broader implications: nanotechnology (i.e., the preparation of materials at the nanoscale) has opened numerous avenues for advancements in optoelectronics, medicine, and renewable energy. In order to play a role in these rapidly emerging areas, plasma technology must overcome significant challenges related to the processing of *nanomaterials*.

Fortunately, there is good news. Although soft chemical methods have made a great deal of progress, plasma technology offers several advantages for nanomaterials processing. Low-temperature plasmas allow chemical processes to be performed near room temperature, an important consideration to lower costs and make processing compatible with device manufacturing. In addition, plasma-based processing is scalable because materials are usually prepared continuously or over large areas. Perhaps the most important driving force for using plasma technology is chemical purity that is unmatched by any other chemical process and the reason why plasmas have been essential to microelectronic device fabrication.

The purpose of this book is to share the exciting and enormous progress that has been made over the last few years to develop plasma technology for nanomaterials synthesis and processing. This includes cutting-edge research from all over the world, including the United States, Belgium, Germany, Slovenia, Turkey, Australia, China, South Korea, and Japan. The diverse range of examples that have been chosen are meant to illustrate the versatility of plasmas and their great potential for applications. The book is structured from two perspectives—material types and length scales, both of which are key issues for plasma processing of nanomaterials (see Figure P.1). The book begins with recent advancements in top-down methods (Part 1) including nanoscale etching and deposition (Chapter 1) and the development of extreme ultraviolet light sources as next-generation photomasks

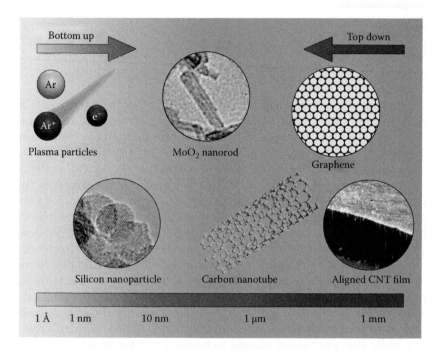

FIGURE P.1

(See color insert.) Plasma processing of materials at different length scales. (Courtesy of Uwe Kortshagen, A. Chandra Bose, Davide Mariotti, and Liming Dai.)

(Chapter 2). Alternatively, nanomaterials must often be prepared from atomic or molecular building blocks (i.e., bottom-up), which requires control at the nanometer scale. Examples of zero-dimensional (0D) nanomaterials, including semiconducting nanocrystals (Chapter 3) and metal or metal oxide nanoparticles (Chapter 4), and one-dimensional (1D) nanomaterials, such as nanowires (Chapter 5) are presented in the subsequent section (Part 2). The next section (Part 3) focuses on carbon-based nanomaterials such as carbon particles (Chapter 6), carbon nanotubes (Chapters 7 and 8), and graphene (Chapter 9), all of which have potential to replace silicon in future electronic devices. The following section (Part 4) introduces modeling efforts (Chapters 10 and 11) and diagnostic methods (Chapter 12) that have paralleled experimental research to gain a basic understanding of plasma processes that are utilized for nanomaterial synthesis. The synthesis and modification of nanomaterials at slightly larger scales is discussed in the next section (Part 5), for example, the fabrication of hybrid structures of organic and inorganic materials (Chapters 13 and 14). The book concludes with a section (Part 6) on an emerging direction for plasma processing, the assembly and organization of nanomaterials into ordered thin films or other hierarchical structures that necessitates processing at macroscopic length scales (Chapter 15). As we look ahead, the ability to process different materials across a wide range of length scales will be essential and determine if plasma technology has a key role in emerging technologies in the 21st century as it has in the past several decades.

R. Mohan Sankaran

Contributors

Annemie Bogaerts
Department of Chemistry
University of Antwerp
Belgium

Jane P. Chang
Department of Chemical and
    Biomolecular Engineering
University of California, Los Angeles
Los Angeles, California

Manish Chhowalla
Rutgers University
Department of Materials Science
    and Engineering
Piscataway, New Jersey

Uros Cvelbar
Jozef Stefan Institute
Ljubljana, Slovenia

Liming Dai
Department of Chemical
    Engineering
Case Western Reserve University
Cleveland, Ohio

Maxie Eckert
Department of Chemistry
University of Antwerp
Belgium

Rikizo Hatakeyama
Department of Electronic
    Engineering
Tohoku University
Sendai, Japan

Toshiro Kaneko
Department of Electronic
    Engineering
Tohoku University
Sendai, Japan

Toshiaki Kato
Department of Electronic
    Engineering
Tohoku University
Sendai, Japan

Holger Kersten
Institute for Experimental and
    Applied Physics
University of Kiel
Kiel, Germany

Uwe Kortshagen
Department of Mechanical
    Engineering
University of Minnesota
Minneapolis, Minnesota

Se Jin Kyung
Department of Materials Science
    and Engineering
Sungkyunkwan University
Suwon, South Korea

Jeong-Soo Lee
Division of IT-Convergence
    Engineering
POSTECH
Pohang, Republic of Korea

Lorenzo Mangolini
Department of Mechanical
    Engineering
University of California, Riverside
Riverside, California

Ming Mao
Department of Chemistry
University of Antwerp
Belgium

Nathan Marchack
Department of Chemical and
    Biomolecular Engineering
University of California, Los Angeles
Los Angeles, California

Davide Mariotti
Nanotechnology and Advanced
    Materials Research Institute
    (NAMRI)
University of Ulster
Newtownabbey, United Kingdom

Horst R. Maurer
Institute for Experimental and
    Applied Physics
University of Kiel
Kiel, Germany

M. Meyyappan
NASA Ames Research Center
Moffett Field, California

Tony Murphy
Plasma Nanoscience Centre
    Australia (PNCA)
CSIRO Materials Science and
    Engineering
Lindfield NSW, Australia

Erik Neyts
Department of Chemistry
University of Antwerp
Belgium

Kostya (Ken) Ostrikov
Plasma Nanoscience Centre
    Australia (PNCA)
CSIRO Materials Science and
    Engineering
Complex Systems, School of Physics
The University of Sydney
Sydney, NSW, Australia

Jae Beom Park
SKKU Advanced Institute of Nano
    Technology (SAINT)
Sungkyunkwan University
Suwon, South Korea

Amanda Evelyn Rider
Plasma Nanoscience Centre
    Australia (PNCA)
CSIRO Materials Science and
    Engineering
Lindfield, NSW, Australia

David N. Ruzic
Department of Nuclear, Plasma, and
    Radiological Engineering
Center for Plasma Materials
    Interactions
University of Illinois at Urbana-
    Champaign
Urbana, Illinois

R. Mohan Sankaran
Department of Chemical
    Engineering
Case Western Reserve University
Cleveland, Ohio

John Sporre
Department of Nuclear, Plasma, and
    Radiological Engineering
Center for Plasma Materials
    Interactions
University of Illinois at
    Urbana–Champaign
Urbana, Illinois

Mahendra K. Sunkara
Department of Chemical
    Engineering
University of Louisville
Louisville, Kentucky

Eugene Tam
Plasma Nanoscience Centre
    Australia (PNCA)
CSIRO Materials Science and
    Engineering
Lindfield NSW, Australia

H. Emrah Unalan
Department of Metallurgical and
    Materials Engineering
Middle East Technical University
Ankara, Turkey

Yuhua Xue
Institute of Advanced Materials for
    Nano-Bio Applications, School of
    Ophthalmology & Optometry
Wenzhou Medical College
Wenzhou, China

Geun Young Yeom
SKKU Advanced Institute of Nano
    Technology (SAINT)
Sungkyunkwan University
Suwon, South Korea
Department of Materials Science
    and Engineering
Sungkyunkwan University
Suwon, South Korea

1

Nanoscale Etching and Deposition

Nathan Marchack
Jane P. Chang

CONTENTS

1.1 Introduction

The vast potential utility offered by nanoscale assemblies and structures due to their remarkable physical, chemical, electrical, and biological properties has led to a burgeoning body of work in recent years.[1] Currently, there are many examples of functional electronic and magnetic devices based on these structures, and nano-array fabrication is also finding much interest in biological applications such as tissue engineering,[2] regenerative medicine,[3] drug delivery,[4] and the study of cell-environment behavior.[5] However, the prevailing difficulty lies in effectively directing the desired assembly of such structures over a large area, which requires reconciling widely disparate length scales: seven to nine orders of magnitude between the nanostructures ($\leq$ nm) to the desired areas of deposition (cm to m).[6] Based on this requirement, it naturally follows that attention should be drawn to the techniques employed by the semiconductor/integrated circuit device industry, which has flourished by overcoming a similar challenge: the patterning of transistor features with critical dimensions (CDs) in the nanometer range over

1

wafers many centimeters in diameter. The method of patterning materials in this industry is a combination of optical lithography and plasma etching, which has been successfully demonstrated for a wide variety of materials such as semiconductors, oxides, and metals. However, attempts to extend this particular means of processing to the creation of well-ordered assemblies of nanostructures have encountered difficulties. This chapter explores the potential use of top-down processing methods for the creation of nanoarrays and highlights limitations that have to be overcome to ensure the success of these processes.

1.2 Bottom-up versus Top-down—Comparing Two Approaches

The creation of nanostructures can be achieved by two different approaches: bottom up and top down, as illustrated in Figure 1.1. Bottom-up methods involve growth of the structures starting at the substrate level, with each successive layer or step building on the last. This nanoscale manipulation was touched upon in Richard Feynman's 1959 lecture entitled "There's Plenty of Room at the Bottom," wherein the concept of physics-based synthesis through arrangement of individual atoms was discussed as a potential means of tapping into the unparalleled potential of quantum mechanics. In top-down methods, the structures are created by selectively removing unwanted material, each time at the layer farthest away from the substrate. Ideally, the creation of nanostructures from the bottom up is desired as this does not require the removal of unwanted nanostructures; the damage created by plasma exposure and its effects on device performance have long been significant issues in the semiconductor industry, even at scale lengths much

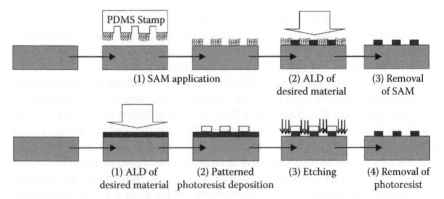

FIGURE 1.1
Comparison between a bottom-up and a top-down process for pattern transfer, with identical starting and final surfaces/structures.

larger than those of current devices.[7] While much literature exists on plasma chemistries that have been used to etch various semiconductors,[8] high-k materials,[9] metals,[10] and composite materials,[11] these studies focus primarily on blanket films, or systems where the etched surface can be represented by "bulk behavior." However, as the length scales continue to decrease toward the size of individual molecules or atoms, it is possible that individual bonds or functional groups present can dictate the type of reactions that occur, thus setting up the possibility of uneven etching or deposition on a surface. Bottom-up methods could potentially overcome this problem as they allow for the manipulation of nanoscale building units beyond the limit of what is obtainable by conventional optical lithography methods, which are largely limited by the wavelength of light being used.[12] Although the primary focus of this chapter is on extending top-down methods to the nanoscale, the bottom-up methods are briefly discussed in the next section to highlight the overall challenge in nanoscale patterning.

In order to controllably synthesize nanomaterials from the bottom up, attention must be paid to several factors. Bottom-up methods are highly dependent on the condition of the growth surface: natural entropic tendencies facilitate the distortion of crystal structures and atomic layouts, while topographic effects such as grain boundary and line dislocations can affect the available reactive area on much larger scales. There has been much study in this field regarding topics such as nanostructure growth on vicinal surfaces[13]—steps can significantly affect the electronic,[14] magnetic,[15] and chemical[16] properties of the surface, opening up the possibility of synthesizing chains or wires along the step structure, as shown in Figure 1.2.

Progressing to a finer scale, the potential reactions that can take place at a particular site also have significant influences on bottom-up growth. Studies on the variability of surface reactivity in the presence of atomic-scale defects or dopants are thus of great value to the emerging field of nanomaterial fabrication. As shown in Figure 1.3, observation and analysis of the surface reactions on oxides such as MgO[17] revealed that species such as CO react with oxide anions in widely disparate ways depending on their location: steps, edges, kinks, and corners lead to the formation of polymeric species through exothermic, nonactivated pathways, while the bulk (100) surface is inert to such reactions. Similarly, the incorporation of impurities such as Ni ions into an MgO matrix for the purpose of N_2O decomposition can lower the barrier for reaction by 0.2 eV, leading to a reaction rate increase by approximately one order of magnitude. Other work focused on transition metal oxide surfaces such as TiO_2 showed the ability of such surfaces to act as either a reducing or oxidizing agent on adsorbed Au particles depending on the presence of oxygen vacancies or impurities such as CO. It was shown that Au is bound to the surface much more strongly on O vacancies than on the bulk surface (~1.8 eV versus 0.5 eV, respectively) and its 6s level is filled and below the Fermi level, thus allowing it to be treated as an Au^- anion. In the presence of CO, the Au 6s electron interacts differently, transferring to the Ti 3d level and

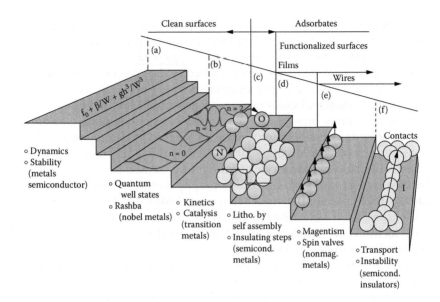

FIGURE 1.2

(See color insert.) Representation of potential functionalizations of vicinal surfaces. (Reprinted from *Journal of Physics: Condensed Matter*, Vol. 21, Tegenkamp, "Vicinal surfaces for functional nanostructures," pp. 013002-2, Copyright (2009) with permission from IOP Publishing.)

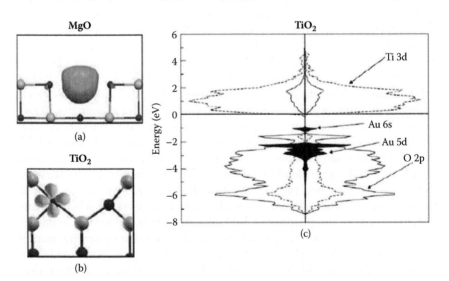

FIGURE 1.3

Charge density plots for an oxygen vacancy created on (a) MgO and (b) TiO_2, and (c) the density of states of an Au atom adsorbed on an oxygen vacancy of the rutile TiO_2 (110) surface. (Reprinted from *Theoretical Chemistry Accounts* Vol. 117, Cinquini et al., "Theory of oxide surfaces, interfaces and supported nano-clusters," pp. 827–845, Copyright (2007) with permission from Springer-Verlag.)

becoming Au[+].[18] Hence, one can envision the creation of nanostructures in an ordered fashion by selectively adding dopants or vacancies to manipulate the reactivity of specific surface areas.

The manipulation of surface reactivity can also be accomplished via external modification (i.e., by the addition of other layers rather than modifying the original surface structure directly). In area-selective atomic layer deposition (ALD) processes, surfaces are passivated as a prelude to creating features. In this method, the substrate is treated with a template such as a self-assembled monolayer (SAM) or protein film. The desired material is then deposited via ALD, and the template film serves as a resist to limit growth to only the exposed areas. Depending upon the choice of SAM or proteins, features with size ranging from 1 to 40 µm down to 9 to 11 nm have been realized for various materials, including ruthenium,[19] titanium dioxide,[20] hafnium dioxide,[21] and platinum.[22] Once the selective ALD growth is completed, the resist can be removed by either ozone plasma treatment[23] or wet etches[24] and rinses,[25] leaving only the desired material patterned on the surface. Several potential issues concerning this technology exist, such as transferring the initial template over a large area with high fidelity and cleanly removing the resist material without compromising the features, and the most significant challenge involves multilayer registry (i.e., successfully superimposing multiple patterns on the surface in order to create three-dimensional (3D) stacked features). Thus far, such complex multilayered structures at nanometer-scale precision have only been created via top-down methods.

Other possible bottom-up approaches for patterning arrays of nanostructures with long-range order have been investigated, including but not limited to combining prepatterned substrates with diblock copolymer coatings[26] and the use of nanosphere lithography to create hole arrays for plasmonic applications.[27] However, these methods involve top-down patterning to a certain degree (e-beam lithographic patterning of the inorganic resist and reactive ion etching [RIE] of the polymers spheres, respectively). Figure 1.4 demonstrates experimental work showcasing the patterning capabilities of diblock copolymers on prepatterned substrates, as well as simulation approaches to understand the formation and assembly of such structures.[28] Three-dimensional features have also been patterned using a probe-based method in conjunction with RIE. Specifically, atomic-scale surface manipulation has been demonstrated with such scanning tips by a variety of means such as local anodic oxidation, field-induced deposition, modification of thermomechanically responsive organic materials, and removal via heating. The largest limitation of this method involves limiting the wear on the tips as patterning areas in the mm[2] range involve the writing of 10^8 to 10^{10} pixels per single tip.[29]

Bottom-up methods have great potential for material patterning at the nanoscale and beyond. Among their key advantages are the potential for self-organization (which allows for simplicity and reduced cost), independence from optical lithography limitations, and low induced damage from minimizing

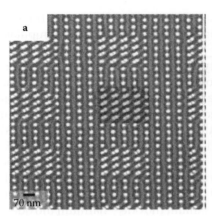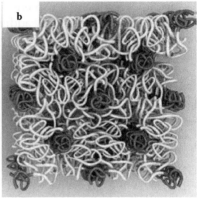

FIGURE 1.4
(a) Interspersed diagonal and vertical structures created using a localized double-post template. (Reprinted from *Nature Nanotechnology*, Vol. 5, Yang et al., "Complex self-assembled patterns using sparse commensurate templates with locally varying motifs," pp. 256–260, Copyright (2010) with permission from Macmillan Publishers Ltd.) (b) Cartoon illustration of proposed chain packing for supramolecular polymer systems based on PEO-b-PS and PS-b-PMMA. (Reprinted from *Science*, Vol. 322, Tang et al., "Evolution of block copolymer lithography to highly ordered square arrays," pp. 429–432, Copyright (2008) with permission from AAAS.)

the need for material removal. However, there are currently many challenges that must be addressed before they become truly viable for high-throughput and industrial integration. Directing and organizing the base units for these structures are divided between two extremes: delivery of the molecules to the surface via a neutral gas can lead to disordered arrays with virtually nonexistent control over the final product,[6] but manipulation of individual molecules via techniques such as scanning probe/dip pen lithography is an extremely tedious and time-consuming endeavor. Additionally, the difficulty of aligning and layering several patterns over each other poses a serious barrier to creating more complex features. For these reasons, top-down patterning methods still remain a useful tool for achieving nanoscale patterning.

Lithography-based top-down methods have been proven to provide high-fidelity, reproducible pattern transfer capabilities; a fact that has sustained their usage for decades in the semiconductor processing industry and other applications in spite of their great expense. As new materials are emerging for various applications with pattern precisions at nanoscales, the ability to preferentially remove specific materials without damaging or etching the other components is critical. Ensuring that etch processes maintain such high degrees of specificity has become even more difficult as research trends toward multicomponent, multifunctional material systems. For example, the chemistries required to etch an organic polymer compound (such as photoresist) as opposed to a crystalline oxide are markedly different. Similarly, the etch disparity between multilayer, multicomponent stacked films as those needed in state-of-the-art memory devices dictates a delicate balance between the desired outcome and the viable etch chemistries.[30] In spite of these challenges, top-down process-ing methods have been used to create multilayered device architectures using semiconductors, oxides, and metals[31] as shown in Figure 1.5, a feat that thus far continues to elude bottom-up patterning techniques.

Given further advances in both the etching process and lithographic tech-niques, it is not inconceivable that top-down processing techniques can serve as a viable candidate for nanoscale patterning. However, there are other potential uses plasmas can fulfill in the field of nanomaterials processing and preparation. Plasma treatment of surfaces has been employed in a great variety of applications, including inducing self-adhesion in polyethylene films,[32] and generating protective,[33] frictional,[34] and hydrophobic[35] coatings. The wide array of reactive species present in the plasma provides the capa-bility for etching and deposition reactions of both the chemical and physical variety. A certain degree of control is also afforded by tailoring controllable parameters such as gas-phase chemistry, source power, pressure, substrate bias, and temperature. In addition to direct exposure, plasmas have been

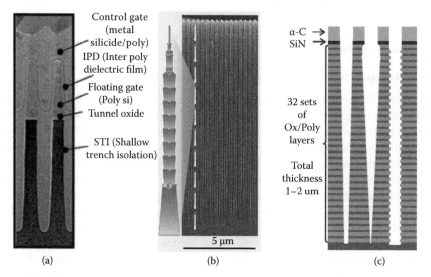

FIGURE 1.5
(a) High resolution transmission electron microscope (HRTEM) image of a 27 nm NAND flash memory with a reported record low 0.00375 μm² unit cell size. (See also Choong-Ho, L.; et al. In *A highly manufacturable integration technology for 2 7nm 2 and 3bit/cell NAND flash memory*, Electron Devices Meeting (IEDM), 2010 IEEE International, 6–8 Dec. 2010; p. 5.1.1.) (b) High aspect ratio trenches in relation to one of the world's tallest freestanding buildings, Taipei 101. (c) High-aspect ratio holes for three-dimensional (3D) flash memory technology showing the complexity of multilayer and material stacks and possible etch profile defects. (Reprinted from *IEDM Conf. Proc.*, Lee et al., "A Highly Manufacturable Integration Technology for 27 nm Multi-Level NAND Flash Memory," pp. 5.1.1–5.1.4, Copyright (2010) with permission from IEEE (part a); *Science* Vol. 319, Lill and Joubert, "Materials Science: The Cutting Edge of Plasma Etching," pp. 1050–1051, Copyright (2008) with permission from AAAS (part b); courtesy of Dr. C. G. N. Lee at Lam Research Corporation (part c).)

indirectly used to treat surfaces using techniques such as plasma electro-lytic deposition, by using microdischarges to mobilize gaseous ionic species that react on a desired surface to form complex compounds.[36] As discussed in the preceding paragraphs concerning bottom-up patterning, such surface modification is of great value in controlling the reactivity and assembly of nanostructures. In this regard, a comprehensive knowledge of plasma–sur-face interactions is essential if one desires to use plasmas to prepare a sur-face for selective nanomaterial growth. Continued study into plasma science could then offer significant utility to the field of nanomaterial patterning even if plasmas are not used directly to etch the structures as they are in lithographic approaches. Therefore, the rest of this chapter features a discus-sion of some of the fundamental scientific concepts behind plasma–surface interactions, the potential applications of plasmas for surface preparation and nanomaterial patterning, and further usage of plasma in other applica-tions such as medicine (e.g., nanotoxicology), waste treatment,[37] and bioma-terial modification.[38]

1.3 Introduction to Plasma Science

A nonequilibrium plasma environment possesses the unique property that the electron temperature is substantially higher than the average ion temper-ature, which itself is greater than the gas temperature ($T_e \gg T_i > T_g$), allowing for reactions to be driven by high-energy electrons rather than intensive and potentially damaging substrate heating. As a result, a large variety of species are generated in plasma: ions, neutrals, radicals, electrons, and photons. In isolation, each of these species is usually insufficient to achieve the desired reaction with the surface to be treated, but when they are supplied simul-taneously to the surface with sufficient fluxes and desirable ratios, a wealth of reactions become attainable. Semiconductors, oxides, nitrides, metals, and polymers have all been etched using various plasma chemistries.[39] As illustrated in Figure 1.6, because of the distinct nature and reactivity of each plasma species and the very different time scales[40] associated with the gas-phase and surface reactions involving these species, the interplay between these species poses great challenges as the dimensions of the features to be transferred approach the size of molecules or even atoms. While it is true that a surface post-ion impact remains reactive for some finite period of time (owing to surface relaxation through reconfiguration and annihilation of dangling bonds), controlling this process window directly is an unlikely proposition, again due to the multitude and diversity of species arriving from the bulk plasma.

In a recently published report by the National Research Council, three main scientific challenges were identified for nonthermal plasmas[41]: (a) the need to

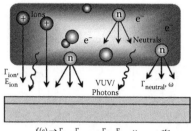

$f(\varepsilon) \rightarrow \Gamma_{ion}, \Gamma_{neutral}, \Gamma_{h\nu}, E_{ion}, \omega_{neutral}, etc.$

(a)

Species	Mass (g)	Gas phase reaction(s)	Surface reaction(s)
Electrons	10^{-28}	10^{-9}	$10^{-16} - 10^{-15}$
Ions	$10^{-24} - 10^{-22}$	—	10^{-12}
Photons	—	$10^{-9} - 10^{-8}$	—
Neutrals	$10^{-24} - 10^{-22}$	$10^{-3} - 10^{-1}$	10^{-12}
Radicals	$10^{-24} - 10^{-22}$	$10^{-5} - 10^{-4}$	$10^{-3} - 10^{-2}$

(Γ = flux, E = energy, ω = angular distribution).

(b) Reactive species in a plasma and their characteristics.

FIGURE 1.6

(a) Plasma showing constituent species and key parameters and (b) a list contrasting disparate mass and time scales for various plasma species.

connect processes that occur on an atomic level with plasma behavior over an area potentially spanning meters; (b) to quantify, characterize, and predict the interactions between reactive plasmas and complex surfaces; and (c) to develop a predictive capability to quantify and advance our understanding of low-temperature plasmas. Much experimental and simulation-based work has been done in an ongoing quest to achieve these goals, but despite much progress, there is still relatively little known about the exact minutiae of plasma–surface interactions, in large part due to the plethora of reactive species and possible reactions that can take place, as well as their interconnected nature that makes control over the process extremely difficult.[42]

The synergistic effects of major plasma species are, arguably, the most intriguing aspect of plasma processing that differentiate it from other processes. Although these deeply interconnected relationships between plasma species and their surface interactions have allowed for the achievement of feats that could not be accomplished by other attempted means, they also present significant challenges to research and study of the field as progress continues toward the nanoscale. The simultaneous arrival of these species to the surface in the magnitude necessary for etching inherently begets an uncontrollable, chaotic process, leading to phenomena such as electron bombardment-induced damage, deviated ion flux as a result of surface charging, and electron tunneling currents, all of which can lead to deviation from ideal feature geometries and device performance.

The synergism between ionic and neutral species necessary to facilitate materials etching was first demonstrated in a seminal paper by Coburn and Winters in 1979,[43] schematically illustrated in Figure 1.7a. The authors demonstrated that a silicon substrate underwent limited etching when physically sputtered by Ar^+ ions or chemically etched by XeF_2 gas. Yet when the two species were combined in an environment simulating a real plasma, the etch rate increased an order of magnitude over the sum of both individual etch rates. The authors' three major hypotheses on this particular system were

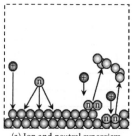

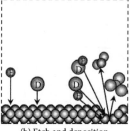

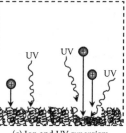

| (a) Ion and neutral synergism for semiconductor etch. | (b) Etch and deposition synergism for oxide etch. | (c) Ion and UV synergism for PR etch. |

FIGURE 1.7

The major synergistic effects in plasmas: (a) ion/neutral, (b) etch/deposition, and (c) ion/ultraviolet (UV).

as follows: the ions initiated the dissociative chemisorption via damage or defect creation, the ions dissociated weakly bonded surface species as a result of direct impact, and the ions facilitated the removal of nonvolatile residues. These hypotheses have been validated over the last 30 years with many other material–plasma combinations, particularly with the halogen chemistries, such as the study of silicon etching with Cl/Cl_2 ions and neutrals.[44] In these processes, the key step involves the breaking of chemical bonds upon ion impact. Etching enhancement then results from the simultaneous exposure and subsequent reaction with radicals and neutrals, leading to the formation of volatile species that desorb either spontaneously or upon ion impact. Other studies revealed complications in deconvoluting reaction mechanisms in the etching of compound semiconductors, carbide/nitride materials, and metals.[45] For example, some degree of control over the dominant neutral radical species could be achieved by tailoring the power supplied or the gas chemistry to favor particular reactions, but this would have to be weighed against damage issues or formation of unwanted residues, respectively.

The second major idea of interest is the balance between ion-/radical-induced etching and neutral stimulated deposition. This is best known in the etching of oxide materials, where the removal of oxygen and nonoxygen elements dictates simultaneous yet different reactions.[46] The earliest and perhaps best known example is SiO_2 etching in a fluorocarbon chemistry, in which silicon reacts with fluorine species while the oxygen reacts with carbon species to produce volatile SiF_x and CO_x, respectively. It was noted in a study by Oehrlein et al. that these discharges were responsible for the deposition of a fluorinated polymer film on an exposed surface. This polymeric film serves to control the etch rate and as such plays a critical role in regard to achieving the desired etching characteristics.[47] More recent studies involving the same general principle have been conducted regarding the etching of metal oxides in BCl_3 chemistries, using the specific example of high-k dielectrics (HfO_2,[9b] ZrO_2[48]). In all of these oxide systems, there is a delicate balance between deposition and etching as halogenated polymer layers are deposited on the surface during the etching process. These layers react slowly with incoming halogen species to generate volatile products, which are predominantly removed by energetic ion bombardment. This type of mechanism allows for the anisotropic etching of various oxides,[49] however, below a certain ion energy threshold net deposition of the halogenated polymer layers occurs. The deposited layers can pose an additional advantage in the context of a multimaterial system: selectivity is enhanced by the preferential passivation of certain layers of a film (e.g., in a Si/oxide system polymer formation, C or B based, on Si far exceeds that on the oxide and as a result selective etching can be achieved).[50]

Third, ion/VUV photon synergism was recently demonstrated through the study of photoresist roughening during plasma exposure.[51] The separate effect of VUV and ions was characterized using Fourier transform infrared (FTIR), which saw the removal of specific functional groups from the polymer as a result of each reactive species (C=O and CH_x losses, respectively). It was demonstrated that simultaneous exposure to VUV and ion beams resulted in far greater roughening than either of the two individual effects. Furthermore, a cross-linked layer generated by ion bombardment was necessary for roughening to occur, but this layer could only be formed on a surface morphology that had been softened by VUV scission. This synergistic effect has direct implications on the formation of line edge roughness (LER)[52] and line width roughness (LWR),[53] both of which become progressively more significant with device downscaling and are discussed in further detail in a later section dealing with lithography challenges.

It should be evident that the synergistic interactions outlined above represent a solid foundation upon which one can expand the field of plasma surface treatment, and ultimately *extend* it into the field of nanomaterials patterning. As discussed earlier, the reactivity and functionalization of the growth surface are of utmost importance when considering the potential growth and organization of nanostructures. Plasmas offer vast possibilities for creating reactive sites and adding reactive functional groups on a surface, thereby setting the stage for etching and deposition of materials at the nanoscale. Both atmospheric and low-pressure plasmas have been used for surface modification.[54] Documented studies in literature include the use of nitrogen-containing plasmas to selectively control amino group density[55] and attachment of various species containing nitrogen, oxygen, and fluorine onto the surface of multiwalled carbon nanotubes.[56] In these cases the plasma environment is responsible for the addition of these additional functional groups: by adjusting plasma and process parameters (e.g., source power, feed gas chemistry, pressure, and exposure time), the degree of incorporation with the surface could be controlled. In other examples, the plasma surface treatment is used as a prerequisite for further functionalization via grafting. Demonstrated cases include acrylic and styrenesulfonic acids, N,N-dimethylacrylamide,[57] and maleic anhydride.[58]

Having delved into the nature of plasma etching, it becomes evident that controlling its outcome is a highly complex problem that is also very costly. To that effect, computation-based modeling has substantially complemented experimental work, contributing tremendously to the understanding of elementary reaction mechanisms involving plasma species that are difficult to examine via physical probes. It also holds promises to help reduce the time and effort associated with time-consuming and expensive design of experiments, by allowing prediction of the process outcome. Based on length scales, these calculations span from reactor (cm), topographical features (μm), down to molecules and atoms on the surface (nm). From the standpoint of time scales, these calculations range from processing time (minutes) to bond breakage (milliseconds) to photostimulated excitation (< nanoseconds), as depicted in Figure 1.8. The large disparity between the length and time scales of the various regimes—reactor scale, film surface, and feature profiles—dictates a variety of techniques (e.g., fluid models,[59] Monte Carlo (MC),[60] active site modeling,[44,61] translating mixed layer (TML) kinetics,[62] molecular dynamics (MD),[63] and a hybrid plasma equipment model (HPEM),[64] as no single method is adequate to address the overall scope of etching kinetics.

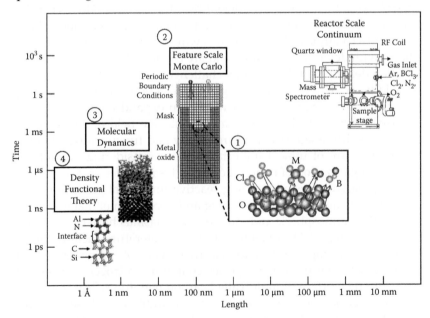

FIGURE 1.8
Length and time scales addressed by several modeling techniques. The lack of overlap suggests that multiple techniques are required to comprehensively handle the simulation of the plasma etch process from the reactor to the individual nanoscale features. The magnified inset shows a representation of a metal oxide surface being etched in a BCl_3 discharge, the modeling of which is used as the starting point for the approach elaborated on in this section.

In order to establish a framework that can be used to predict bulk surface etching behavior (labeled as 1 in Figure 1.8), it is necessary to combine experimental and theoretical approaches. Diagnostic tools are of great utility in obtaining data about the etch process. Optical emission enabled actinometry and absorption spectroscopy can be used to measure and quantify the concentrations of gas-phase species in the plasma, and mass sensitive analysis (MS[65], quartz crystal microbalance [QCM]), spectroscopy[66] (photon, electron, or ion based) and profilometry (atomic force microscopy) provide an assessment of the growth/etch kinetics and determine the sticking and reaction coefficients and the composition and morphology of an evolving surface.[67] Surface-related analytical tools such laser-induced fluorescence and desorption (LIF/LID) can be used to provide in situ chemical information about the surface and desorbed species that would be otherwise very difficult to obtain.[68] For example, a phenomenological model based on data from such probes was recently demonstrated in a surface site balance based form that accounted for synergistic etching and deposition by ionic and neutral species.[69] It was formulated to accurately describe the etching of composite oxide films in complex plasma chemistries involving competing deposition and etching mechanisms, explaining the etch-rate (Rt) dependence on key plasma parameters including plasma chemistry/conditions, neutral-to-ion flux ratio, and ion energy, as well as the film composition, yielding the rate equation depicted in Figure 1.9.

$$R_t = \frac{J_e^2 v_{es} S_{es} v_{ep} S_{ep} - J_d^2 v_{ds} S_{ds} v_{dp} S_{dp}}{\left\{ J_e v_{ep} S_{ep} + \dfrac{J_d^2 v_{ds} S_{ds} v_{dp} S_{dp}}{D_p} + \dfrac{J_d J_e v_{ds} S_{ds} v_{ep} S_{ep}}{J_i C_p \left(E_{ion}^{1/2} - E_{th,p}^{1/2}\right)} + J_d v_{ds} S_{ds} + \dfrac{J_d J_e v_{ds} S_{ds} v_{ep} S_{ep}}{D_s} + \dfrac{J_e^2 v_{es} S_{es} v_{ep} S_{ep}}{J_i \left[A_s \left(E_{ion}^{1/2} - E_{ion}^{1/2}\right) + B_s \left(E_{ion}^{1/2} - E_{tr,s}^{1/2}\right)\right]} \right\}}$$

$(J_d = 0)$

$(J_e = 0)$

$$R_t = \frac{J_i \left[A_s \left(E_{ion}^{1/2} - E_{th,s}^{1/2}\right) + B_s \left(E_{ion}^{1/2} - E_{tr,s}^{1/2}\right)\right]}{1 + J_i \left[A_s \left(E_{ion}^{1/2} - E_{th,s}^{1/2}\right) + B_s \left(E_{ion}^{1/2} - E_{tr,s}^{1/2}\right)\right]/J_e v_{es} S_{es}}$$

$$R_t = -\frac{J_d v_{dp} S_{dp}}{1 + J_d v_{dp} S_{dp}/D_p}$$

FIGURE 1.9
A viable model delineating the dominant effect in plasma etching of complex materials, where *J* terms are fluxes of ionic, etching, and depositing species; *S* terms are sticking probabilities on the substrate/polymer layer for etching and depositing species; *D* terms are deposition rate on the substrate/polymer layer; *A, B, C* terms refer to the volume of substrate/polymer removed as a function of ion energy; *v* terms refer to the volume of substrate/polymer removed or deposited as a function of species flux; and *E* terms are energy values (e.g., incoming ion energy, threshold energy, th, transition energy, tr). (Reprinted from *JVST A*, Vol. 27, Martin and Chang, "Plasma etching of Hf-based high-k thin films. Part III. Modeling the reaction mechanisms," pp. 224–229, Copyright (2009) with permission from JVST.)

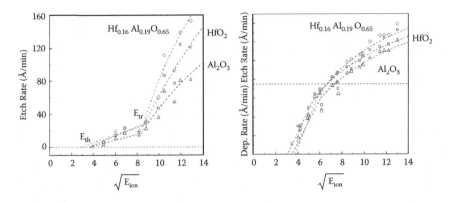

FIGURE 1.10
Phenomenological model fits for the cases of HfO_2, Al_2O_3, and $Hf_{0.16}Al_{0.19}O_{0.65}$ films in Cl_2 (left) and BCl_3 (right) plasmas. Plots show etch/deposition rate as a function of the square root of ion energy. The dashed horizontal line marks the zero value of the y-axis, showing clearly the deposition and etching regimes for BCl_3 plasmas.

Preliminary results show agreement between experiment and simulation for the base system of high-k oxide film (HfO_2, Al_2O_3, and HfAlO) etching in Cl_2 and BCl_3 plasmas, as shown in Figure 1.10. This model has the potential to become the platform for predicting the etching of even complex materials, including elements that form less-volatile reaction products such as those projected to be implemented in memory devices (e.g., phase change materials [PCMs] and multiferroics used in magnetic random access memory [MRAM] and ferroelectric random access memory [FeRAM] such as $YMnO_3$).[70]

However, the ultimate impact of this model depends on if it can be utilized in a feature profile simulator such as one based on Monte Carlo or similar methods, by properly parameterizing the rate coefficients of reaction kinetics. One approach is to convert such a model into a translating mixed layer (TML) format. This model uses a defined set of reaction equations to predict the etch yield of a process and can be used to take into account the effect of variety of species present in the plasma, as well as different material removal pathways such as sputtering and ion-enhanced etching. It was previously applied to the etching of SiO_2 in fluorocarbon chemistry, demonstrating agreement with experimental data.[62b] A comparison of the phenomenological and TML models is shown in Figure 1.11. In order to verify the accuracy of the TML model, its results must be used as input for a Monte Carlo (MC)–based feature profile simulator (labeled as 2 in Figure 1.8) and the simulated etch profiles compared to experimentally obtained ones.

With the capability and flexibility to include large amounts of kinetic data (such as chemical reaction sets) that improve the accuracy of the model in characterizing specific plasma processes, MC methods can be used to simulate profile evolution in patterned surfaces exposed to plasma.[71] Specifically,

$DR_s = D_s \theta_{ds} = J_d \nu_{ds} S_{ds} \theta_1$ $DR_p = D_p \theta_{dp} = J_p \nu_{dp} S_{dp} \theta_2$	$Cl_{(g)} \rightarrow Cl_{(s)}$ $r_{A1} = S_{Cl\_on\_M} R_{Cl} \times M_{\_for\_Cl}$ $Cl^+_{(g)} \rightarrow Cl_{(s)}$ $r_{A2} = S_{Cl+} R_{Cl+}$
$ER_s = J_t A_s (E_{ion}^{1/2} - E_{ths}^{1/2})\, \theta_{es} + J_t B_s (E_{ion}^{1/2} - E_{trs}^{1/2})\, \theta_{es} = J_e \nu_{es} S_{es} \theta_1$ $ER_p = J_i C_p (E_{ion}^{1/2} - E_{thp}^{1/2})\, \theta_{ep} = J_e \nu_{es} S_{es} \theta_2$	$Cl_{(s)} + O_{(s)} \rightarrow ClO_{(g)}$ $r_{E1} = \beta_{ClO}(J_{Cl-O})$ $BCl_{(s)} + O_{(s)} \rightarrow BOCl_{(g)}$ $r_{E2} = \beta_{BOCl}(J_{B-Cl})(J_{B-O})$ $M_{(s)} + 4Cl_{(s)} \rightarrow BCl_{4(g)}$ $r_{E3} = \beta_{MCl_4}(J_{M-Cl})^4$
$\theta_{es} + \theta_{ds} + \theta_{ep} + \theta_{dp} + \theta_1 + \theta_2 = 1$	$r_M = r_{A1} + r_{A2} - 2r_{E1} - 3r_{E2} - 5r_{E3}$

FIGURE 1.11

Comparison of the formulation of the phenomenological and the translating mixed layer (TML) models. The top row describes the reactions with incoming plasma species, the middle row illustrates the etching reactions, and the bottom row depicts the balance of the above, using metal oxide etching in a BCl_3 chemistry as an example.

the species population in question is represented by a collection of particles followed in space and time. Events involving the particles take place at random times but with an average frequency close to that of the real (measured) frequency. This method utilizes relatively elementary concepts such as Newton's laws of motion and reaction probabilities. A large number of particles are sampled in order to obtain a decent representation of the configuration and velocity space being studied. The MC method is often combined with other modeling techniques such as fluid models or particle-in-cell methods in a "hybrid approach" to obtain the best possible results in systems that contain physical properties that differ by orders of magnitude, such as plasmas containing ionic and electronic transport.[72]

If a predictive model for features approaching the monolayer regime based on rate equations and reaction probabilities is to be formulated, it is necessary to obtain a basic comprehension of the competing mechanisms involved when a surface is exposed to plasma.[73] Given the limitations of experimental probes in comprehensively observing these reactions in real time, an atomistic simulation of what is happening on a surface would thus be quite suitable for this field. It is for these reasons that use of molecular dynamics (MD) (labeled as 3 in Figure 1.8) becomes essential for studying the limits of plasma etching. The work of Harrison was the first to use MD to study the case of physical sputtering,[74] with subsequent studies dealing with such topics as silicon etching (Figure 1.12a), silicon surface rippling after exposure to ion beams,[75] and the nickel-catalyzed formation of carbon nanotubes.[76] MD simulations allow for manipulation and observation of events happening in the pico- and femto-second time scales, far shorter than what can be controlled in an experimental setting. However, while these time scales are suitable for reactions such as the ion-neutral synergism collision cascade in chemical and physical sputtering, it also means that MD simulations cannot be used for longer time scale reactions such as the desorption of weakly bound species from a plasma treated surface.[77] The problem has often been addressed in the field by coupling MD with other models that are equipped to handle the time scale difference, but there has been work that discussed

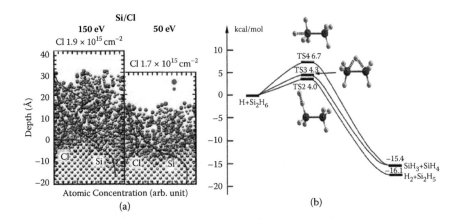

FIGURE 1.12
(a) Molecular dynamics (MD) simulations of Si bombarded with Cl ions from the top of the cell. (Reprinted from *Thin Solid Films*, Vol. 518 (13), Ono et al., "Plasma–surface interactions for advanced plasma etching processes in nanoscale ULSI device fabrication: A numerical and experimental study," pp. 3461–3468, Copyright (2010) with permission from Elsevier.) (b) An energy diagram of a possible H + Si$_2$H$_6$ surface reaction. (Reprinted from *J. Phys. Chem. A*, Vol. 114 (1), Wu et al., "Ab Initio Chemical Kinetic Study for Reactions of H Atoms with SiH$_4$ and Si$_2$H$_6$: Comparison of Theory and Experiment," pp. 633–639, Copyright (2009) with permission from American Chemical Society.) (See also Wu, S. Y.; Raghunath, P.; Wu, J. S.; Lin, M. C., Ab initio chemical kinetic study for reactions of H atoms with SiH$_4$ and Si$_2$H$_6$: Comparison of theory and experiment. *J. Phys. Chem. A* 2009, 114 (1), 633.)

the incorporation of transition state theory (TST)[78] as another potential solution to allow for the extension of the time scales in MD simulations into the microsecond regime while maintaining the level of detail.

Although MD simulations allow for modeling at an atomistic level, they are not without limitations. The accuracy of the calculation requires reliable interatomic potential values. One such way of obtaining these values is by combining MD with *ab initio* calculations[79] (labeled as 4 in Figure 1.8), in which the electronic structure is calculated at each step in the simulation, allowing for real-time adjustment of the interatomic potentials. This method allows for greater accuracy, but the computing cost is so high that it is often considered infeasible.[80] However, through careful use of this technique for specific chemical interactions, it is possible to utilize these calculations to model surface processes, as shown in Figure 1.12b.[81] *Ab initio* modeling allows for the prediction of such information as reaction pathways, the effect of energy distribution and molecular orientation, and the influence of surface defects on dynamics and reactivity.[82] These types of predictions are obviously very important for the study of plasma etching at an atomic level and would provide a significant augmentation to experiments. As a surface is exposed to plasma, it undergoes changes in chemical composition and structure that affect later reactions; thus, the ability of *ab initio* modeling to reassess changes at the atomic level as reactions occur is required in order to depict these processes accurately. Unfortunately, this method has its own limitations. Its quantitative accuracy is primarily limited by approximations for the exchange correlation (XC) functionals involved.[83] Lowering the computational intensity of the model used also results in the least accurate prediction of parameters such as adsorption barriers, and even proposed correctional functions (such as gradient corrections) have inherent limitations in producing systematic changes. Additionally, there is an issue of modeling ions versus neutral species, which requires the use of multiple sets of interatomic potentials and the modeling of charge-exchange events and other electronic transitions.[84]

Further integration of MD and *ab initio* methods into these types of model can allow for more accurate depictions of surface processes during plasma etching, especially at smaller length scales. In some ways, the synergism of these various simulation methods to achieve a goal mimics the synergism of different species that populate plasmas and allow for reactions to take place. Just as one isolated type of species is not effective in realizing the end goal of material removal, a single modeling method is not suitable to simulate the holistic plasma etching process.

1.4 Overcoming Damage Issues—Neutral Beams and Atomic Layer Etching

The use of plasma in top-down processes as an etching tool to pattern materials has played a significant role in the microelectronics industry and allowed for the extension of Moore's Law over two decades. However, as the scale of integrated circuit (IC) devices continues to shrink to the 22 nm node and beyond, the fundamental limits of plasma etching are fast approaching. The damage induced by radiation and charge buildup is among the most pertinent issues facing IC device manufacturing that could also pose serious challenges to nanomaterials processing. At nanometer-length scales, it is imperative that precise control be exhibited over the etch process as even the slightest deviation in feature geometry could produce severely compromised or unusable structures. This problem is largely due to the wide variety of species that are produced in a plasma (e.g., ions and electrons can induce charge buildup on surfaces which greatly alters the trajectories of incident charged particles, while exposure to UV/VUV photons creates crystal defects). Besides physical damage, these phenomena can have deleterious effects on other characteristics, such as electrical properties. Among the possible solutions that have been proposed for this issue include the use of neutral-beam etching,[85] in which low energy (~10 eV) neutrals can be generated from negatively charged ions passing through apertures in a carbon plate. Although standard plasma etching has been demonstrated as viable for pattern transfer in the removal

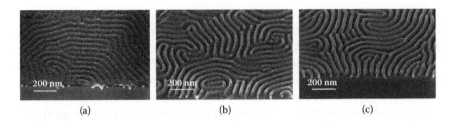

(a)   (b)   (c)

FIGURE 1.13
Illustration of etched PS-b-PMMA block copolymer samples showing damage/roughness after plasma etching in (a) and (b) and smooth surface in (c). (Reprinted from *Journal of Vacuum Science and Technology B*, Vol. 26, Ting et al., "Plasma etch removal of poly(methyl methacrylate) in block copolymer lithography," pp. 1684–1689, Copyright (2008) with permission from American Vacuum Society.)

of soft materials such as the PS-b-PMMA diblock copolymer system[5] (shown in Figure 1.13), the absence of charged species and high bombardment energies in the neutral-beam technique leads to a much gentler process that has been successfully applied to highly delicate masks (e.g., nanotemplate/nanoparticle fabrication using a protein film).

Another possible avenue to explore for precise control over the amount of material removed in an etching process is atomic layer etching (ALE).[86] The difference between ALE and a conventional "bulk etching" process can be thought of as an analogy between atomic layer deposition (ALD) and chemical vapor deposition (CVD) processes, as shown in Figure 1.14. In the conventional plasma etch/CVD processes, the synergism of species in the gas phase as well as on the surface is fully embraced and a multitude of reactions take place, resulting in a somewhat uncontrollable, nonconformal etch or deposition. Thus, ALE can be considered as analogous to ALD in that it involves reproducibly removing single atomic layers of a surface through a series of self-limiting etching steps. The four main steps in an ALE/ALD process can be thought of as (1) adsorption (chemisorption) of one species onto the surface which self-terminates upon consumption of the available surface site, (2) evacuation of the reactor to prevent reaction in the gas-phase, (3) exposure of the surface to a second species to induce chemical reaction with the adsorbed species, leading to atomic layer etch or deposition, and (4) evacuation of the reactor to remove the reaction products. It is important to note that the most important factor in achieving atomic scale control is that at least one of the steps, (1) or (3), should be self-limiting.

To achieve atomic scale control of etching processes, the most critical component to control is the ion energy and its distribution, as they determine the degree and thickness of plasma damage,[87] which translates to degradation in materials property and device functionality.[88] Low-temperature plasma sources (such as microwave and pulsed inductively coupled plasma [ICP]) have been examined to provide better control of the electron temperature and ion energy distribution). It has been shown that

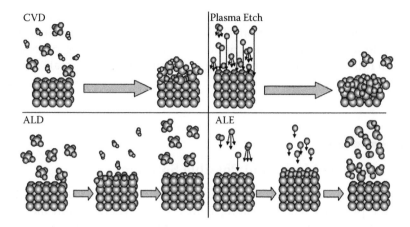

FIGURE 1.14
Comparison of a traditional plasma etch process with an ALE process, in comparison to chemical vapor deposition (CVD) and atomic layer deposition (ALD).

a synchronous DC bias during the afterglow of a pulsed plasma allows better control of the ion energy distribution and the relative fluxes of ion/neutral/photon to the surface, over a die and the entire wafer.[89] Being able to control the energy and flux of ions is the first step toward achieving layer-by-layer control in plasma etching.

In a plausible ALE process, ions can be the enabling species to form and remove a volatile product. The energy of the ions should be low enough to prevent excessive sputtering but higher than the threshold energy for etching. This general concept has been demonstrated for etching silicon,[90] germanium,[91] and gallium arsenide,[92] with only a few minor differences between the three cases. There has also been some simulation work done on this research topic,[93] but more thorough experimental work, particularly with regard to different material systems is in great need. The following half reactions describe the deposition and etching of metal oxide (MO_2) at one atomic layer precision.

Half Reactions for ALD of MO_2:

    (1) $M\text{-}(OH)_2(s)+MCl_4(g) \rightarrow M\text{-}O\text{-}M\text{-}Cl_2(s)+2HCl(g)$

    (2) $M\text{-}O\text{-}M\text{-}Cl_2(s)+2H_2O(g)\rightarrow M\text{-}O\text{-}M\text{-}(OH)_2(s)+2HCl(g)$

Half Reactions for ALE of MO_2:

    (1) $MO_2(s)+Cl^+(g)\rightarrow MO_2^*(s)$

    (2) $MO_2^*(s)+ 6Cl(g)\rightarrow MCl_4(g)+2OCl\ (g)$

Granted, the aforementioned ALE process is not without difficulties. Among them include selecting the right combination of species to ensure the process is carried out as close to ideally as possible, and measuring the etch products produced, which typically have to be in reasonably high concentrations for in situ observation. However, the potential of ALE to provide insight into a surface-based description of the plasma etching process, as well as to fulfill a role in the processing of features at the nanoscale regime, is undoubtedly great.

1.5 Lithography Limits and Possible Solutions

The issues of line edge/width roughness (LER/LWR) were called to attention in the previous section dealing with UV/ion synergisms in the etching of photoresist. Excessive roughness can influence characteristics such as capacitance, light scattering, fluid flow, wettability and adsorption. As stated earlier, roughness is an especially significant issue for structures and devices at the nanoscale because it does not necessarily scale down with feature size. As an example, a LWR of 2.5 nm for a 36 nm DRAM half-pitch produced in 2012 was called for by the 2009 International Technology Roadmap for Semiconductors. Given that the radius of gyration of the polymers used in photoresist is on the order of 2 to 5 nm, it is clear that this is a difficult obstacle to surmount. This has led to research to suppress LER/LWR both during and after the lithographic processing[94] and to model both the patterning[95] as well as the deleterious effects on electrical performance.[96]

Controlling the etch characteristics is only one component of the patterning process, however. Current critical dimensions (CDs) of features patterned by top-down processes are largely limited by the minimum possible resolution of optical lithography. As we contemplate the use of lithography for patterning nanostructures at even smaller dimensions than the existing state of the art, perhaps the two most pressing obstacles are ensuring high fidelity of pattern transfers, and overcoming the inherent limitations put forth by the Rayleigh equation for half-pitch critical dimension,

$$CD = \frac{k_1 \cdot \lambda}{NA},$$

where k_1 is the process aggressiveness factor, λ is the wavelength of the light used, and NA is the numerical aperture. A two-pronged approach including physics-based quantitative assessment of phenomena contributing to transfer defects such as mask edge effects, defect-feature interactions, and stress-generated variations[97] and techniques such as optical proximity correction (OPC) and off-axis illumination (OAI) have been demonstrated as an effective engineering solution to achieve improved fidelity in pattern transfer.[98] In order to lower the feature half-pitch CD, either k_1 must be lowered, or NA must be increased. Given that improving NA requires significant concurrent advances in lens materials, fluids, and resists, reducing k_1 appears to be the best solution for achieving this goal. Although the minimum value for k_1 with a single exposure is 0.25, this limit can be surmounted by using double-patterning lithography (DPL) methods such as double exposure-single etch (DESE) and double exposure double etch (DE2). Both methods can be employed with existing exposure tools (thus requiring minimal additional capital investment), but at the cost of reduced throughput as a result of the additional process steps. Additionally, DPL techniques are particularly susceptible to overlay errors due to the multiple exposure steps required, and in the case of DE2, the need to remove the wafer from the chuck prior to the second exposure step.[99] Nonetheless, DPL offers a viable means of reducing printed critical dimensions beyond that of existing single exposure methods (Figure 1.15). By combining advanced lithographic techniques with the feedback from projected outcomes from plasma etching, intricate patterning that would be impossible by either of them in isolation can be carried out.

There are other top-down techniques discussed in the literature to extend lithographic patterning beyond the diffraction limit. Among these are nanoimprint lithography (NIL),[101] quantum interferometric optical lithography (QIOL),[102] surface plasmon resonant interference nanolithography (SPRINT),[103] and metamaterial-based "superlenses."[104] The NIL method utilizes mechanical molding of polymer materials to create features, which theoretically could overcome the diffraction limit; however, plasma etching is still deeply involved in the process—from the creation of the template molds to the removal of residual polymer on the substrate. However, this approach offers the possibility of hybridization; by implanting metal pads

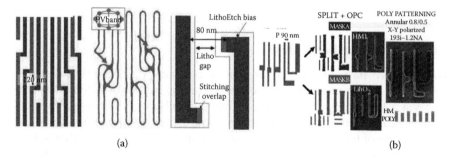

(a)   (b)

FIGURE 1.15

(a) Desired feature versus obtained feature due to plasma-lithography interplay. (Reprinted from *IEEE Transactions on Circuits and Systems I*, Vol. 56, Ronse et al., "Lithography Options for the 32 nm Half Pitch Node and Beyond," pp. 1883–1890, Copyright (2009) with permission from IEEE.) (b) Top view of double-patterning lithography (DPL) approach to patterning a feature with critical dimension (CD) lower than that of single exposure. (See also Ronse, K.; et al., Lithography options for the 32 nm half pitch node and beyond. *IEEE Trans. Circuits Syst. I—Regular Papers* 2009, 56 (8), 1884–1891.) (IMEC Scientific Report, "Immersion lithography with double patterning," Copyright (2009) with permission from IMEC.)

into the template mold it can be used as both a conventional lithographic mask as well as an NIL tool. QIOL is based on the concept of nonclassical entangled photon-number states and theoretically allows for the patterning of features with a minimum CD smaller than the diffraction limit by a factor of N, where N in this case is the number of photons entangled at a time, and also the number absorbed by the substrate. Entangled photon pairs can be generated by spontaneous parametric down-conversion and allow for lithography below the typical limits of diffraction.[102] SPRINT relies on the principle that illumination light can be guided with a prism to couple with surface plasmons to obtain a new state with a much shorter wavelength and higher field intensity than that of the illumination light. The resulting enhanced optical field close to the metal mask can then cause localized exposure of a thin resist layer below the mask.[103] The superlens concept hinges heavily on "negative index media" (NIM) (i.e., materials with a negative index of refraction). Engineering NIM allows for the enhancement of evanescent waves, which carry fine details about the object but are confined to the near field and subsequently become lost by conventional glass lenses. When utilized in conjunction with a coupling element, the enhanced evanescent waves can be coupled into propagating waves, which makes far-field detection possible. A variation of this idea is the hyperlens, which uses an artificial metamaterial to transfer deep subwavelength information into the far field by a two-stage process. First, evanescent waves are enhanced through surface resonance, followed by conversion into a propagation wave at the exit surface by means of a designed surface scatter.[104] Such novel techniques provide examples of expanding the possibilities of plasma etching.

1.6 Plasma in other Roles?

Thus far, we have examined the use of plasmas for patterning features defined by either premanufactured lithographic masks or a template created via a bottom-up process (e.g., diblock copolymers). Recently, Vourdas et al. postulated that plasma exposure can also be used to initially create the template of features on a surface layer that acts as a mask, so that subsequent plasma treatment with varied chemistry can then remove the material and pattern the surface.[105] Linked to this idea is the fact that simultaneous etching and deposition occur in a typical plasma. In theory, if exact control over the various species in a discharge could be exhibited in a manner based on the ALE process described earlier, plasmas could be used to fabricate structures from both bottom-up and top-down perspectives. Currently, the problem lies in controlling the features that are created via the plasma process. Experimental work in the aforementioned paper showed the creation of arrays of nanodots; however, only limited control was exhibited over the periodicity and size of the structures formed. To manufacture simple arrays, this could be a viable approach, bolstered by its inexpensive nature. However, for more complicated designs it is unlikely that this approach is feasible, making a combination of methods more likely to succeed in effectively patterning these types of features over larger scales.

Outside of patterning, another role that plasmas have been applied to is that of direct material synthesis. Among the reviewed topics in literature include synthesis of nanomaterials via plasma spray synthesis,[106] metal plasma immersion ion implantation,[107] thermal plasma processing,[108] and plasma- or radical-enhanced atomic layer deposition (PE- or RE-ALD).[109] Plasmas play highly varied roles in these methods: from merely providing a high energy source (certain instances of thermal plasma processing) to generating reactive species that facilitate reaction pathways at lower temperatures to forming films from direct condensation. The overall efficacy of these processes is greatly enhanced through the controlled use of plasma. For example, in conventional ALD processes, the disparate process temperature windows for different precursors make it difficult to deposit multicomponent films. The use of RE-ALD lowers the requisite deposition temperature for all elements and thus allows for the synthesis of complex films as depicted in Figure 1.16. For example, RE-ALD has been used to deposit co-doped rare earth ion doped oxide films with controlled atomic percentages and spatial distribution based on the deposition cycle parameters.[110]

The plasma spray synthesis method combines elements of these approaches, using a high-temperature plasma jet to both melt and disperse the reactants. In this case, the distinctive properties of the plasma (e.g., high

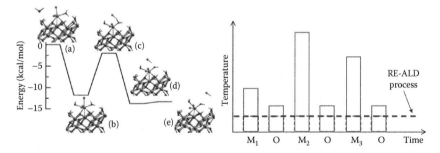

FIGURE 1.16

(Left) Energy diagram for ligand exchange reaction between gaseous water and a metal precursor in a conventional atomic layer deposition (ALD) process. (Right) Representation of radical-enhanced atomic layer deposition (RE-ALD) used to lower process temperature windows for deposition of multicomponent metal (M) oxide films. (See also Mukhopadhyay, A. B.; Musgrave, C. B.; Sanz, J. F., Atomic layer deposition of hafnium oxide from hafnium chloride and water. *J. Am. Chem. Soc.* 2008, *130* (36), 11996–12006.) (Reprinted from *Journal of the American Chemical Society*, Vol. 130, Mukhopadhyay et al., "Atomic layer deposition of hafnium oxide from hafnium chloride and water," pp. 11996–12006, Copyright (2008) with permission from ACS.)

enthalpy density, temperature, velocity, and heating/cooling) facilitate techniques such as particle spheroidization and rapid solidification processing. A large assortment of films and materials have been grown using the methods described above, from co-doped rare earth oxide films, to metal alloys, and to ceramic powders for biological applications.[112] Given the subject material, of key interest is the usage of plasmas to create nanomaterial powders and deposits. A particularly recent and relevant example of novel adaptation of plasmas to achieve this goal was seen in the use of a microplasma reactor to synthesize nanoparticles with precise control over the size distribution and composition. [113]

On a slight tangent, there exist many other potential avenues for plasmas regarding nanostructures and materials research. Plasmas have recently begun to gain more attention from the medical community, with examples ranging from the treatment of surfaces for tissue grafting, to direct exposure of tissue for sterilization, and wounds for regeneration.[114] All of these cases involve concepts discussed earlier in the chapter (e.g., the creation of reactive species in the plasma allow for physical and chemical modification of the grafting surface, making the process more favorable, while reactive species potentially play a role in wound regeneration by accelerating the pathways responsible for the healing process to take place). It is possible that selective generation of certain species could be utilized to combat other deleterious elements present in the body or tissue samples. Interestingly enough, nanostructures are also finding themselves under more detailed investigation in the medical field, but under much less auspicious circumstances: there have been several reviews in literature focused on evaluating the potential toxic effects of nanoparticles on individuals, species, and entire ecosystems.[115] Even though nanostructures have an extensively documented record of existing in nature (caused by phenomena such as volcanic eruptions), such naturally occurring examples are often ephemeral, easily susceptible to dissolution or aggregation. In contrast, the treatment of manufactured nanostructures to increase their stability means these particles could possibly exist for much longer periods of time in the natural environment. Additional modification to enhance catalytic or other properties could result in increased toxicity over naturally formed nanoparticles; thus, further research is required into ways of treating these particles to ensure safe disposal. Plasma jets have been effectively used in waste treatment facilities for the removal of greenhouse gases, organic material, and other contaminants,[116] and thus could possibly be extended to nanomaterials by combining this topic with the study of nanoparticle nucleation in plasmas.[117] This work also ties in with nanoparticle synthesis outlined in the preceding paragraph. By analyzing growth mechanisms and particle-plasma interaction phenomena in discharges, an effective means of "coagulating" these particles together to assist in their removal could be realized. These topics are beyond the scope of this chapter but are mentioned to motivate potential new directions for plasma science.

1.7 Conclusion

This chapter has examined the potential role that plasmas can play in the creation and processing of nanostructures and nanomaterials. In conjunction with top-down lithographic patterning, plasma etching has already met success in creating features with an average half-pitch of <22 nm as per the current technology node. As a synthesis method, groups have reported the use of plasma-assisted techniques to create a wide variety of nanomaterials and nanostructured thin films, such as nanoparticles with mean diameters <10 nm.[113] Plasma treatment has also been used to functionalize nanostructures with a variety of organic and inorganic surface groups. Even though future fabrication standards pose a set of unique challenges that may hinder the application of conventional lithographic processing methods, there are new techniques in the developmental stage that may yet overcome some of the fundamental limitations. For example, the use of a metamaterial-based superlens has demonstrated the possibility of printing features with resolution as low as an order of magnitude below the diffraction limit of the light used.[118] Additionally, top-down approaches utilizing plasmas can be used to indirectly influence the synthesis and patterning of nanomaterials, through the ability to manipulate substrate reactivity and functionalization. The research path may be extended even further into multidisciplinary endeavors (e.g., studying plasma–nanoparticle interactions may yield benefits in both the medical and waste treatment fields as more light is shed on the consequences of nanoparticle production in those disciplines). As we progress further into the realm of nanometer-scale technologies, it would appear that top-down plasma processing methods will continue to play a highly important role.

References

1. (a) Cui, D. X., Advances and prospects on biomolecules functionalized carbon nanotubes. *J. Nanosci. Nanotechnol.* 2007, 7 (4–5), 1298–1314; (b) Cui, D. X.; Gao, H. J., Advance and prospect of bionanomaterials. *Biotechnol. Progress* 2003, *19* (3), 683–692; (c) Wang, K. L., Issues of nanoelectronics: A possible roadmap. *J. Nanosci. Nanotechnol.* 2002, 2 (3–4), 235–266; (d) Balandin, A. A., Nanophononics: Phonon engineering in nanostructures and nanodevices. *J. Nanosci. Nanotechnol.* 2005, 5 (7), 1015–1022.
2. (a) Murugan, R.; Ramakrishna, S., Nano-featured scaffolds for tissue engineering: A review of spinning methodologies. *Tissue Engineering* 2006, 12 (3), 435–447; (b) Ma, J.; Wong, H. F.; Kong, L. B.; Peng, K. W., Biomimetic processing of nanocrystallite bioactive apatite coating on titanium. *Nanotechnology* 2003, 14 (6), 619–623.

3. Thomas, V.; Dean, D. R.; Vohra, Y. K., Nanostructured biomaterials for regenerative medicine. *Current Nanoscience* 2006, 2 (3), 155–177.
4. Otsuka, H.; Nagasaki, Y.; Kataoka, K., Self-assembly of poly(ethylene glycol)-based block copolymers for biomedical applications. *Curr. Opinion Colloid Interface Sci.* 2001, 6 (1), 3–10.
5. Ting, Y. -H.; Park, S. -M.; Liu, C. -C.; Liu, X.; Himpsel, F. J.; Nealey, P. F.; Wendt, A. E., Plasma etch removal of poly(methyl methacrylate) in block copolymer lithography. *J. Vacuum Sci. Technol. B: Microelectronics Nanometer Struct.* 2008, 26 (5), 1684.
6. Ostrikov, K., Plasma Nanoscience: From nature's mastery to deterministic plasma-aided nanofabrication. *Plasma Sci., IEEE Trans.* 2007, 35 (2), 127.
7. (a) Gottscho, R. A.; Jurgensen, C. W.; Vitkavage, D. J., Microscopic uniformity in plasma-etching. *J. Vacuum Sci. Technol. B* 1992, 10 (5), 2133–2147; (b) Cao, X. A.; Pearton, S. J.; Zhang, A. P.; Dang, G. T.; Ren, F.; Shul, R. J.; Zhang, L.; Hickman, R.; Van Hove, J. M., Electrical effects of plasma damage in p-GaN. *Appl. Phys. Lett.* 1999, 75 (17), 2569–2571; (c) Gabriel, C. T.; McVittie, J. P., How plasma-etching damages thin gate oxides. *Solid State Technol.* 1992, 35 (6), 81–87.
8. (a) Oehrlein, G. S.; Bestwick, T. D.; Jones, P. L.; Jaso, M. A.; Lindstrom, J. L., Selective dry etching of germanium with respect to silicon and vice-versa. *J. Electrochem. Soc.* 1991, 138 (5), 1443–1452; (b) Paul, D. J.; Law, V. J.; Jones, G. A. C., Si1-xGex pulsed plasma etching using CHF3 and H-2. *J. Vacuum Sci. Technol. B* 1995, 13 (6), 2234–2237.
9. (a) Efremov, A.; Min, N. K.; Jin, S.; Kwon, K. H., Effect of gas mixing ratio on etch behavior of ZrO_2 thin films in Cl^{-2}-based inductively coupled plasmas. *J. Vacuum Sci. Technol. A* 2008, 26 (6), 1480–1486; (b) Sha, L.; Puthenkovilakam, R.; Lin, Y. S.; Chang, J. P., Ion-enhanced chemical etching of HfO_2 for integration in metal-oxide-semiconductor field effect transistors. *J. Vacuum Sci. Technol. B* 2003, 21 (6), 2420–2427.
10. (a) Kim, H. W., Characteristics of Ru etching using ICP and helicon O^{-2}/Cl^{-2} plasmas. *Thin Solid Films* 2005, 475 (1–2), 32–35; (b) Morel, T.; Bamola, S.; Ramos, R.; Beaurain, A.; Pargon, E.; Joubert, O., Tungsten metal gate etching in Cl^{-2}/O^{-2} inductively coupled high density plasmas. *J. Vacuum Sci. Technol. B* 2008, 26 (6), 1875–1882.
11. (a) Noda, S.; Ozaki, T.; Samukawa, S., Damage-free metal-oxide-semiconductor gate electrode patterning on thin HfSiON film using neutral beam etching. *J. Vacuum Sci. Technol. A* 2006, 24 (4), 1414–1420; (b) Martin, R. M.; Blom, H.-O.; Chang, J. P., Plasma etching of Hf-based high-k thin films. Part II. Ion-enhanced surface reaction mechanisms. *J. Vacuum Sci. Technol. A: Vacuum, Surfaces, and Films* 2009, 27 (2), 217.
12. (a) Liu, Z. W.; Durant, S.; Lee, H.; Pikus, Y.; Fang, N.; Xiong, Y.; Sun, C.; Zhang, X., Far-field optical superlens. *Nano Lett.* 2007, 7 (2), 403–408; (b) Brueck, S. R. J., Optical and interferometric lithography—nanotechnology enablers. *Proc. IEEE* 2005, 93 (10), 1704–1721.
13. (a) Tegenkamp, C., Vicinal surfaces for functional nanostructures. *J. Phys.—Condes. Matter* 2009, 21 (1); (b) Jeong, H. -C.; Williams, E. D., Steps on surfaces: Experiment and theory. *Surf. Sci. Reports* 1999, 34 (6–8), 171.
14. Mugarza, A.; Ortega, J. E., Electronic states at vicinal surfaces. *J. Phys.—Condes. Matter* 2003, 15 (47), S3281–S3310.

15. Hahlin, A.; Dunn, J. H.; Karis, O.; Poulopoulos, P.; Nunthel, R.; Lindner, J.; Arvanitis, D., Ultrathin Co films on flat and vicinal Cu(111) surfaces: Per atom determination of orbital and spin moments. *J. Phys.—Condes. Matter* 2003, *15* (5), S573–S586.

16. Zambelli, T.; Trost, J.; Wintterlin, J.; Ertl, G., Diffusion and atomic hopping of N atoms on Ru(0001) studied by scanning tunneling microscopy. *Phys. Rev. Lett.* 1996, *76* (5), 795–798.

17. (a) Spoto, G.; Gribov, E. N.; Ricchiardi, G.; Damin, A.; Scarano, D.; Bordiga, S.; Lamberti, C.; Zecchina, A., Carbon monoxide MgO from dispersed solids to single crystals: A review and new advances. *Prog. Surf. Sci.* 2004, *76* (3–5), 71–146; (b) Izumi, Y.; Shimizu, T.; Kobayashi, T.; Aika, K., Nitrous oxide decomposition active site on Ni-MgO catalysts characterized by X-ray absorption fine structure spectroscopy. *Chem. Commun.* 2000, (12), 1053–1054.

18. Cinquini, F.; Di Valentin, C.; Finazzi, E.; Giordano, L.; Pacchioni, G., Theory of oxides surfaces, interfaces and supported nano-clusters. *Theor. Chem. Acc.* 2007, *117* (5–6), 827–845.

19. Park, K. J.; Doub, J. M.; Gougousi, T.; Parsons, G. N., Microcontact patterning of ruthenium gate electrodes by selective area atomic layer deposition. *Appl. Phys. Lett.* 2005, *86* (5), 051903.

20. Park, M. H.; Jang, Y. J.; Sung-Suh, H. M.; Sung, M. M., Selective atomic layer deposition of titanium oxide on patterned self-assembled monolayers formed by microcontact printing. *Langmuir* 2004, *20* (6), 2257–2260.

21. Liu, J.; Mao, Y.; Lan, E.; Banatao, D. R.; Forse, G. J.; Lu, J.; Blom, H. -O.; Yeates, T. O.; Dunn, B.; Chang, J. P., Generation of oxide nanopatterns by combining self-assembly of S-layer proteins and area-selective atomic layer deposition. *J. Am. Chem. Soc.* 2008, *130* (50), 16908–16913.

22. Jiang, X. R.; Chen, R.; Bent, S. F., Spatial control over atomic layer deposition using microcontact-printed resists. *Surf. Coatings Technol.* 2007, *201* (22–23), 8799–8807.

23. Jiang, X.; Huang, H.; Prinz, F. B.; Bent, S. F., Application of atomic layer deposition of platinum to solid oxide fuel cells. *Chem. Mater.* 2008, *20* (12), 3897–3905.

24. Campina, J. M.; Martins, A.; Silva, F., A new cleaning methodology for efficient Au-SAM removal. *Electrochim. Acta* 2008, *53* (26), 7681–7689.

25. Mullings, M. N.; Lee, H. B. R.; Marchack, N.; Jiang, X. R.; Chen, Z. B.; Gorlin, Y.; Lin, K. P.; Bent, S. F., Area selective atomic layer deposition by microcontact printing with a water-soluble polymer. *J. Electrochem. Soc.* 157 (12), D600–D604.

26. (a) Cheng, J. Y.; Ross, C. A.; Thomas, E. L.; Smith, H. I.; Vancso, G. J., Fabrication of nanostructures with long-range order using block copolymer lithography. *Appl. Phys. Lett.* 2002, *81* (19), 3657-3659; (b) Yang, J. K. W.; Jung, Y. S.; Chang, J. B.; Mickiewicz, R. A.; Alexander-Katz, A.; Ross, C. A.; Berggren, K. K., Complex self-assembled patterns using sparse commensurate templates with locally varying motifs. *Nature Nanotechnol.* 2010, *5* (4), 256–260.

27. Klein, M. J. K.; Guillaumee, M.; Wenger, B.; Dunbar, L. A.; Brugger, J.; Heinzelmann, H.; Pugin, R., Inexpensive and fast wafer-scale fabrication of nanohole arrays in thin gold films for plasmonics. *Nanotechnology* 2010, *21* (20), 205301, 1–7.

28. Tang, C. B.; Lennon, E. M.; Fredrickson, G. H.; Kramer, E. J.; Hawker, C. J., Evolution of block copolymer lithography to highly ordered square arrays. *Science* 2008, *322* (5900), 429–432.

29. Knoll, A. W.; Pires, D.; Coulembier, O.; Dubois, P.; Hedrick, J. L.; Frommer, J.; Duerig, U., Probe-based 3-D nano lithography using self-amplified depolymerization polymers. *Adv. Mater.* 2010, *22* (31), 3361–3365.

30. Jung, K. B.; Cho, H.; Hahn, Y. B.; Lambers, E. S.; Onishi, S.; Johnson, D.; Hurst, A. T.; Childress, J. R.; Park, Y. D.; Pearton, S. J., Relative merits of Cl^{-2} and CO/NH_3 plasma chemistries for dry etching of magnetic random access memory device elements. *J. Appl. Phys.* 1999, *85* (8), 4788–4790.

31. Choong-Ho, L.; Suk-Kang, S.; Donghoon, J.; Sehoon, L.; Seungwook, C.; Jonghyuk, K.; Sejun, P.; Minsung, S.; Hyun-Chul, B.; Eungjin, A.; Jinhyun, S.; Kwangshik, S.; Kyunghoon, M.; Sung-Soon, C.; Chang-Jin, K.; Jungdal, C.; Keonsoo, K.; Jeong-Hyuk, C.; Kang-Deog, S.; Tae-Sung, J. In *A highly manufacturable integration technology for 2 7nm 2 and 3bit/cell NAND flash memory*, Electron Devices Meeting (IEDM), 2010 IEEE International, 6–8 Dec. 2010; p. 5.1.1.

32. Owens, D. K., Mechanism of corona-induced self-adhesion of polyethylene film. *J. Appl. Polym. Sci.* 1975, *19* (1), 265–271.

33. Sun, Y.; Bell, T., Plasma surface engineering of low-alloy steel. *Mater. Sci. Eng. A—Struct. Mater. Prop. Microstruct. Process.* 1991, *140*, 419–434.

34. Yerokhin, A. L.; Voevodin, A. A.; Lyubimov, V. V.; Zabinski, J.; Donley, M., Plasma electrolytic fabrication of oxide ceramic surface layers for tribotechnical purposes on aluminium alloys. *Surf. Coatings Technol.* 1998, *110* (3), 140.

35. Teare, D. O. H.; Spanos, C. G.; Ridley, P.; Kinmond, E. J.; Roucoules, V.; Badyal, J. P. S.; Brewer, S. A.; Coulson, S.; Willis, C., Pulsed plasma deposition of super-hydrophobic nanospheres. *Chem. Mat.* 2002, *14* —(11), 4566–4571.

36. Yerokhin, A. L.; Nie, X.; Leyland, A.; Matthews, A.; Dowey, S. J., Plasma electrolysis for surface engineering. *Surf. Coat. Technol.* 1999, *122* (2–3), 73–93.

37. Kogelschatz, U., Atmospheric-pressure plasma technology. *Plasma Phys. Control. Fusion* 2004, *46*, B63–B75.

38. Chu, P. K.; Chen, J. Y.; Wang, L. P.; Huang, N., Plasma-surface modification of biomaterials. *Mater. Sci. Eng. R-Rep.* 2002, *36* (5–6), 143–206.

39. Kastenmeier, B. E. E.; Matsuo, P. J.; Beulens, J. J.; Oehrlein, G. S., Chemical dry etching of silicon nitride and silicon dioxide using $CF4/O^{-2}/N^{-2}$ gas mixtures. *J. Vac. Sci. Technol. A—Vac. Surf. Films* 1996, *14* (5), 2802–2813.

40. (a) Barone, M. E.; Graves, D. B., Molecular dynamics simulations of plasma-surface chemistry. *Plasma Sources Sci. Technol.* 1996, *5* (2), 187–192; (b) Lymberopoulos, D. P.; Economou, D. J., Fluid simulations of radio-frequency glow-discharges—2-dimensional argon discharge including metastables. *Appl. Phys. Lett.* 1993, *63* (18), 2478–2480; (c) Frenklach, M., Simulation of surface reactions. *Pure Appl. Chem.* 1998, *70* (2), 477–484; (d) Veprek, S., Statistical-model of chemical reactions in nonisothermal low-pressure plasma. *J. Chem. Phys.* 1972, *57* (2), 952–959.

41. Plasma 2010 Committee, N. R. C., Plasma Science: Advancing Knowledge in the National Interest. In *The National Academies Press*, 2010; pp 41–48.

42. Kota, G. P.; Coburn, J. W.; Graves, D. B., The recombination of chlorine atoms at surfaces. *J. Vac. Sci. Technol. A—Vac. Surf. Films* 1998, *16* (1), 270–277.

43. Coburn, J. W.; Winters, H. F., Ion-assisted and electron-assisted gas-surface chemistry—Important effect in plasma-etching. *J. Appl. Phys.* 1979, *50* (5), 3189–3196.

44. Chang, J. P.; Sawin, H. H., Kinetic study of low energy ion-enhanced polysilicon etching using Cl, Cl^{-2}, and Cl$^+$ beam scattering. *J. Vac. Sci. Technol. A-Vac. Surf. Films* 1997, *15* (3), 610–615.

45. (a) Pearton, S. J., Dry-etching techniques and chemistries for III-V semiconductors. *Mater. Sci. Eng.: B* 1991, *10* (3), 187; (b) Wang, J. J.; Lambers, E. S.; Pearton, S. J.; Ostling, M.; Zetterling, C. M.; Grow, J. M.; Ren, F.; Shul, R. J., Inductively coupled plasma etching of bulk 6H-SiC and thin-film SiCN in NF3 chemistries. *J. Vac. Sci. Technol. A—Vac. Surf. Films* 1998, *16* (4), 2204–2209; (c) Eddy, C. R.; Leonhardt, D.; Shamamian, V. A.; Meyer, J. R.; Hoffman, C. A.; Butler, J. E., Characterization of the $CH_4/H^{-2}/Ar$ high density plasma etching process for HgCdTe. *J. Electron. Mater.* 1999, *28* (4), 347–354; (d) Pearton, S. J., High ion density dry etching of compound semiconductors. *Mater. Sci. Eng. B—Solid State Mater. Adv. Technol.* 1996, *40* (2–3), 101–118; (e) Yonts, O. C.; Harrison, D. E., Surface cleaning by cathode sputtering. *J. Appl. Phys.* 1960, *31* (9), 1583–1584.

46. (a) Mogab, C. J.; Adams, A. C.; Flamm, D. L., Plasma etching of Si and SiO_2—Effect of oxygen additions to Cf_4 Plasmas. *J. Appl. Phys.* 1978, *49* (7), 3796–3803; (b) Dagostino, R.; Cramarossa, F.; Debenedictis, S.; Ferraro, G., Spectroscopic diagnostics of Cf_4-O_2 plasmas during Si and SiO_2 etching processes. *J. Appl. Phys.* 1981, *52* (3), 1259–1265; (c) Steinbruchel, C., Langmuir probe measurements on Chf_3 and Cf_4 plasmas—The role of ions in the reactive sputter etching of SiO_2 and Si. *J. Electrochem. Soc.* 1983, *130* (3), 648–655.

47. Oehrlein, G. S.; Williams, H. L., Silicon etching mechanisms in a Cf_4/H_2 glow-discharge. *J. Appl. Phys.* 1987, *62* (2), 662–672.

48. Sha, L.; Cho, B. O.; Chang, J. P., Ion-enhanced chemical etching of ZrO_2 in a chlorine discharge. *J. Vac. Sci. Technol. A—Vac. Surf. Films* 2002, *20* (5), 1525–1531.

49. Donnelly, V. M.; Flamm, D. L.; Dautremontsmith, W. C.; Werder, D. J., Anisotropic etching of SiO_2 In Low-Frequency Cf_4/O_2 and Nf_3/Ar plasmas. *J. Appl. Phys.* 1984, *55* (1), 242–252.

50. Oehrlein, G. S.; Zhang, Y.; Vender, D.; Joubert, O., Fluorocarbon high-density plasmas. 2. Silicon dioxide and silicon etching using Cf_4 and Chf_3. *J. Vac. Sci. Technol. A—Vac. Surf. Films* 1994, *12* (2), 333–344.

51. (a) Titus, M. J.; Nest, D. G.; Chung, T. Y.; Graves, D. B., Comparing 193 nm photoresist roughening in an inductively coupled plasma system and vacuum beam system. *J. Physics D—Appl. Phys.* 2009, *42* (24), 245205, 1–13; (b) Titus, M. J.; Nest, D. G.; Graves, D. B., Modelling vacuum ultraviolet photon penetration depth and C = O bond depletion in 193 nm photoresist. *J. Phys. D—Appl. Phys.* 2009, *42* (15), 152001, 1–4; (c) Titus, M. J.; Nest, D.; Graves, D. B., Absolute vacuum ultraviolet flux in inductively coupled plasmas and chemical modifications of 193 nm photoresist. *Appl. Phys. Lett.* 2009, *94* (17), 171501, 1–3.

52. Rau, N.; Stratton, F.; Fields, C.; Ogawa, T.; Neureuther, A.; Kubena, R.; Willson, G., Shot-noise and edge roughness effects in resists patterned at 10 nm exposure. *J. Vac. Sci. Technol. B* 1998, *16* (6), 3784–3788.

53. Shimizu, D.; Maruyama, K.; Saitou, A.; Kai, T.; Shimokawa, T.; Fujiwara, K.; Kikuchi, Y.; Nishiyama, I., Progress in EUV resist development. *J. Photopolym Sci. Technol.* 2007, *20* (3), 423–428.

54. Shenton, M. J.; Stevens, G. C., Surface modification of polymer surfaces: Atmospheric plasma versus vacuum plasma treatments. *J. Phys. D—Appl. Phys.* 2001, *34* (18), 2761–2768.

55. Meyer-Plath, A. A.; Schroder, K.; Finke, B.; Ohl, A., Current trends in biomaterial surface functionalization—Nitrogen-containing plasma assisted processes with enhanced selectivity. *Vacuum* 2003, *71* (3), 391–406.

56. Felten, A.; Bittencourt, C.; Pireaux, J. J.; Van Lier, G.; Charlier, J. C., Radio-frequency plasma functionalization of carbon nanotubes surface O^{-2}, NH_3, and CF_4 treatments. *J. Appl. Phys.* 2005, *98* (7).

57. Kang, E. T.; Tan, K. L.; Kato, K.; Uyama, Y.; Ikada, Y., Surface modification and functionalization of polytetrafluoroethylene films. *Macromolecules* 1996, *29* (21), 6872–6879.

58. Tseng, C. H.; Wang, C. C.; Chen, C. Y., Functionalizing carbon nanotubes by plasma modification for the preparation of covalent-integrated epoxy composites. *Chem. Mat.* 2007, *19* (2), 308–315.

59. (a) Hsu, C. C.; Coburn, J. W.; Graves, D. B., Etching of ruthenium coatings in O_{-2}- and Cl_{-2}-containing plasmas. *J. Vacuum Sci. Technol. A* 2006, *24* (1), 1–8; (b) Hsu, C. C.; Titus, M. J.; Graves, D. B., Measurement and modeling of time- and spatial-resolved wafer surface temperature in inductively coupled plasmas. *J. Vacuum Sci. Technol. A* 2007, *25* (3), 607–614.

60. (a) Hsu, C. C.; Hoang, J.; Le, V.; Chang, J. P., Feature profile evolution during shallow trench isolation etch in chlorine-based plasmas. II. Coupling reactor and feature scale models. *J. Vacuum Sci. Technol. B* 2008, *26* (6), 1919–1925; (b) Hoang, J.; Hsu, C. C.; Chang, J. P., Feature profile evolution during shallow trench isolation etch in chlorine-based plasmas. I. Feature scale modeling. *J. Vacuum Sci. Technol. B* 2008, *26* (6), 1911–1918.

61. (a) Greer, F.; Coburn, J. W.; Graves, D. B., Vacuum beam studies of photoresist etching kinetics. *J. Vac. Sci. Technol. A—Vac. Surf. Films* 2000, *18* (5), 2288–2294; (b) Levinson, J. A.; Shaqfeh, E. S. G.; Balooch, M.; Hamza, A. V., Ion-assisted etching and profile development of silicon in molecular and atomic chlorine. *J. Vacuum Sci. Technol. B* 2000, *18* (1), 172–190.

62. (a) Kwon, O.; Sawin, H. H., Surface kinetics modeling of silicon and silicon oxide plasma etching. II. Plasma etching surface kinetics modeling using translating mixed-layer representation. *J. Vacuum Sci. Technol. A* 2006, *24* (5), 1914–1919; (b) Kwon, O.; Bai, B.; Sawin, H. H., Surface kinetics modeling of silicon and silicon oxide plasma etching. III. Modeling of silicon oxide etching in fluorocarbon chemistry using translating mixed-layer representation. *J. Vacuum Sci. Technol. A* 2006, *24* (5), 1920–1927.

63. (a) Helmer, B. A.; Graves, D. B., Molecular dynamics simulations of Ar^+ and Cl^+ impacts onto silicon surfaces: Distributions of reflected energies and angles. *J. Vac. Sci. Technol. A—Vac. Surf. Films* 1998, *16* (6), 3502–3514; (b) Rauf, S.; Sparks, T.; Ventzek, P. L. G.; Smirnov, V. V.; Stengach, A. V.; Gaynullin, K. G.; Pavlovsky, V. A., A molecular dynamics investigation of fluorocarbon based layer-by-layer etching of silicon and SiO_2. *J. Appl. Phys.* 2007, *101* (3); (c) Barone, M. E.; Robinson, T. O.; Graves, D. B., Molecular dynamics simulations of direct reactive ion etching: Surface roughening of silicon by chlorine. *IEEE Trans. Plasma Sci.* 1996, *24* (1), 77–78.

64. Zhang, D.; Kushner, M. J., Surface kinetics and plasma equipment model for Si etching by fluorocarbon plasmas. *J. Appl. Phys.* 2000, *87* (3), 1060–1069.

65. Martin, R. M.; Chang, J. P., Plasma etching of Hf-based high-k thin films. Part I. Effect of complex ions and radicals on the surface reactions. *J. Vacuum Sci. Technol. A: Vacuum, Surf., Films* 2009, *27* (2), 209.

66. Layadi, N.; Donnelly, V. M.; Lee, J. T. C., Cl^{-2} plasma etching of Si(100): Nature of the chlorinated surface layer studied by angle-resolved x-ray photoelectron spectroscopy. *J. Appl. Phys.* 1997, *81* (10), 6738–6748.

67. Simpson, W. C.; Yarmoff, J. A., Fundamental studies of halogen reactions with III-V semiconductor surfaces. *Annu. Rev. Phys. Chem.* 1996, *47*, 527–554.

68. Herman, I. P.; Donnelly, V. M.; Guinn, K. V.; Cheng, C. C., Laser-induced thermal-desorption as an in-situ surface probe during plasma processing. *Phys. Rev. Lett.* 1994, *72* (17), 2801–2804.

69. Martin, R. M.; Chang, J. P., Plasma etching of Hf-based high-k thin films. Part III. Modeling the reaction mechanisms. *J. Vacuum Sci. Technol. A: Vacuum, Surf., Films* 2009, *27* (2), 224.

70. Marchack, N. P.; Pham, C.; Hoang, J.; Chang, J. P. In *Predicting the Surface Response upon Simultaneous Plasma Etching and Depostion*, American Vacuum Society 57th International Symposium Albuquerque, New Mexico, 2010.

71. (a) Ono, K.; Ohta, H.; Eriguchi, K., Plasma-surface interactions for advanced plasma etching processes in nanoscale ULSI device fabrication: A numerical and experimental study. *Thin Solid Films* 2010, *518* (13), 3461-3468; (b) Shoeb, J.; Kushner, M. J., Mechanisms for plasma etching of HfO_2 gate stacks with Si selectivity and photoresist trimming. *J. Vac. Sci. Technol. A* 2009, *27* (6), 1289–1302.

72. van Dijk, J.; Kroesen, G. M. W.; Bogaerts, A., Plasma modelling and numerical simulation. *J. Phys. D—Appl. Phys.* 2009, *42* (19), 19301, 1–14.

73. Abrams, C. F.; Graves, D. B., Molecular dynamics simulations of Si etching by energetic CF^{3+}. *J. Appl. Phys.* 1999, *86* (11), 5938–5948.

74. (a) Harrison, D. E.; Levy, N. S.; Johnson, J. P.; Effron, H. M., Computer simulation of sputtering. *J. Appl. Phys.* 1968, *39* (8), 3742–3761; (b) Harrison, D. E., Application of molecular-dynamics simulations to the study of ion-bombarded metal-surfaces. *CRC Crit. Rev. Solid State Mater. Sci.* 1988, *14*, S1–S78.

75. Vegh, J. J.; Graves, D. B., Molecular dynamics simulations of sub-10 nm wavelength surface rippling by CF^{3+} ion beams. *Plasma Sources Sci. Technol.* 2010, *19* (4).

76. Bogaerts, A.; De Bie, C.; Eckert, M.; Georgieva, V.; Martens, T.; Neyts, E.; Tinck, S., Modeling of the plasma chemistry and plasma-surface interactions in reactive plasmas. *Pure Appl. Chem.* 2010, *82* (6), 1283.

77. Barone, M. E.; Graves, D. B., Chemical and physical sputtering of fluorinated silicon. *J. Appl. Phys.* 1995, *77* (3), 1263–1274.

78. (a) Voter, A. F., A method for accelerating the molecular dynamics simulation of infrequent events. *J. Chem. Phys.* 1997, *106* (11), 4665–4677; (b) Voter, A. F.; Montalenti, F.; Germann, T. C., Extending the time scale in atomistic simulation of materials. *Ann. Rev. Mater. Res.* 2002, *32*, 321–346.

79. Wu, S. Y.; Raghunath, P.; Wu, J. S.; Lin, M. C., Ab initio chemical kinetic study for reactions of H atoms with SiH_4 and Si_2H_6: Comparison of theory and experiment. *J. Phys. Chem. A* 2009, *114* (1), 633.

80. Graves, D. B.; Brault, P., Molecular dynamics for low temperature plasma-surface interaction studies. *J. Phys. D—Appl. Phys.* 2009, *42* (19), 194011, 1–27.

81. Sriraman, S.; Agarwal, S.; Aydil, E. S.; Maroudas, D., Mechanism of hydrogen-induced crystallization of amorphous silicon. *Nature* 2002, *418* (6893), 62–65.

82. Radeke, M. R.; Carter, E. A., Ab initio dynamics of surface chemistry. *Annu. Rev. Phys. Chem.* 1997, *48*, 243–270.

83. Huang, P.; Carter, E. A., Advances in correlated electronic structure methods for solids, surfaces, and nanostructures. *Annu. Rev. Phys. Chem.* 2008, *59*, 261–290.

84. Helmer, B. A.; Graves, D. B., Molecular dynamics simulations of fluorosilyl species impacting fluorinated silicon surfaces with energies from 0.1 to 100 eV. *J. Vac. Sci. Technol. A—Vac. Surf. Films* 1997, *15* (4), 2252–2261.

85. Samukawa, S., Ultimate top-down etching processes for future nanoscale devices: Advanced neutral-beam etching. *Jpn. J. Appl. Phys. Part 1—Regular Papers Brief Commun. Rev. Papers* 2006, *45* (4A), 2395–2407.

86. Athavale, S. D.; Economou, D. J. In *Molecular dynamics simulation of atomic layer etching of silicon*, Proceedings of the 41st National Symposium of the American Vacuum Society, Denver, CO, AVS: Denver, CO, 1995; p. 966.

87. Iwakawa, A.; Nagaoka, T.; Ohta, H.; Eriguchi, K.; Ono, K., Molecular dynamics simulation of Si etching by off-normal Cl^+ bombardment at high neutral-to-ion flux ratios. *Jpn. J. Appl. Phys.* 2008, *47* (11), 8560–8564.

88. Eriguchi, K.; Matsuda, A.; Nakakubo, Y.; Kamei, M.; Ohta, H.; Ono, K., Effects of plasma-induced Si recess structure on n-MOSFET performance degradation. *IEEE Electron Device Lett.* 2009, *30* (7), 712–714.

89. Xu, L.; Economou, D. J.; Donnelly, V. M.; Ruchhoeft, P., Extraction of a nearly monoenergetic ion beam using a pulsed plasma. *Appl. Phys. Lett.* 2005, *87* (4), 041502, 1–3.

90. Athavale, S. D.; Economou, D. J., Realization of atomic layer etching of silicon. *J. Vacuum Sci. Technol. B* 1996, *14* (6), 3702–3705.

91. Sugiyama, T.; Matsuura, T.; Murota, J., Atomic-layer etching of Ge using an ultraclean ECR plasma. *Appl. Surf. Sci.* 1997, *112*, 187–190.

92. Lim, W. S.; Park, S. D.; Park, B. J.; Yeom, G. Y., Atomic layer etching of (100)/(111) GaAs with chlorine and low angle forward reflected Ne neutral beam. *Surf. Coat. Technol.* 2008, *202* (22–23), 5701–5704.

93. Agarwal, A.; Kushner, M. J., Plasma atomic layer etching using conventional plasma equipment. *J. Vacuum Sci. Technol. A* 2009, *27* (1), 37–50.

94. (a) Kim, S. H.; Hiroshima, H.; Komuro, M., Photo-nanoimprint lithography combined with thermal treatment to improve resist pattern line-edge roughness. *Nanotechnology* 2006, *17* (9), 2219–2222; (b) Padmanaban, M.; Rentkiewicz, D.; Hong, C.; Lee, D.; Rahman, D.; Sakamuri, R.; Dammel, R. R., Possible origins and some methods to minimize LER. *J. Photopolym Sci. Technol.* 2005, *18* (4), 451–456.

95. (a) Oldham, W. G.; Nandgaonkar, S. N.; Neureuther, A. R.; Otoole, M., General simulator for VLSI lithography and etching processes. 1. Application to projection lithography. *IEEE Trans. Electron Dev.* 1979, *26* (4), 717–722; (b) Oldham, W. G.; Neureuther, A. R.; Sung, C.; Reynolds, J. L.; Nandgaonkar, S. N., A General simulator for VLSI lithography and etching processes. 2. Application to deposition and etching. *IEEE Trans. Electron Dev.* 1980, *27* (8), 1455–1459; (c) Wong, A. K.; Neureuther, A. R., Rigorous 3-dimensional time-domain finite-difference electromagnetic simulation for photolithographic applications. *IEEE Trans. Semiconductor Manufact.* 1995, *8* (4), 419–431.

96. (a) Deng, Y. F.; Neureuther, A. R., Electromagnetic characterization of nanoimprint mold inspection. *J. Vac. Sci. Technol. B* 2003, *21* (1), 130–134; (b) Diaz, C. H.; Tao, H. J.; Ku, Y. C.; Yen, A.; Young, K., An experimentally validated analytical model for gate line-edge roughness (LER) effects on technology scaling. *IEEE Electron Device Lett.* 2001, *22* (6), 287–289.

97. Neureuther, A., In *5th Bi-Annual Workshop: Measuring, Understand and Controlling Variability in Deep Sub-Micron Patterning*, Integrated Modeling, Process and Computation for Technology, San Jose, CA, 2010.

98. (a) Gupta, P.; Kahng, A. B.; Park, C. H., Detailed placement for enhanced control of resist and etch CDs. *IEEE Trans. Computer-Aided Design Integrated Circuits Syst.* 2007, *26* (12), 2144–2157; (b) Gupta, P.; Kahng, A. B.; Park, C. H.; Samadi, K.; Xu, X., Wafer topography-aware optical proximity correction. *IEEE Trans. Computer-Aided Design Integrated Circuits Syst.* 2006, *25* (12), 2747–2756.

99. Byers, J.; Lee, S.; Jeri, K.; Zimmerman, P.; Turr, N. J.; Willson, C. G., Double exposure materials: Simulation study of feasibility. *J. Photopolym. Sci. Technol.* 2007, *20* (5), 707–717.

100. Ronse, K.; Jansen, P.; Gronheid, R.; Hendrickx, E.; Maenhoudt, M.; Wiaux, V.; Goethals, A. M.; Jonckheere, R.; Vandenberghe, G., Lithography options for the 32 nm half pitch node and beyond. *IEEE Trans. Circuits Syst. I—Regular Papers* 2009, *56* (8), 1884–1891.

101. (a) Guo, L. J., Nanoimprint lithography: Methods and material requirements. *Adv. Mater.* 2007, *19* (4), 495–513; (b) Harrer, S.; Yang, J. K. W.; Salvatore, G. A.; Berggren, K. K.; Ilievski, F.; Ross, C. A., Pattern generation by using multistep room-temperature nanoimprint lithography. *IEEE Trans. Nanotechnol.* 2007, *6* (6), 639–644.

102. Boto, A. N.; Kok, P.; Abrams, D. S.; Braunstein, S. L.; Williams, C. P.; Dowling, J. P., Quantum interferometric optical lithography: Exploiting entanglement to beat the diffraction limit. *Phys. Rev. Lett.* 2000, *85* (13), 2733–2736.

103. Luo, X. G.; Ishihara, T., Surface plasmon resonant interference nanolithography technique. *Appl. Phys. Lett.* 2004, *84* (23), 4780–4782.

104. Zhang, X.; Liu, Z. W., Superlenses to overcome the diffraction limit. *Nat. Mater.* 2008, *7* (6), 435–441.

105. Vourdas, N.; et al., Plasma directed assembly and organization: bottom-up nanopatterning using top-down technology. *Nanotechnology* 2010, *21* (8), 085302.

106. Karthikeyan, J.; Berndt, C. C.; Tikkanen, J.; Reddy, S.; Herman, H., Plasma spray synthesis of nanomaterial powders and deposits. *Mater. Sci. Eng. A-Struct. Mater. Prop. Microstruct. Process.* 1997, *238* (2), 275–286.

107. Brown, I. G.; Anders, A.; Anders, S.; Dickinson, M. R.; Ivanov, I. C.; Macgill, R. A.; Yao, X. Y.; Yu, K. M., Plasma synthesis of metallic and composite thin-films with atomically mixed substrate bonding. *Nucl. Instrum. Methods Phys. Res. Sect. B—Beam Interact. Mater. Atoms* 1993, *80-1*, 1281–1287.

108. Boulos, M. I., Thermal plasma processing. *IEEE Trans. Plasma Sci.* 1991, *19* (6), 1078–1089.

109. Van, T. T.; Hoang, J.; Ostroumov, R.; Wang, K. L.; Bargar, J. R.; Lu, J.; Blom, H. O.; Chang, J. P., Nanostructure and temperature-dependent photoluminescence of Er-doped Y2O3 thin films for micro-optoelectronic integrated circuits. *J. Appl. Phys.* 2006, *100* (7).

110. Tu Van, T.; Chang, J. P., Controlled erbium incorporation and photolumines-
     cence of Er-doped Y_2O_3. *Appl. Phys. Lett.* 2005, *87* (1), 011907.
111. Mukhopadhyay, A. B.; Musgrave, C. B.; Sanz, J. F., Atomic layer deposition of
     hafnium oxide from hafnium chloride and water. *J. Am. Chem. Soc.* 2008, *130*
     (36), 11996–12006.
112. Xu, S. Y.; Long, J. D.; Sim, L. N.; Diong, C. H.; Ostrikov, K., RF plasma sputtering
     deposition of hydroxyapatite bioceramics: Synthesis, performance, and biocom-
     patibility. *Plasma Process. Polym.* 2005, *2* (5), 373–390.
113. Chiang, W.-H.; Sankaran, R. M., Microplasma synthesis of metal nanoparticles
     for gas-phase studies of catalyzed carbon nanotube growth. *Appl. Phys. Lett.*
     2007, *91* (12), 121503.
114. (a) Fridman, G.; Peddinghaus, M.; Balasubramanian, M.; Ayan, H.; Fridman,
     A.; Gutsol, A.; Brooks, A., Blood coagulation and living tissue sterilization
     by floating-electrode dielectric barrier discharge in air. *Plasma Chem. Plasma
     Process.* 2006, *26* (4), 425; (b) Sensenig, R.; Kalghatgi, S.; Cerchar, E.; Fridman, G.;
     Shereshevsky, A.; Torabi, B.; Arjunan, K.; Podolsky, E.; Fridman, A.; Friedman,
     G.; Azizkhan-Clifford, J.; Brooks, A., Non-thermal plasma induces apoptosis in
     melanoma cells via production of intracellular reactive oxygen species. *Ann.
     Biomed. Eng.* 2010, *39* (2), 674; (c) Dobrynin, D.; Fridman, G.; Mukhin, Y. V.;
     Wynosky-Dolfi, M. A.; Rieger, J.; Rest, R. F.; Gutsol, A. F.; Fridman, A., Cold
     Plasma Inactivation of *Bacillus cereus* and *Bacillus anthracis* (Anthrax) spores.
     Plasma Sci., IEEE Trans. 38 (8), 1878.
115. (a) Handy, R.; Owen, R.; Valsami-Jones, E., The ecotoxicology of nanoparticles
     and nanomaterials: Current status, knowledge gaps, challenges, and future
     needs. *Ecotoxicology* 2008, *17* (5), 315; (b) Buzea, C.; Pacheco, I. I.; Robbie, K.,
     Nanomaterials and nanoparticles: Sources and toxicity. *Biointerphases* 2007, *2* (4),
     MR17; (c) Powers, K. W.; Brown, S. C.; Krishna, V. B.; Wasdo, S. C.; Moudgil,
     B. M.; Roberts, S. M., Research strategies for safety evaluation of nanomateri-
     als. Part VI. Characterization of nanoscale particles for toxicological evaluation.
     Toxicol. Sci. 2006, *90* (2), 296–303.
116. (a) Heberlein, J.; Murphy, A. B., Thermal plasma waste treatment. *J. Phys. D—Appl.
     Phys.* 2008, *41* (5); (b) Nishioka, H.; Saito, H.; Watanabe, T., Decomposition mech-
     anism of organic compounds by DC water plasmas at atmospheric pressure.
     Thin Solid Films 2009, *518* (3), 924–928; (c) Nezu, A.; Morishima, T.; Watanabe,
     T., Thermal plasma treatment of waste ion-exchange resins doped with metals.
     Thin Solid Films 2003, *435* (1–2), 335–339.
117. Ravi, L.; Girshick, S. L., Coagulation of nanoparticles in a plasma. *Phys. Rev. E*
     2009, *79* (2), 1–9.
118. Taubner, T.; Korobkin, D.; Urzhumov, Y.; Shvets, G.; Hillenbrand, R., Near-field
     microscopy through a SiC superlens. *Science* 2006, *313* (5793), 1595–1595.

2

Extreme Ultraviolet Light Lithography for Producing Nanofeatures in Next-Generation Semiconductor Processing

John Sporre
David N. Ruzic

CONTENTS

2.1 Introduction

The invention of the integrated circuit in the late 1950s eventually led to the development of the computer chip and laid the foundation for the semiconductor industry as we know it today. This industry, largely driven by technological advancements and demand, has resulted in the miniaturization of integrated circuit components, a progression forecasted by Gordon Moore in 1965. Moore stated, in an effort to emphasize the future role of integrated circuits in the world, that the number of transistors on a die would need to double approximately every 2 years [1]. To meet this goal, the size of a transistor has been reduced, in accordance with Moore's prescient prediction, as shown in Figure 2.1. This feat has largely been accomplished by a combination of plasma processing and optical lithography. Here, focus on optical lithography.

The physical limit of shrinking a transistor is approximately 1.5 nm—beyond this point, quantum tunneling becomes important and adversely impacts the functional feasibility of a computer chip. Before this limit is reached, however, there are several hurdles that need to be overcome. One such issue is the fact that at 1.5 nm, roughly 3.5×10^6 W/cm^2 of thermal removal is required. This is an insurmountable limit considering the goal for 2016 is to achieve 93 W/cm^2 [3]. Nevertheless, before even these hurdles present a problem, many challenges remain in the realm of optical lithography.

Although solutions currently exist to go beyond 32 nm on the International Technology Roadmap for Semiconductors, it is clear that there is a need for a more cost-effective, long-term solution that can be extended to create the sub-32 nm features. Currently, excimer lasers coupled with double patterning and immersion lithography is one possibility, but the cost effectiveness of this technique makes it evident that smaller wavelength lithography sources are required. One of the possible next-generation sources that will enable reduction in wavelength is extreme ultraviolet (EUV) light. With its origins beginning in the 1980s, EUV lithography has been widely researched and presents a possible solution to the problems that exist for expanding to sub-10 nm features [4].

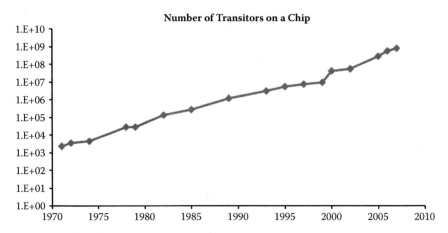

FIGURE 2.1
The number of transistors on a die has doubled every 18 to 24 months. This trend will not continue forever, but until the physical limits imposed by silicon are reached, efforts are being made in finding technology to extend this trend into the future. (Data obtained from *60 Years of the Transistor: 1947–2007.* (cited 2010 February 22); Available from: http://www.intel.com/technology/timeline.pdf.)

Unfortunately, simply implementing a new wavelength of radiation is not as simple as replacing a light source. The large reduction of wavelength from the currently used deep ultraviolet (DUV) to EUV region requires the development of new reflective optics, reflective masks, as well as new photoresists, to name only a few current obstacles. One issue of great importance to the application of EUV is the fact that a simple transparent pellicle cannot be used to inhibit the transportation of debris (anything that is not in band EUV radiation) created by the dense hot plasmas used to create EUV light. As such, in order to provide an EUV source that does not have an extravagant cost of ownership, great care needs to be taken to mitigate the harmful effects of energetic debris. In this chapter, the use of EUV lithography is discussed as a possible tool for the next generation of creating nanoscale features in the production of computer chips.

2.2 Optical Lithography

2.2.1 Current State of Lithography

Lithography is the process by which a designed pattern is transferred onto a wafer using radiation and a radiation-sensitive resist that changes its chemical properties upon exposure to the radiation. Optical lithography has been the driving force behind the technological advancements within the computer chip industry since the creation of the integrated circuit in 1958 [5]. Many varieties of lithography exist, but the most compelling method for high volume manufacturing of computer chips is optical lithography. EUV lithography, a small subset of optical lithography, will be primarily highlighted as a means to achieve nano-features in the future of computer chip manufacturing, but in order to provide an understanding of EUV lithography, traditional lithographic approaches will first be examined. The conventional optical lithography system with nonreflective optics is shown in Figure 2.2. A typical system contains five different components: the source of radiation, the condenser optics, a mask, the projection optics, and a substrate [6]. The radiation source can be any type of light-emitting source including, but not limited to, high-pressure mercury arc lamps, excimer lasers, and even tabletop X-ray sources [7–9]. In order to achieve smaller feature sizes that can be printed on a wafer, the radiation source wavelength has been progressively reduced from 436 nm (g-line) to 365 nm (i-line), and eventually down to 248 nm (KrF excimer lasers) and 193 nm (ArF excimer lasers) [5,10]. After the required photons are created, the condenser optics are used to create a plane wave that is either passed through or diffracted off of the mask. The mask is a critical component of the optical lithography chain, as it is the component that contains the information that is to be printed onto the wafer.

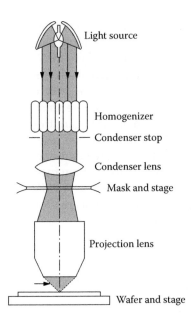

FIGURE 2.2
A generic optical lithography system using lenses, as opposed to reflective optics. The system consists of a homogenizer, which creates a plane wave, a condenser lens, which focuses the photons onto the mask, a mask that contains the information to be printed, and projection lenses that reduce the size of the printed image onto the wafer. (Reprinted with permission from Levenson, M.D., Wavefront engineering for photolithography. *Physics Today*, 1993. 46(7): 28. Copyright 1993, American Institute of Physics.)

Because of the high absorption of EUV photons in any material, the mask for such a system must be made reflective; areas on the mask that are not to be printed are coated in an absorbing material that will not reflect the photons. The projection lens then focuses the photons, containing the information to be printed, onto the wafer with a reduction in size of a factor of four to five times [11].

Wafers are exposed to the projected image using either a stepper or a scanner. A stepper operates by illuminating the entire mask and projecting the whole image with 4 to 5× reduction onto the wafer. After exposure, the wafer is stepped to the next location to be printed, and the process is repeated. A scanner only illuminates a small portion of the mask. The wafer and the mask are moved simultaneously in opposite directions in order to print the entire image. The wafer is then stepped to the next location and the process is repeated. In both cases, the light that is projected onto the wafer causes a chemical change in the photoresist where it has been exposed. Two types of photoresists exist: positive and negative. A positive resist will harden where it has been exposed, and a negative resist will become weaker. The weaker areas of the photoresist are then chemically washed off of the wafer, leaving a pattern of resist that mimics the design on the mask. The remaining material serves as a buffer for the etching and deposition steps in the creation of a computer chip. It is the fundamental size of the features printed onto the resist using the optical lithography process that determine the critical features of a computer chip. Ultimately the ability to create nano-features on a computer chip, for a given lithography tool, is determined by a trade-off between resolution and depth of focus.

2.2.2 Resolution and Depth of Focus

Resolution and depth of focus are the confining terms of an optical lithography system that account for the trade-off between the minimum feature size that is printable and the quality of the image being printed on the photoresist. Resolution, logically, defines the smallest feature size that can be printed onto the wafer. Mathematically this term is defined as shown in Equation (2.1), where R is the smallest resolvable half-pitch feature, k_1 is a constant that is defined by the optical train, λ is the wavelength of light being used, and NA is the numerical aperture of the lens system within the optical lithography tool [12]. Numerical aperture is defined in Equation (2.2), where n is the refractive index of the medium between the final lens and the wafer, and θ is the half-angle formed by the rays that are incident on the wafer.

$$R = k_1 \frac{\lambda}{NA} \qquad (2.1)$$

$$NA = n\sin(\theta) \qquad (2.2)$$

By increasing numerical aperture, reducing the wavelength of the radiation, or reducing k_1, printed feature sizes can be reduced. Changing numerical aperture requires changing the optics used for focusing the image onto the surface, changing the wavelength requires an alteration in the optical radiation utilized, and changing k_1 requires improvements in photoresist materials. The numerical aperture of lithography systems can range anywhere from 0.3 to 1.35, while k_1 is typically larger than 0.25 [13,14]. Although it seems that it would be simple to reduce the feature sizes being printed by simply increasing the numerical aperture of the system, an issue arises in the confinement of depth of focus. The depth of focus (DOF) is defined in Equation (2.3), with k_2 being another system-based constant, and NA and λ being similarly defined as in Equation (2.2).

$$DOF = k_2 \frac{\lambda}{NA^2} \qquad (2.3)$$

Depth of focus is defined as the distance over which a projected image remains in focus. When considering the need to print highly anisotropic features, a low depth of focus will result in the bottom of the photoresist not being exposed the same as the top. This results in a lack of the ideal anisotropic exposure that is required for nanometer features. Typically, the numerical aperture of a system is increased before retooling an entire lithography line with a new source of radiation due to cost effectiveness. Increasing numerical aperture, however, can only continue for so long before it is more cost effective to decrease the wavelength being used.

2.3 Extreme Ultraviolet Light Lithography

2.3.1 The Drive for Smaller Wavelength Lithography Sources

As it stands in early 2011, industry is utilizing extensions of excimer laser technology to achieve the 32 nm node. Excimer lasers, which were first developed in 1975, create 193 nm light that is then reduced through modifications to numerical aperture and printing techniques [15]. Immersion lithography, double patterning, phase shift masks, and off-axis illumination are each techniques being utilized to create sub-193 nm features. Immersion lithography, which places a thin layer of water between the last lens and the wafer, increases the numerical aperture. Unfortunately, in order to reach the 32 nm node by 2011, as suggested by the International Technology Roadmap for Semiconductors, materials with a higher index of refraction than water, are required to increase the NA up to 1.6. It has been shown that this technique would not be performed rapidly enough to meet the node in time, and as such the current infrastructure within industry has to be abandoned [16]. The second technique, double patterning, utilizes multiple exposure processes that are stitched together in increments smaller than the lowest resolution that is printable, in effect lowering the k_1 term in resolution. One advantage of this process is that there does not need to be a great deal of infrastructure investment. Furthermore, there is no detrimental effect on the depth of focus [16–18]. Phase shift masks employ changes in mask thickness to change the phase of light from certain parts of the mask. This phase-shifted light coupled with the unaltered light serves to reduce the printable feature size. Its applicability as a mainstream technique is limited, however, by the complexity of the mask designs [19–21]. Last, off-axis illumination collects the zeroth and first-order light, which in effect reduces the wavelength of light being transmitted to the surface because of the reduction in size between the outer orders. This technique also improves depth of focus [22,23]. Despite the ability to extend the current use of excimer lasers, it is inevitable that a new form of radiation be required for further continuation of Moore's law. As such, EUV lithography is currently being eyed for high-volume manufacturing at the 22 nm node and beyond. A simple economic analysis, as shown in Figure 2.3 reveals that EUV simply becomes more cost effective than continuing with old technology [24].

EUV lithography utilizes 13.5 nm wavelength light to pattern features on a wafer, roughly an order of magnitude smaller than the 193 nm excimer laser i-line processes. Switching to the lower wavelength improves the resolution of the lithography too, while not sacrificing the depth of focus that is associated with changing the numerical aperture of a system. Simply switching to a new wavelength is not as easily performed as stated, for one issue with EUV light is that the photoelectric effect causes it to be readily absorbed into almost all material [25]. As such, EUV lithography needs to be performed solely under vacuum and only reflective optics can be used because EUV light is absorbed in the material of lenses. Another issue of concern is the photoresist. Photoresists are typically designed to be only reactive to the designated wavelength of choice and, therefore, current resists cannot be used in continuation with the change in technology. One of the most critical problems with switching to EUV lithography, however, is the fact that the plasma used to create the 13.5 nm emission line generates energetic debris. This

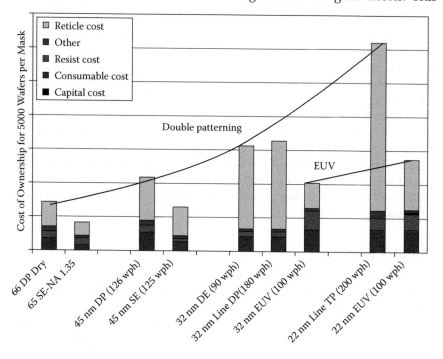

FIGURE 2.3

The cost of continuing double patterning with currently used technology becomes prohibitive at lower feature sizes. It is this economic strain that makes extreme ultraviolet (EUV) lithography likely the successor of excimer laser lithography. (Figure taken from Semiconductor Industry Association. The International Technology Roadmap for Semiconductors, 2009 Edition. International SEMATECH: Austin, TX, 2009. With permission.)

debris, if not mitigated, can cause degradation of the collector optics and dramatically increase the cost of ownership of the EUV tool. Traditionally these collector optics have been protected using a transparent pellicle placed between the light source and the optics, but because these pellicles absorb EUV light, they are no longer applicable and new methods of debris mitigation are required.

2.3.2 Extreme Ultraviolet (EUV) Emitting Plasma

In order to create EUV light, a considerable amount of energy is required to excite the fuel atoms to an appropriate energy state. There have been three different fuels actively pursued as possible EUV-emitting fuels: Li, Xe, and Sn. Of these three, Sn has proven to be the most likely candidate, as it has the highest conversion efficiency. In order to produce EUV light using Sn, the atom must be ionized to the 8+ to 12+ ionizations states [26]. The energy required to achieve these ionization states also has the effect of creating highly energetic ions and neutral atoms that pose a threat to the collector optics. These plasmas have electron temperatures of approximately 30 eV and densities of approximately 10^{18} to 10^{19} cm^{-3} [27,28]. In addition to the desired 13.5 nm emission, out-of-band radiation is also emitted from the plasma. This out-of-band radiation has a deleterious effect of increasing the thermal loads on the plasma facing components, yet another challenge that must be solved before EUV lithography can be implemented in a manufacturing setting.

EUV light is created by either of two methods: laser-produced plasma, or rotating disk discharge plasma. In the first method, plasma is formed by irradiating a liquid Sn droplet with a high-power laser (>10^{10} W/cm^2) [29]. In the second method, a laser is used to liberate Sn atoms from a rotating drum covered in molten Sn. The droplets then form a bridge from one electrode to another and a current is dropped across the bridge. This current heats the Sn atoms and causes plasma to develop.

2.3.3 EUV Fuels

Over the last decade, three different EUV fuels have been investigated as viable methods for generating EUV light: Li, Xe, and Sn. These fuels each have their own advantages and disadvantages. The first fuel investigated was Li because of its tight bandwidth around 13.5 nm for the 2p to 1s transition of Li^{2+} as shown in Figure 2.4 [30,31]. In addition, Li is also ideal because of its low mass, which is likely to cause less impact damage than something as heavy as either Sn or Xe. It also has a conversion efficiency up to 2.5%, which is competitive with the other two fuels. Conversion efficiency is the fraction of light in a 2% bandwidth around 13.5 nm relative to the amount of power put into the system. The main deterrent for Li use, however, is that it is highly condensable (meaning that it will readily coat the surface of a

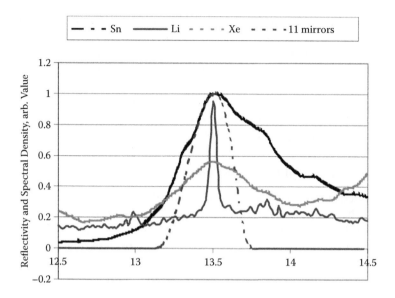

FIGURE 2.4

(See color insert.) The spectral density plots of Li, Sn, and Xe are presented. Each plot is normalized. These plots reveal why Li is ideal as a mono-energetic light emitter, and why Xe and Sn are more ideal for their output. The fourth plot, 11 mirrors, shows the normalized reflectivity of 11 mirrors each made of Sn/Mo bilayer Bragg reflectors. (Figure reproduced from Banine, V., and R. Moors, Plasma sources for EUV lithography exposure tools. *Journal of Physics D: Applied Physics*, 2004. 37(23): p. 3207–3212. With permission.)

collector optic) and highly reactive [31–33]. Li is also attractive; because there is very little out-of-band radiation; however, less consideration needs to be taken for spectral purity filters, which are filters that limit the light emanating from the EUV light source to 13.5 nm. Condensation of the metal onto the collector optics causes EUV reflection degradation and leads to higher cost of ownership.

In light of the condensability and reactivity of Li, the second fuel proposed was Xe. Unfortunately, the high mass of the Xe atom is much more capable of causing damage to components. Xe also suffers from the fact that its conversion efficiency is less than 1% [34,35]. Unlike Sn, which has multiple ionization states contributing to the emission of EUV light, only the 10+ ionization state is active for the Xe atom. Even though it is still possible to create large outputs of radiation power, it requires more input energy, which in turn leads to a larger amount of out-of-band thermal radiation. These additional thermal loads, if unabated, can lead to more rapid diffusion within the mirror surface as well as surface warping due to the temperature gradients induced [36]. In addition, Xe is quite expensive, and Xe reclamations systems add to the already considerable price of an EUV source.

The final fuel mentioned, and the one currently being used for EUV light production is Sn. Sn has the advantage of utilizing 8+ through 12+ ionization states for creating EUV light. These additional ionization states allow the conversion efficiency of Sn plasmas to theoretically approach 6%, though current practice is limited to less than 4% [37–39]. Although Sn is more efficient than Xe, it suffers from the same condensability issues as Li, though it is a less reactive species. In order to deal with the issue of condensation, research is ongoing to address methods by which to clean the mirror surface [40]. Despite these additional issues, Sn is the only fuel currently viewed as viable for high-volume manufacturing.

2.3.4 EUV Sources and Collector Optics

As mentioned previously, there are currently two different competing designs for high-volume manufacturing EUV lithography plasma sources: the laser-produced plasma, and the rotating disk electrode plasma. Although the consequent result of both of these methods is the same, clean EUV photons on the wafer surface, they each approach the generation and transport of the photons differently.

The first method, laser-produced plasma, is performed by focusing a CO_2 laser onto a molten Sn droplet. The laser irradiance liberates and accelerates electrons from the outer edge of the droplet into the core. These electrons cause ionization and assist in the heating of the core material, resulting in ionization levels required for EUV light emission. The kinetic energy of the electrons is proportional to the energy deposited by the laser, and as such it is required to have greater than 10^{10} W/cm² fluence. A droplet generator, as shown in Figure 2.5, produces the molten Sn droplets that are ejected with diameters of 10 to 150 µm [41]. As shown in Figure 2.6, the laser is introduced from behind the collector optic through an orifice. The created EUV light (as well as out-of-band radiation) is isotropically radiated from the droplet location and is collected with a normal incidence mirror [42]. Normal incidence mirrors are used with laser-produced plasma systems because they allow the largest solid angle collection of light. The small size of the Sn droplet is an intentional consequence of mass limitation. By not using more Sn than is necessary to create the maximum amount of EUV light output, there is less extraneous mass that can be accelerated and deposited onto the mirror surfaces. A technique using a prepulse laser burst can also be implemented to increase conversion efficiency. This laser burst causes the Sn droplet to expand and reduce its density before it is hit with the primary laser burst. The increase in conversion efficiency is due to the fact that at higher densities, the droplet is optically dense for EUV light to be emitted, and as such it is often a trade-off between how much fuel should be included in each shot and the ability to extract as much EUV as possible [43]. The current state-of-the-art laser-produced plasma source, the HVM I tool produced by CYMER, Inc., is capable of producing 20 W of EUV light at the intermediate focus (after reflection off of the collector optics). It is projected that by the third quarter of 2011, this result will be improved to 105 W [44].

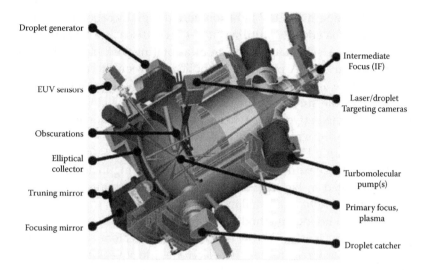

FIGURE 2.5
The various components of a Cymer laser-produced plasma assembly. A liquid drop generator ejects molten Sn droplets into the chamber where a laser is used to create an extreme ultraviolet (EUV) plasma from behind the collector optic. The light is collected, using a normal incidence mirror and projected onto the intermediate focus. (Reprinted from Farrar, N.R., et al., EUV laser produced plasma source development. *Microelectronic Engineering*, 2009. 86(4–6): 509–512. With permission from Elsevier.)

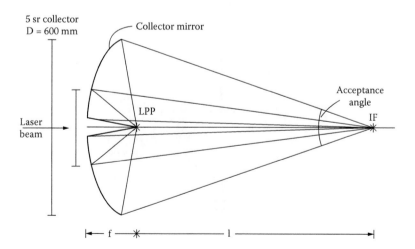

FIGURE 2.6
How laser-produced plasma collector optics collect extreme ultraviolet (EUV) light. Laser-produced plasma mirrors are normal incidence and placed directly behind the EUV source. A laser is fired through a hole in the collector to allow for the production of EUV light from a Sn droplet. (Figure taken from Feigl, T., et al. Enhanced reflectivity and stability of high-temperature LPP collector mirrors, *Advances in X-Ray/EUV Optics and Components III*, 2008. 7077. San Diego, CA: SPIE. With permission.)

The second competitive EUV light source, a design incorporating rotating disk electrodes coated in Sn, utilizes an electric discharge to perform the ionization of Sn. A schematic diagram of a rotating disk electrode system is shown in Figure 2.7. A potential is applied between the two electrodes, which are rotating and constantly being replenished with molten Sn. A laser is fired to ablate the surface of the molten Sn, creating a cloud of Sn atoms that bridges the gap between the two electrodes. Current then flows through the Sn atoms, creating a pinch effect that heats and compresses the formed Sn plasma, thus creating EUV light [45]. Because the plasma is formed between two electrodes, these sources require the use of grazing incidence collector optics as illustrated in Figure 2.8. These mirrors, which are not capable of collecting as much light as normal incidence mirrors in a similar size, collect light using a two bounce method [46]. Unfortunately, even though electric discharge is the most efficient way of ionizing the Sn atoms to their desired levels, the presence of plasma near the electrodes introduces extra debris that is detrimental to the collector optics. This can cause a great deal of problems

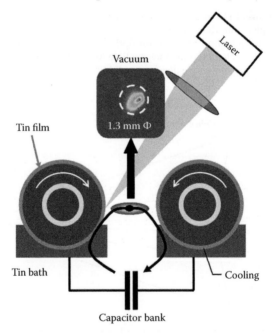

FIGURE 2.7
(See color insert.) A rotating disk electrode source is one of a few different types of gas discharge produced plasmas. The two disks rotate, and are coated in a thin Sn layer, which replenishes the fuel. A laser is fired externally that causes a plume of Sn gas to gap the two electrodes. The potential on the electrodes is then discharged across the gap creating extreme ultraviolet (EUV) light. (Figure taken from Yoshioka, M. *Tin DPP Source Collector Module (SoCoMo): Status of beta products and HVM developments*, Extreme Ultraviolet (EUV) Lithography, 2010. 7636. San Diego, CA:SPIE. With permission.)

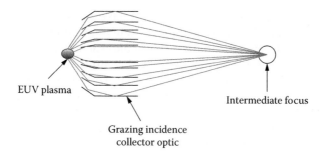

FIGURE 2.8
The gas discharge produced plasma source grazing-incidence collector optic. Unlike laser-produced plasma normal incidence mirrors, grazing incidence mirrors collect light between the source and the intermediate focus. Grazing incidence mirrors often utilize a two-bounce collection technique in order to increase the amount of light collected while being able to reduce the physical footprint of the collector.

with the grazing incident mirrors used to collect the light. The current leader in the rotating disk electrode plasma EUV source, Xtreme Technologies NXE 3100, is capable of producing 15 W at the intermediate focus, with projected power emission reaching 107 W by the second quarter of 2011 [47].

In order to enable reflection of EUV light off of normal incidence mirror, the mirror's outer coating must consist of multiple layers of reflective materials. The chosen materials for the multilayer mirror structure are typically alternating layers of high and low mass. This design maximizes the constructive interference of EUV light being reflected off of the surface when used with layers approximately half the wavelength. These materials are also chosen for their high contrast and low absorption of EUV light. The segmented length of 6.9 nm is chosen because it is nearly half the wavelength of 13.5 nm light, thus optimizing Bragg diffraction [25,49,50]. The optimum number of Mo/Si bilayers is between 40 and 50 because additional bilayers do not increase the reflection and only add to the fabrication costs. In order to prevent oxidation of the outermost surface of the mirror, an approximately 2 nm thick layer of Ru is deposited on the outermost surface. Ru is chosen for its high reflectivity of EUV and high oxidation resistance [51]. Diffusion among the layers is also a concern, and much research has been invested in Gibbsean segregation alloys and other diffusion barriers [52,53].

Grazing incidence mirrors do not have the same reflectivity issues as normal incidence mirrors, and as such do not require multiple layers to reflect EUV light. Much like the normal incidence mirrors, however, an outer capping layer of Ru is deposited onto Si to provide reflectivity and oxidation resistance [51]. Multilayer mirror stacking can be employed by grazing incidence mirror manufacturers, though the purpose is not to increase reflectivity but instead to provide wear resistance against degradation of the surface due to energetic species bombardment [54].

Although both methods of collecting light have the same goal of collecting light and transporting it to the intermediate focus, there are several differences between the two techniques that make each challenging. Because normal incidence mirrors require the use of a multilayer structure, diffusion-induced interlayer mixing is of grave concern to the reflectivity of the mirror. This diffusion is a result of the mirror being too hot, a consequence of out-of-band light thermal heating, as well as energetic debris causing outermost layers to be damaged. Energetic debris is also detrimental to grazing incidence mirrors due to the fact that surface roughness of these mirrors can lead to a drop in reflectivity. Multilayer mirror production is also a task in upon itself because of the requirement to maintain minimal (<1 nm) surface roughness while depositing sub-10 nm layers up to 50 times on a curved surface. The biggest difference between the two mirror types is the amount of light that can be collected. Despite the difficulties in fabricating and maintaining the mirror surface of a normal incidence mirror, it is able to collect more light in a given area than grazing incidence mirrors. This is because the grazing incidence mirrors only collect a small portion of light with each shell. As such there are many shells required to collect the EUV light. Cooling of the inner shells, and structural issues are also difficult in these types of mirrors. For both mirrors, however, deposition of the condensable Sn fuel on the surface of the mirrors can rapidly lead to loss of reflection and, as such, debris mitigation and in situ cleaning methods are required.

2.3.5 Extreme Ultraviolet Light Source Debris Considerations

In the previous section it was mentioned that out-of-band radiation, the condensability of the fuel, and energetic debris emitted from the plasma are of grave concern to the EUV collector optic. The lifetime of the collector optic directly correlates to the cost of ownership of the EUV lithography tool. If the collector optic is damaged due to deposited debris or interlayer diffusion, not only does the million dollar collector optic need to be replaced, but there is incredible cost incurred from the tool downtime required to replace the optic. Because the mirrors are so close to the EUV-emitting plasma, and their lifetime is so important, considerable effort has been placed into studying how to implement debris (anything that is not an EUV photon) mitigation [55].

The first concern for the collector optics, thermal issues, is a consequence of the out-of-band radiation created by the EUV-emitting plasma. The process of creating EUV light has the deleterious effect of also creating infrared light, ultraviolet light, as well as light in the visible spectrum. Because EUV collector optics are only designed to reflect EUV photons, the out-of-band radiation will be partially absorbed. This partial absorption means that as the output power of the source is increased, so is the amount of out-of-band power that is absorbed in the collector optic. This absorbed energy can lead to inter-layer diffusion in multilayer mirrors and can result in a loss of transmission of EUV light. Of even greater consequence is the light that is reflected by the collector optics (>90%). The out-of-band radiation can also heat and damage components beyond the intermediate focus, such as the illumination optics and the reflective mask. To this end, various methods of spectral purity filters have been designed. These range from thin films of Zr, Si, or Mo to gas jets that resonantly absorb non-EUV emission before it is transmitted to the intermediate focus [56]. The use of plasma windows is also being explored, where the plasma is used to not only restrict the transmittance of infrared light but also the transmission of fuel species [57].

The second influence on collector optic lifetime, deposited contamination of EUV fuels, is of considerable concern in light of industry's shift to condensable Sn fuel. The materials chosen to reflect EUV light off of the mirror surface are chosen because they have electrical properties that are ideal for reflecting a certain bandwidth of light. Sn, however, strongly absorbs EUV light and is detrimental to reflectivity if the fuel is allowed to condense onto the optics. In an effort to mitigate this issue, hydrogen radicals have been investigated as a means for removing Sn from the collector optic surface while maintaining the integrity of the mirror. It has been shown that, when coupled with a thin layer of Si_3N_4 added between the Ru capping layer and the first layer of the multilayer mirror, complete reflectivity of a ~14 nm coated mirror can be restored in 20 seconds of cleaning [58].

The last influence on collector optic lifetime, and also one of the most difficult to address, is the effect of energetic debris flux. EUV-light-emitting plasmas can create energetic neutral and ionized fuel debris with energies on the order of 10s of keVs [59]. This energetic debris can be implanted into the mirror or sputter the surface layer of the mirror. Implantation into the mirror can create substrate-doping defects that affect the optical parameters of the mirror, cause interlayer scattering that reduces the effectiveness of Bragg diffraction, and can even cause EUV absorption that leads to thermal concerns. Surface sputtering leads to an increase in surface roughness, which can reduce the reflectivity of the mirror and increase cost of ownership [60]. Thermal issues can be addressed with external cooling lines, and EUV fuel contamination can be addressed with in situ cleaning, but energetic debris is very difficult to mitigate. Part of the issue with mitigating debris created by the plasmas is the fact that, in order to collect a large solid angle of light and reduce the cost of the collector, the collector optic needs to be located very close to the plasma. As shown in Figure 2.9, which is a flux measurement of the Xtreme XTS 13-35 gas discharge Xe-fueled EUV light source, there is a considerable amount of high-energy debris that needs to be addressed.

Approaches to mitigating debris emanating from EUV sources are somewhat dependent on the type of source being considered. Laser-produced plasma sources approach debris mitigation through the use of buffer gasses and magnetic fields that confine the energetic charged debris to a location that is not detrimental to the EUV source. The buffer gas of choice is typically hydrogen based on its low EUV light absorption, its ability to clean Sn, as well as its ability to reduce not only the flux of emanating debris but also

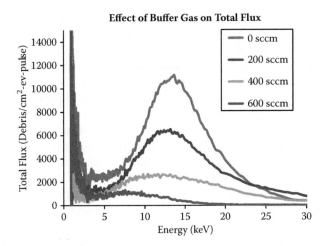

FIGURE 2.9

(See color insert.) Measurements of total ion and neutral flux from an XTS 13-35 extreme ultra-violet (EUV) light source from 25° off axis are shown as a function of buffer gas flow rate. The buffer gas is injected between the EUV source and a foil trap as a method for mitigating the debris reaching the collector optics. The measured flux is predominately composed of Xe gas atoms. (Figure taken from Sporre, J., Detection of energetic neutral flux emanating from extreme ultraviolet light lithography sources, in *Department of Nuclear, Plasma, and Radiological Engineering*. 2010, University of Illinois at Urbana-Champaign. p. 126. With permission.)

to reduce its energy [62]. Rotating disk electrodes also implement the use of a buffer gas, although the species used is Ar rather than H_2. This debris mitigation technique is frequently coupled with the use of a collimated foil trap that is composed of tungsten. The buffer gas causes scattering events that cause the energetic flux to propagate at angles that are not radial from the EUV light source. These scattered fuel species then impact the walls of the foil trap, depositing their energy into the foil trap instead of the collector optic. Sn condensation is also addressed by the fact that the collimated foil trap cannot stop all fuel species entirely. The argon species that are accelerated by the plasma serve as a sputtering source that removes the Sn as it is being deposited [63].

2.4 Conclusion

In this chapter, it was shown that one approach to extending Moore's law for the production of nanoscale features is to develop new lithography tools. The current 193 nm i-line lithography tool has already been stretched beyond its economic limits, and as such it is necessary to change the wavelength of radiation used for exposing wafers in computer chip manufacturing. Although there exist alternative solutions to fulfill this need, EUV lithography presents the most likely technology to achieve sub-20 nm nodes on the International Technology Roadmap for Semiconductors. This light source utilizes 13.5 nm light generated by either laser-produced plasmas or rotating disk electrode plasmas. These ~30 eV, 10^{19}cm^{-3} plasmas liberate upwards of 10 electrons off of Sn atoms to generate the EUV light. Unfortunately, because EUV light is readily absorbed in almost all materials, considerations must be made to protect the collector optics used to focus EUV photons onto the intermediate focus. Utilizing various debris mitigation methods such as buffer gas injection, collimated foil traps, and applied magnetic fields, source suppliers are able to protect the collector optics and maintain tool cost efficiency. Thermal consideration, a consequence of the out-of-band radiation emitted by the plasmas, is also addressed using cooling and spectral purity filters. Ultimately, however, as these issues are being addressed it becomes clear that EUV lithography is a viable method for creating the nanostructures to be utilized for next-generation computer technology.

References

1. Moore, G., Cramming more components onto integrated circuits. *Proceedings of the IEEE*, 1998. 86(1): 82–85.
2. *60 Years of the Transistor: 1947–2007*. (cited 2010 February 22); Available from: http://www.intel.com/technology/timeline.pdf.
3. Zhirnov, V.V., et al., Limits to binary logic switch scaling—A gedanken model. *Proceedings of the IEEE*, 2003. 91(11): 1934–1939.
4. Hawryluk, A.M., and L.G. Seppala, Soft x-ray projection lithography using an x-ray reduction camera. *Journal of Vacuum Science & Technology B: Microelectronics and Nanometer Structures*, 1988. 6(6): 2162–2166.
5. Wu, B., and A. Kumar, Extreme ultraviolet lithography: A review. *Journal of Vacuum Science and Technology B: Microelectronics and Nanometer Structures*, 2007. 25(6): 1743–1761.
6. Levenson, M.D., Wavefront engineering for photolithography. *Physics Today*, 1993. 46(7): 28.
7. Nazmov, V., J. Mohr, and E. Reznikova, Visualization of the development process in deep X-ray lithography. *Nuclear Instruments and Methods in Physics Research Section A: Accelerators, Spectrometers, Detectors and Associated Equipment*, 2009. 603(1–2): 153–156.
8. Wakana, K., et al. *Optical Performance of Laser Light Source for ArF Immersion Double Patterning Lithography Tool*. 2009. San Jose, CA: SPIE.
9. Owen, G., et al. *1/8 mu m Optical Lithography*. 1992. Orlando, FL: AVS.
10. Jain, K., *Excimer Laser Lithography*. 1990, Bellingham: SPIE Optical Engineering Press.

11. Levinson, H.J., *Principles of Lithography*. 2001, Bellingham: SPIE—The International Society for Optical Engineering.
12. Liebmann, L.W. *Resolution Enhancement Techniques in Optical Lithography: It's Not Just a Mask Problem*. 2001. Kanagawa, Japan: SPIE.
13. Lin, B.J., The ending of optical lithography and the prospects of its successors. *Microelectronic Engineering*, 2006. 83(4–9): 604–613.
14. Torres, J.A., O. Otto, and F.G. Pikus. *Challenges for the 28 nm Half Node: Is the Optical Shrink Dead?* 2009. Monterey, CA: SPIE.
15. Basting, D., K. Pippert, and U. Stamm. History and future prospects of excimer laser technology. In *Focused on Second International Symposium on Laser Precision Microfabrication*. 2002.
16. Ronse, K., et al., Lithography options for the 32 nm half pitch node and beyond. *Circuits and Systems I: Regular Papers, IEEE Transactions on*, 2009. 56(8): 1884–1891.
17. Hazelton, A.J., et al., Double-patterning requirements for optical lithography and prospects for optical extension without double patterning. *Journal of Micro/Nanolithography, MEMS and MOEMS*, 2009. 8(1): 011003–11.
18. Dusa, M., et al. *Pitch Doubling through Dual-Patterning Lithography Challenges in Integration and Litho Budgets*. 2007. San Jose, CA: SPIE.
19. Levenson, M.D., et al., The phase-shifting mask II: Imaging simulations and submicrometer resist exposures. *Electron Devices, IEEE Transactions on*, 1984. 31(6): 753–763.
20. Noguchi, M., et al. *Subhalf-Micron Lithography System with Phase-Shifting Effect*. 1992. San Jose, CA: SPIE.
21. Brunner, T.A. *Rim Phase-Shift Mask Combined with Off-Axis Illumination: A Path to .5 Lambda/NA Geometries*. 1993. San Jose, CA: SPIE.
22. Matthew, I., et al. *Design Restrictions for Patterning with Off-Axis Illumination*. 2004. San Jose, CA: SPIE.
23. Rigolli, P., et al. *High-Order Distortion Effects Induced by Extreme Off-Axis Illuminations at Hyper NA Lithography*. 2009. San Jose, CA: SPIE.
24. Semiconductor Industry Association. The International Technology Roadmap for Semiconductors, 2009 Edition. International SEMATECH: Austin, TX, 2009.
25. Attwood, D.T., *Soft X-rays and Extreme Ultraviolet Radiation*. 1999, Cambridge University Press: Cambridge. 1–21.
26. Borisov, V.M., EUV sources using Xe and Sn discharge plasmas. *Journal of Physics. D, Applied Physics*, 2004. 37(23): 3254–3265.
27. Nastoyashchii, A.F. *Optimal Physical Conditions for Extreme UV Generation*. 2004. Taos, NM: SPIE.
28. Masnavi, M., et al., Estimation of optimum density and temperature for maximu efficiency of tin ions in Z discharge extreme ultraviolet sources. *Journal of Applied Physics*, 2007. 101(033306).
29. Hassanein, A., et al. *Effects of Plasma Spatial Profile on Conversion Efficiency of Laser Produced Plasma Sources for EUV Lithography*. 2009. San Jose, CA: SPIE.
30. Banine, V., and R. Moors, Plasma sources for EUV lithography exposure tools. *Journal of Physics D: Applied Physics*, 2004. 37(23): 3207–3212.
31. Masnavi, M., et al., Potential of discharge-based lithium plasma as an extreme ultraviolet source. *Applied Physics Letters*, 2006. 89(3): 031503-3.

32. Neumann, M.J., et al., Plasma cleaning of lithium off of collector optics material for use in extreme ultraviolet lithography applications. *Journal of Micro/ Nanolithography, MEMS and MOEMS*, 2007. 6(2): 023005-6.

33. Nagano, A., *Extreme ultraviolet source using laser-produced Li plasma. Transactions of the Institute of Electrical Engineers of Japan, Part C*, 2009. 129(2): 249–252.

34. Ter-Avetisyan, S., et al., Efficient extreme ultraviolet emission from xenon-cluster jet targets at high repetition rate laser illumination. *Journal of Applied Physics*, 2003. 94(9): 5489–5496.

35. Komori, H., EUV radiation characteristics of a CO_2 laser produced Xe plasma. *Applied Physics. B, Lasers and Optics*, 2006. 83(2): 213–218.

36. Bianucci, G., et al. *Thermal Management Design and Verification of Collector Optics into High-Power EUV Source Systems*. 2007. San Jose, CA: SPIE.

37. Cummings, A., et al., Conversion efficiency of a laser-produced Sn plasma at 13.5 nm, simulated with a one-dimensional hydrodynamic model and treated as a multi-component blackbody. *Journal of Physics. D, Applied Physics*, 2005. 38(4): 604–616.

38. Shimada, Y., et al., Characterization of extreme ultraviolet emission from laser-produced spherical tin plasma generated with multiple laser beams. *Applied Physics Letters*, 2005. 86(5): 051501-3.

39. Krucken, T., Fundamentals and limits for the EUV emission of pinch plasma sources for EUV lithography. *Journal of Physics. D, Applied Physics*, 2004. 37(23): 3213–3224.

40. Shin, H., R. Raju, and D.N. Ruzic. *Remote Plasma Cleaning of Sn from an EUV Collector Mirror*. 2009. San Jose, CA: SPIE.

41. Farrar, N.R., et al., EUV laser produced plasma source development. *Microelectronic Engineering*, 2009. 86(4–6): 509–512.

42. Feigl, T., et al. Enhanced reflectivity and stability of high-temperature LPP collector mirrors, *Advances in X-Ray/EUV Optics and Components III*, 2008. 7077. San Diego, CA: SPIE.

43. Tao, Y., Mass-limited Sn target irradiated by dual laser pulses for an extreme ultraviolet lithography source. *Optics Letters*, 2007. 32(10): 1338–1340.

44. Brandt, D.C., et. al. LPP EUV source production for HVM. in *International Symposium on EUVL*. 2010. Kobe, Japan: SEMATECH.

45. Borisov, V.M., et al., *Xenon and Tin Pinch Discharge Sources*, in *EUV Sources for Lithography*, V. Bakshi, Editor. 2006, SPIE—The International Society for Optical Engineering: Bellingham.

46. Bianucci, G., et al. *Design and Fabrication Considerations of EUVL Collectors for HVM*. 2009. San Jose, CA: SPIE.

47. Corthout, M., Y. Teramoto, M. Yoshioka. First Tin Beta SoCoMo ready for wafer exposure. In *International Symposium on EUVL*. 2010. Kobe, Japan: SEMATECH.

48. Yoshioka, M. *Tin DPP Source Collector Module (SoCoMo): Status of beta products and HVM developments*, Extreme Ultraviolet (EUV) Lithography, 2010. 7636. San Diego, CA:SPIE.

49. Maury, H., et al., Non-destructive X-ray study of the interphases in Mo/Si and Mo/B4C/Si/B4C multilayers. *Thin Solid Films*, 2006. 514(1–2): 278–286.

50. Hecquet, C., et al. *Design, Conception, and Metrology of EUV Mirrors for Aggressive Environments*. 2007. Prague, Czech Republic: SPIE.

51. Bajt, S., Oxidation resistance of ru-capped EUV multilayers. *Proceedings of SPIE—The International Society for Optical Engineering*, 2005. 5751(1): 137–146.

52. Bajt, S., et al., Improved reflectance and stability of Mo-Si multilayers. *Optical Engineering*, 2002. 41(8): 1797–1804.
53. Qiu, H., Time exposure performance of Mo-Au Gibbsian segregating alloys for extreme ultraviolet collector optics. *Applied Optics*, 2008. 47(13): 2443–2451.
54. Braic, V., M. Balaceanu, and M. Braic. Grazing incidence mirrors for EUV lithography. in *Semiconductor Conference, 2008. CAS 2008. International*. 2008.
55. Ruzic, D.N., *Origin of Debris in EUV Sources and Its Mitigation*, in *EUV Sources for Lithography*, V. Bakshi, Editor. 2006, SPIE—The International Society for Optical Engineering: Bellingham. pp. 957–991.
56. Chimaobi Mbanaso, G.D., Alin Antohe, Horace Bull, Frank Goodwin, Ady Hershcovitch. Development of a spectral purity filter for CO_2 laser produced plasma based on magnetized plasma confinement of absorbing gases. in *International Symposium on EUVL*. 2010. Kobe, Japan: SEMATECH.
57. Pinkoski, B.T., X-ray transmission through a plasma window. *Review of Scientific Instruments*, 2001. 72(3): 1677–1679.
58. van Herpen, M.M.J.W., Sn etching with hydrogen radicals to clean EUV optics. *Chemical Physics Letters*, 2010. 484(4): 197–199.
59. Ruzic, D.N., Srivastava, S. N., Normal incidence (multilayer) collector contamination, in *EUV Lithography*, V. Bakshi, Editor. 2008, SPIE: Bellingham. pp. 285–318.
60. Thompson, K.C., et al., Experimental test chamber design for optics exposure testing and debris characterization of a xenon discharge produced plasma source for extreme ultraviolet lithography. *Microelectronic Engineering*, 2006. 83(3): 476–484.
61. Sporre, J., Detection of energetic neutral flux emanating from extreme ultraviolet light lithography sources, in Department of Nuclear, Plasma, and Radiological Engineering. 2010, University of Illinois at Urbana-Champaign. p. 126.
62. Fomenkov, I.V., Ershov, A.I., Partlo, W.N., Myers, D.W., Sandstrom, R.L., Bowering, N.R., Vaschenko, G.O., Khodykin, O.V., Bykanov, A.N., Srivastava, S.N., Ahmad, I., Rajyaguru, C., Golich, D.J., De Dea, S., Hou, R.R., O'Brien, K.M., Dunstan, W.J., Brandt, D.C., Laser-produced plasma light source for EUVL. *Proceedings of SPIE—The International Society for Optical Engineering*, 2010. 7636: 6.
63. M. Corthout, Y.T., M. Yoshioka. First Tin Beta SoCoMo ready for wafer exposure. in *International Symposium on EUVL*. 2010. Kobe, Japan: SEMATECH.

3

Nonthermal Plasma Synthesis
of Semiconductor Nanocrystals

Uwe Kortshagen
Lorenzo Mangolini

CONTENTS

3.1 Introduction

Nonthermal plasmas are a common fabrication tool in the semiconductor industry. Low-pressure, partially ionized gases are a convenient source of radicals and are of great use for a variety of applications, such as for the deposition of thin films, or for the etching of features with high aspect ratio for microelectronic applications. The formation of particles in processing plasmas is a well-known phenomenon, but it has always been considered by the semiconductor industry as a source of contamination and a problem to eliminate [1]. The initial studies on particle formation in plasmas, and in particular in silane-containing discharges, focused on quenching the nucleation process and avoiding particle transport onto the substrate. Soon, however, several research groups realized that nanocrystals could be of great interest for a variety of applications, such as optoelectronic devices, photovoltaic devices, bio-related applications, and so forth. The group of Roca i Cabarrocas was the first to demonstrate that the inclusion of small silicon nanocrystals into an amorphous thin film grown by plasma-enhanced chemical vapor deposition (PECVD) would result in a film with improved optoelectronic properties [2]. This new silicon material, called polymorphous silicon, or pm-Si:H, showed an improved stability with respect to the Staebler–Wronski effect [3]—that is, the light-induced creation of defects in amorphous silicon. Following these reports, the interest in *dusty plasmas* grew significantly.

At about the same time, interest in free-standing semiconductor nanocrystals grew rapidly, following reports from Bawendi's group at the Massachusetts Institute of Technology (MIT) [4,5] that introduced a simple route for the synthesis of CdSe nanocrystals small enough to show quantum confinement effects. The band gap of small semiconductor nanocrystals is a size-dependent property [6,7]; an increase in the optical band gap is observed when the crystal size approaches the Bohr exciton radius, which is typically less than 10 nm for most semiconductors. This leads to size-dependent optical emission, as shown in Figure 3.1 for the case of plasma synthesized silicon nanocrystals. In fact, many material properties exhibit strong deviations from bulk values when the crystal size reaches down to a few nanometers,

FIGURE 3.1
(See color insert.) Photoluminescence of silicon nanocrystals that were synthesized in a two-stage plasma process: a first synthesis step and a subsequent etching step using a CF₄-etch chemistry. (From Pi, X. D., R. W. Liptak, J. D. Nowak et al. 2008b. Air-stable full-visible-spectrum emission from silicon nanocrystals synthesized by an all-gas-phase plasma approach. *Nanotechnology* 19: 245603. With permission.)

such as the rate of optical emission in indirect band gap semiconductors [8], the melting point temperature [9], and the hardness of the nanocrystal material compared to bulk materials [10].

The interest in group II-VI materials such as cadmium selenide (CdSe), cadmium sulfide (CdS), and cadmium telluride (CdTe) and group IV-VI compounds such as lead selenide (PbSe) and lead sulfide (PbS) has been motivated by their excellent optical and electronic properties, and by the fact that liquid-phase synthetic approaches work well for this material system, allowing excellent control over particle size and dispersity [5]. Several groups have reported the development of nanocrystal devices based on group II-VI semiconductor quantum dots, such as light-emitting devices [11,12] and photovoltaic cells [13–15]. Group II-VI semiconductor quantum dots are also being proposed as fluorescent tags for biomedical applications [16]. Despite their many attractive properties, solution processable semiconductors face many challenges. Group II-VI and IV-VI semiconductors contain toxic heavy metals, which is a major concern for all applications involving large-scale production, including photovoltaic devices. Some of the constituent elements are not abundant in the earth's crust, raising concerns about the long-term economic viability of their use. Finally, batch, liquid-phase synthesis is not easily scalable to industrial levels and utilizes large amounts of solvents, which are expensive to process as waste and also pose some safety concerns.

In contrast, the group IV materials carbon (C), silicon (Si), and germanium (Ge) are nontoxic in their bulk form, environmentally benign and abundant, with silicon being the second most abundant element in the earth's crust. This makes group IV materials interesting for large-scale applications. Producing small nanocrystals of group IV materials, however, presents some challenges. Similar to III-V compound semiconductors, group IV semiconductors have relatively high melting points and as a consequence have higher crystallization temperatures. For instance, the crystallization temperatures of silicon nanoparticles with diameters of 4, 6, 8, and 10 nm were found to be around 773, 1073, 1173, and 1273 K, respectively [17]. Yet, the formation of high-quality crystallites is mandatory to take advantage of many of the novel properties of quantum confined nanocrystals. Amorphous nanoparticles have higher densities of defects, compromising their optical and electronic properties. The higher temperatures required to process group IV nanocrystals are incompatible with liquid-phase techniques, which are limited by the decomposition temperature of solvents.

Plasmas and other gas phase approaches easily lend themselves to high-temperature synthesis. Techniques such as high-temperature thermal reactions (pyrolysis) in furnace flow reactors [18,19], laser-induced reactions (photolysis) [20], and laser pyrolysis using high-power infrared lasers [21–22] were explored for the synthesis of small silicon nanocrystals. However, nanoparticles produced in such systems have very high diffusivity, which results in high losses to the reactor walls and low production rate. Moreover, as particles in these approaches remain neutral, high production rates lead to particle agglomeration, which results in a poor control over the particle size and broad size distributions compared to particles produced in the liquid phase. In contrast, nonthermal plasmas have a series of unique characteristics that differentiate them from other gas-phase synthesis techniques:

1. *High production rate*: In partially ionized gases, free electrons acquire energies of a few eV, where 1 eV corresponds to a temperature of ~11,000 K. Collisions between electrons and the background gas lead to the efficient production of radicals and, in the case of silane-containing discharges, to the chemical nucleation of nanoparticles. The high production rate is of interest for industrial-scale applications.

2. *Particle charging*: Electrons in a nonthermal plasma have a temperature that largely exceeds that of heavy particles. The ion temperature is typically close to the background gas-temperature, which is near room temperature for nonthermal plasmas. The imbalance in the energy of electrons and ions leads to an initial imbalance of electron and ion fluxes to any surface exposed to the plasma. As a consequence, the surface acquires a negative charge to balance the ion and electron currents to the surface. Hence, in nonthermal plasmas, both reactor walls and the surfaces of entrained nanoparticles charge negatively with respect to the plasma [23,24]. The unipolar negative nanoparticle charge induces electrostatic repulsion between the particles, reducing nanoparticle agglomeration rates and growth due to coalescence [25,26]. This enables the synthesis of smaller particle sizes with a narrower size distribution compared to other gas-phase techniques. Similarly, the electrostatic repulsion between nanoparticles and reactor walls slows the diffusion loss of particles. These are two significant advantages compared to other gas phase processes.

3. *Selective nanoparticle heating*: In low-pressure nonthermal plasmas, energetic surface reactions such as electron-ion recombination and surface chemical reactions heat the nanoparticles to temperatures that exceed the temperature of the surrounding gas by several hundreds of Kelvin [27,28]. Cooling processes through convection/conduction are relatively inefficient at low pressure due to the low collision rate with neutral gas atoms. The selective heating of the nanoparticles in plasmas is believed to play an important role in the formation of high-quality nanocrystals while the surrounding gas is close to room temperature.

This chapter is organized as follows: Section 2 discusses the particular advantages and mechanisms of nanocrystal synthesis in nonthermal plasmas. In Section 3 we present different nonthermal plasma reactors that have been proposed for nanocrystal synthesis. We also describe approaches for doping and in-flight functionalization of nanoparticles. In Section 4 we review development efforts for various devices based on plasma-produced nanoparticles. Conclusions will be presented in Section 5.

Most of this chapter focuses on silicon because the vast majority of the research has focused on this material. However, the production of silicon nanocrystals may be viewed as a model for the production of other materials using nonthermal plasmas, because the mechanisms discussed in this chapter are general and not limited to any particular materials system.

3.2 Mechanisms Controlling the Nucleation and Growth of Particles in Plasmas

3.2.1 Nanoparticle Nucleation in Nonthermal Plasmas

Even though silicon is the most studied material with regard to particle nucleation and growth in nonthermal plasmas, other material systems have also been investigated such as silicon oxide, carbon, germanium, and so forth. There is no general model that describes the nucleation of particles of various materials in nonthermal plasmas, because the underlying chemistries are different for each material. The energetic electrons in a nonthermal plasma, with temperatures typically between 2 and 5 eV, are very efficient at dissociating the precursor gas molecule, resulting in the formation of various neutral and charged radicals that are crucial for the formation of nanoparticles. Nanoparticle formation can be described as a series of chemical clustering events, determined by the kinetics of chemical reactions and by the densities and lifetimes of the important radicals, which in turn are coupled to the charging of radicals and possibly diffusion losses to the reactor walls. These processes are strongly material specific.

Even for silicon nanoparticles, the most widely studied materials system, the underlying processes are not yet fully understood, and competing scenarios for particle formation have been proposed. We briefly review here the different scenarios for silicon particle formation in plasmas and illustrate the possible mechanism that may also play a role in other materials systems. For more details the reader is referred to a recent review by Watanabe [29].

For silicon nanoparticle formation from silane (SiH_4), it is important to note that small unsaturated silane clusters Si_nH_m have positive electron affinities and thus easily form negative ions because of electron attachment [30]. These negative clusters are electrostatically trapped in the plasma and act as seeds for particle nucleation. Hollenstein and coworkers [31–33] provided evidence supporting this theory in mass spectrometric studies, in which the authors found that negatively charged clusters (anions) grew to larger sizes than positive (cation) or neutral clusters, as shown in Figure 3.2a [32]. The

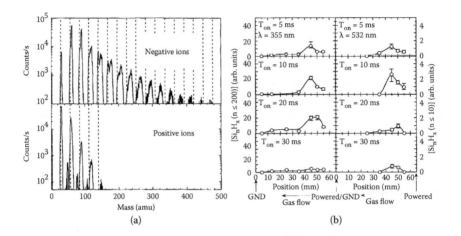

(a)

(b)

FIGURE 3.2
(a) Mass spectra of silicon hydride cluster anions and cations in a radio-frequency silane plasma, for pure silane at 76 mTorr. (From Howling, A. A., L. Sansonnens, J.-L. Dorier, and C. Hollenstein. 1994. Time-resolved measurements of highly polymerized negative ions in radio frequency silane plasma deposition experiments. *J. Appl. Phys.* 75: 1340–1353. With permission.) (b) Spatial profiles of short-lived silane radicals and the distribution of small silicon clusters in a capacitively coupled radio-frequency discharge. (Reproduced from Watanabe, Y., M. Shiratani, T. Fukuzawa et al. 1996. Contribution of short lifetime radicals to the growth of particles in SiH$_4$ high frequency discharges and the effects of particles in deposited films. *J. Vac. Sci. Technol. A* 14: 995. With permission.)

authors proposed that cations and neutrals radicals would tend to diffuse to the walls before growing to large sizes and that anionic clusters, being electrostatically trapped, would have time to grow.

Following these results, a number of groups proposed that a likely path to particle nucleation was through anion-molecule reactions, for example, through sequential hydrogen elimination reactions such as the following: $Si_nH_{2n+1}^- + SiH_4 \rightarrow Si_{n+1}H_{2n+3}^- + H_2$ [34]. Such ion-molecule reactions are expected to be faster than neutral chemistry. A number of theoretical studies supported this hypothesis. Fridman and coworkers [35] proposed that the reaction between negative silyl radicals ($Si_nH_{2n+1}^-$) with vibrationally excited silane monomers is a key reaction pathway. The author's model predicted that the reaction probability of vibrationally excited silane with silicon clusters decreased with increasing cluster size due to the more efficient vibrational deexcitation provided by larger clusters, and thus provided a possible explanation for the size of the smallest "primary" particles of about 2 nm, which was often observed in experiments [36]. Gallagher [37] developed a model that included cluster growth and charging by several generic processes including attachment of radicals, silane, electrons, and cations; detachment of electrons; and diffusive loss. He concluded that that SiH$_3$ is the key particle growth species [37,38], and that electron attachment to SiH$_3$ to form negative silyl radicals is an important process in initiating clustering.

A rather comprehensive chemical kinetics model was presented in the literature [39–41]. This model included over one hundred neutral and charged silicon-hydride species and several hundreds of reversible reactions including radical-induced and ion-induced reactions. The results of this model pointed to the importance of two ion-neutral clustering pathways involving negative silyl ($Si_nH_{2n+1}^-$) and silylene ($Si_nH_{2n}^-$) species.

Watanabe and Shiratani [42,43] developed an alternative view of particle formation in plasmas. In a sequence of sophisticated studies, these authors showed a close correlation between the spatial profiles of the short-lived neutral radicals SiH_2, SiH, and Si and the spatial distribution of the first smallest silicon clusters, see Figure 3.2b. The authors interpreted these findings as an indication that a neutral cluster path through short-lived radicals is the dominant growth mechanism and that negative ion-induced clustering did not play an important role for their discharge conditions. More recently, Watanabe proposed an explanation for these apparent differences in interpretation [29]. He argues that the dominant growth mechanism depends on the typical timescales of cluster growth and on the gas residence in the reactor: if the residence time is small compared to the cluster growth time, a negative-ion induced pathway must be present, because negative ions remain electrostatically confined enabling particles to grow even when neutral species are transported out of the reactor in times shorter than the cluster growth time. Neutral clustering is likely to prevail if the residence time of neutral species in the reactor is longer than the cluster growth time.

This discussion for silicon nanoparticles, the most widely explored nanoparticle-in-plasma system, exemplifies that the details of the particle nucleation and growth will depend on the chemistry as well as on the particular reactor properties considered.

3.2.2 Nanoparticle Charging

The charging of nanoparticles in nonthermal plasmas is an important attribute of nanoparticle synthesis with plasmas. The charging and mutual electrostatic repulsion between the particles results in the reduction of the agglomeration rate, which enables the production of small particles with a narrow size distribution [26]. Charging is also directly related to the heating of nanoparticles to temperatures exceeding that of the background gas, which will be described in the next section. Nanoparticle charging is often described using the orbital motion limited (OML) theory [44,45]. Within this model, the electron and ion currents approaching a negatively charged particle are modeled according to the following relations. For the electron current,

$$I_e = \frac{1}{4} e n_e S \sqrt{\frac{8 k_B T_e}{\pi m_e}} \exp\left(-\frac{e|\Phi|}{k_B T_e} \right), \quad \Phi < 0 \tag{3.1}$$

I_e is the electron current, e the elementary charge, S the particle surface area, k_B the Boltzmann constant, T_e the electron temperature, m_e the electron mass, and Φ the particle potential. This corresponds to the electron flux diminished by the Boltzmann factor for electrons with a Maxwell-Boltzmann velocity distribution. For the ion current,

$$I_i = \frac{1}{4} e n_i S \sqrt{\frac{8k_B T_i}{\pi m_i}} \exp\left(1 + \frac{e|\Phi|}{k_B T_i}\right), \quad \Phi < 0 \tag{3.2}$$

Here I_i denotes the ion current, T_i the ion temperature, and m_i the ion mass. This assumes that ions follow a collisionless trajectory around the particle, and that only the fraction of ions whose angular momentum, referenced to the particle, is below a certain threshold will be collected [44,45]. By balancing the electron and ion current, one can obtain the particle potential Φ and the average particle charge Q from

$$Q = 4\pi\varepsilon_0 R_p \Phi \tag{3.3}$$

with R_p the particle radius.

Knowledge of the average charge alone is not very meaningful, because charging is a stochastic process and particles can hold only an integer number of elementary charges. It is more meaningful to use Equations (3.1) and (3.2) to define charging frequencies ($v_{e,i} = I_{e,i}/e$), and then use the principle of detailed balancing to derive the particle charge distribution. Using this approach, Matsoukas et al. [24,46] derived the charge distribution for nanometer-sized particles in nonthermal plasmas, as shown in Figure 3.3.

Following the approach proposed by Matsoukas et al. [24,46], Schweigert and Schweigert [25] developed a model to predict the influence of particle charging on the coagulation rate. They predicted coagulation rates that were significantly smaller than those experimentally measured [36], casting doubts on the validity of the OML theory. In Kortshagen and Bhandarkar [26], the authors showed that coagulation rates that are consistent with experiments can be obtained if the effect of the particle charging on the plasma parameters is taken into account. "Quasineutrality" predicts that the overall densities of positive and negative charges within the plasma volume are equal; under high particle density conditions, there are not enough electrons present to charge all particles negatively. This situation is frequently encountered at the initial phases of particle nucleation in plasmas. Particle coagulation at this stage is not retarded by the unipolar particle charge distribution; in fact particles at the initial phase of nucleation can have a bipolar charge distribution that increases the coagulation rate compared to a neutral aerosol. After the initial, fast agglomeration phase, the particle density decreases to the point that enough electrons are present to charge negatively all particles, reducing the agglomeration rate. The dynamics of particle charging is shown in Figure 3.4 from Kortshagen and

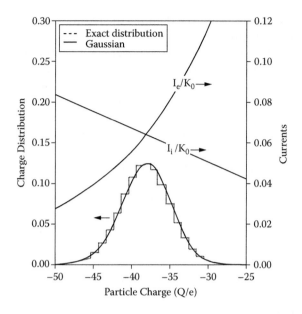

FIGURE 3.3

Charge distribution of monodisperse nanoparticles in a nonthermal plasma. (From Matsoukas, T., M. Russel, and M. Smith. 1996. Stochastic charge fluctuations in dusty plasmas. *J. Vac. Sci. Technol. A* 14: 624. With permission.)

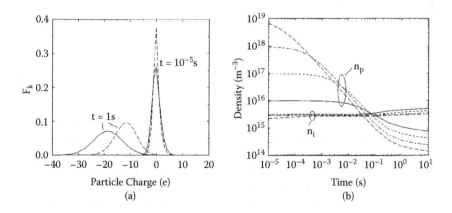

FIGURE 3.4

(a) Particle charge distribution for particles with an initial density of 10^{12} cm^{-3} ($t = 10^{-5}$ s) and a final density of 3×10^8 cm^{-3} ($t = 1$ s) in a plasma with a positive ion density of 3×10^9 cm^{-3}. The dashed on solid lines refer to different charging mechanisms. (See Kortshagen, U., and U. Bhandarkar. 1999. Modeling of particulate coagulation in low pressure plasmas. *Phys. Rev. E* 60: 887.) (b) Evolution of the particle density for different initial values of 2 nm–diameter particles: n_p is the particle density and n_i the self-consistent ion density. (See Kortshagen, U., and U. Bhandarkar. 1999. Modeling of particulate coagulation in low pressure plasmas. *Phys. Rev. E* 60: 887, for details.)

Bhandarkar [26]. The parameter that delimits the regimes of particle densities in which particles are either neutral or mostly negatively charged is the ion density: particles are expected to have on average a neutral average charge when their density exceeds the ion density and are expected to have a negative charge distribution when the density is lower than the ion density.

In addition to the effect of particle density in the plasma, the effect of collisions needs to be accounted for to obtain a more accurate description of charging of particles in plasma. In fact, recent experiments with micron-sized particles confirmed that the OML theory often overpredicts the particle charge by a factor of 2 to 3 [47,48]. In particular, charge exchange collisions between ions and neutrals can lead to the formation of ions that are trapped in the potential well around the particle, effectively increasing the ion current to the particle and decreasing the particle negative charge. Molecular dynamics simulations by Zobnin et al. [49] confirmed that collisions may lead to significant modifications of the particle charge compared to predictions of the OML theory. An analytical model developed by Lampe and coworkers [50,51] confirmed the results predicted by Zobnin.

Khrapak et al. [48] proposed an analytical model based on the idea of a capture radius, defined as the distance at which the total energy of an ion created after a charge exchange collision is smaller than $-k_B T_g$, with T_g the gas temperature. Below this distance, additional collisions with gas atoms cannot raise the ion energy to $E_{tot} > 0$, which is required for detrapping, and the ion is bound to be collected by the particle. Gatti and Kortshagen followed up on this idea and developed an analytical model that describes the particle charge over a wide range of particle diameters and gas pressures [52]. For particles smaller than about 100 nm, the capture radius R_0 can be approximated as

$$R_0(E_{kin}) = \frac{e|\Phi|}{E_{kin}} R_p \tag{3.4}$$

with E_{kin} the ion kinetic energy. The model describes the ion current as being composed of three components: a strictly collisionless OML component, a transitional "collision enhanced current" component, and a hydrodynamic component to describe the strongly collisional limit. These components are weighted with the respective probabilities of performing zero, one, and more than one collision in a sphere around the particle with radius R_0. Figure 3.5 shows the dependence of the particle normalized potential on the dimensionless capture radius Knudsen number, defined as $Kn_{R_0} = \lambda_i / 2\alpha R_0$, where λ_i is the ion mean free path and α a factor that accounts for an average over a Maxwellian ion distribution function. Large Kn_{R_0} correspond to the collisionless OML regime, while small Kn_{R_0} describe the hydrodynamic limit. Small enhancements in the ion collection probability are predicted already for Knudsen numbers $Kn_{R_0} \approx 10^4$—that is, $\lambda_i \approx O(10^6)R_p$, with R_p the particle radius. For $Kn_{R_0} \approx 1$ the OML theory overpredicts the particle potential and

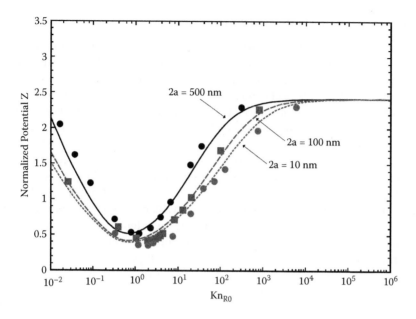

FIGURE 3.5

Comparison of the analytical model (lines) developed in Marco Gatti and Kortshagen (2008) with molecular dynamics simulations (symbols). Shown is the normalized particle potential $z = e|\Phi|/k_bT_e$ as a function of the ion capture radius Knudsen number $Kn_{R_0} = \lambda_i/2aR_0$, in which a is used to denote the particle radius. (Figure reproduced from Marco Gatti, and U. Kortshagen. 2008. Analytical model of particle charging in plasmas over a wide range of collisionality. *Phys. Rev. E* 78: 046402. With permission.)

charge by a factor of ~5. For this value, the OML theory underpredicts the ion and electron current to the particle by a factor of e^2 : 7.4. This large deviation from the OML theory has important consequences for the particle temperature in the plasma, as will be discussed later. At Knudsen numbers $Kn_{R_0} < 1$, the particle potential increases again, because the ion current to the particle gets more and more impeded by an increasing number of collisions.

For typical conditions in low-pressure plasmas, for example, a particle of $R_p = 10$ nm in an argon plasma at 1 Torr (133 Pa), one finds $Kn_{R_0} \approx 10^2$, for which the OML theory would overpredict the particle potential and charge by a factor of 1.66 and underpredict electron and ion currents to the particle by a factor of ~3. For smaller particles, the errors of the OML theory would be less severe. Moreover, in laboratory dusty plasmas the particle density may be high and most of the negative charges may reside on the particles, resulting in a reduced free electron density. As shown in Marco Gatti and Kortshagen [52], this leads to an overall lower particle charge and potential and less severe errors of the OML theory. Other effects may lead to the reduction of the negative particle charge as well such as ultraviolet (UV) photodetachment [26], relaxation of excited molecules at the surface, and tunneling emission from surface states with positive total energy [37].

3.2.3 Nanoparticle Heating in Plasmas

The selective heating of nanoparticles in plasmas remains an underappreciated aspect of plasma synthesis of nanocrystals. It is precisely this feature that makes plasmas attractive for the synthesis of covalently bonded semiconductor materials such as group IV and III-V semiconductors, which require high temperatures for crystal formation. Until recently, the formation of crystals in plasmas had posed a persistent puzzle [28,53,54]. Short-time sintering studies of silicon nanoparticles found crystallization temperatures of 773, 1073, 1173, and 1273 K for particles with diameters of 4, 6, 8, and 10 nm, respectively [17]. Yet, the background gas temperature in nonthermal plasmas is close to room temperature. For instance, Figure 3.6 shows silicon nanocrystals that were synthesized in a nonthermal plasma in which the gas temperature was measured to be <500 K and the nanoparticles residence time in the plasma is ~6 ms [28]. These results suggested that nanoparticles in plasmas can be at temperatures significantly higher than the gas temperature.

The fact that nanoparticles can reach temperatures higher than the gas temperature is well-known. Daugherty and Graves [55] reported particle temperatures exceeding the gas temperature by 75 K for 0.5 to 3 µm using spectroscopic fluorescence measurements. They also proposed an energy balance for the particles taking into account particle heating through electron-ion recombination and cooling through radiation and conduction. However, it was only recently that the magnitude of the nanoparticle heating in nonthermal plasma was understood. In Bapat et al. [27] the authors suggested that the particle temperature of ~35 nm–sized nanoparticles can

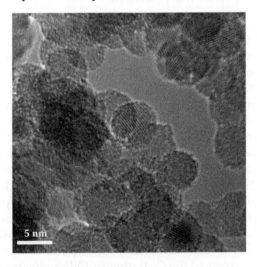

FIGURE 3.6
Silicon nanocrystals formed in a nonthermal plasma during a residence time of ~6 ms. (See Kortshagen, U., R. Anthony, R. Gresback et al. 2008. Plasma synthesis of group IV quantum dots for luminescence and photovoltaic applications. *Pure Appl. Chem.* 80: 1901.)

exceed the gas temperature by hundreds of Kelvin. In a study by Mangolini and Kortshagen [56], we extended previous investigations by examining the temperature history of nanoparticles with diameters <10 nm, such as the ones shown in Figure 3.6. The time-dependent energy balance was solved to predict the particle temperature history:

$$\frac{4}{3}\pi R_p^3 \rho C \frac{dT_p}{dt} = G - S \tag{3.5}$$

ρ is the density of the materials, silicon, C the specific heat, and T_p represents the particle temperature. On the right-hand side, the particle temperature is affected by the heat release term G and the heat loss term S. The loss term includes heat loss by conduction/convection. Radiation is not included because it is not important in the range of particle temperatures considered here [57]. The loss term is treated as a continuous term, because collision with the gas atoms occurs at a very high rate. Surface reactions, such as reaction with radicals and recombination of ions at the surface, lead to the particle heating through the heat release term G. In Mangolini and Kortshagen [56], we considered the case of silicon particles entrained in an argon-hydrogen plasma. We assumed that argon is the dominant ion and that the ionization energy of argon of 15.6 eV is released to the particle in every recombination event. The electron and ion capture frequencies were found using the traditional OML theory as we focused on very small particles. Hence, according to our discussion above, results shown here may even underestimate the effect of particle heating, because collisional effects would lead to a further increase of the ion and electron capture rates. Other surface reactions that were included in the model are reactions with hydrogen atoms, such as attachment to dangling bonds, hydrogen-induced abstraction through the Eley-Rideal mechanism [58], and hydrogen thermal desorption through the Langmuir-Hinshelwood mechanism. The energy released as a consequence of each of these events was carefully accounted for, and all surface reactions were treated stochastically using a Monte Carlo model.

The simulation was performed for plasma conditions that were measured in Mangolini et al. [28]. In Figure 3.7a, we show the excess temperature of small particles in a plasma (i.e., the difference between the particle temperature and the gas temperature, which was assumed to be 300 K in this simulation). The particle temperature of small particles is highly unsteady, with temperatures spikes due to individual exothermic surface reactions. Obviously, the smaller the particles the larger the extent of the temperature spikes, because a smaller particle has a smaller number of degrees of freedom to absorb the energy released. It is important to note that for the realistic plasma conditions in this simulation, the instantaneous particle temperature can easily exceed the crystallization temperature for small silicon particles that were reported in Hirasawa et al. [17].

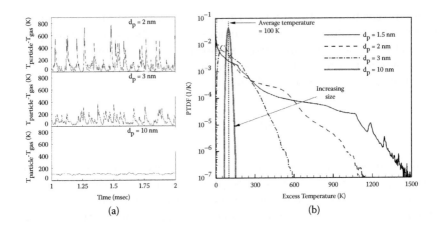

FIGURE 3.7
(a) Simulated temperature history of small nanoparticles in a nonthermal plasma. The plasma density is 5×10^{10} cm^{-3}, the gas temperature is 300 K. (b) Particle temperature distribution function as a function of particle size. The average excess temperature does not depend on the size and is equal to +100 K. (Reproduced from Mangolini, L., and U. Kortshagen. 2009. Selective nanoparticle heating: Another form of nonequilibrium in dusty plasmas. *Phys. Rev. E* 79: 026405. With permission.)

Given the unsteady temperature of small particles, we introduced the concept of a particle temperature distribution function (PTDF) that represents the probability of finding a single particle at a certain temperature. A typical PTDF for different-sized particles is shown in Figure 3.7b. Large particles have a narrow PTDF around the average excess temperature, while small particles have a broad PTDF because of the large temperature fluctuations.

Based on these results, we suggest that the strong heating of nanoparticles through energetic surface reactions, coupled to the slow cooling through conduction and convection at low pressure is likely responsible for the ability of nonthermal plasmas to generate nanocrystals of high crystallization temperature materials. Other mechanisms, such as hydrogen-induced crystallization [59], may play a role as well, and their importance remains to be evaluated.

3.3 Nonthermal Plasma Synthesis of Nanocrystals with Controlled Properties

3.3.1 Nanocrystal Synthesis

In this section, we will focus on processes that are capable of producing free-standing (i.e., nonembedded and nonagglomerated) nanocrystals. For the production of free-standing nanocrystals, it is critical to control the residence time of the particle in the discharge. Various approaches have been proposed to achieve this. Oda's group [60,61] developed a plasma reactor based on an ultra-high frequency capacitively coupled discharge. A schematic of the reactor is shown in Figure 3.8a. Nanoparticles are produced in a capacitively coupled plasma and extracted using periodic pulses of hydrogen gas. The high excitation frequency of 144 MHz used in these studies likely contributed to a higher plasma density than excitation at 13.56 MHz, which in turn probably contributed to the successful synthesis of nanocrystals. Significantly more narrowly dispersed size distributions were demonstrated than using a continuous discharge in silane.

Studies using pulsed capacitively coupled plasmas were reported [62,63]. In these approaches, the plasma is turned off in order to allow negatively charged nanoparticles that during the active plasma are confined by the ambipolar electric field to be extracted from the reactor. Both pulsed power and pulsed gas flow approaches appear to be more complicated than approaches that allow for continuous operation and extraction of particles.

Continuous plasma operation is enabled by plasma systems in which a strong gas flow is used to continuously extract particles from the plasma volume. The plasma excitation can be achieved through different modes of power coupling, including inductive, capacitive, or microwave excitation. Gorla et al. [64] used an inductively coupled plasma system, shown in Figure 3.8b, to produce a high-density plasma at a pressure of 200 mTorr (27 Pa) and a gas flow rate of ~1000 sccm (standard cubic centimeter per minute) to produce both silicon and germanium nanocrystals with sizes between 5 and 200 nm. The authors observed spherical particles with a modest degree of agglomeration, and also reported the observation of partially sintered, not fully coalesced particles. This suggests that the nanoparticles leave the plasma while they are still at fairly high temperatures, and then cool rapidly and remain "frozen" in their current state.

Wiggers and coworkers described a flow-through microwave sustained plasma reactor that is capable of producing silicon nanocrystals at high rates [65,66]. A schematic is shown in Figure 3.8c. The authors use a simple plasma model to predict the gas temperature, which they estimate to be as high as 6000 K. Nucleation of particles in such a hot gas is impossible because small clusters would evaporate, and the silicon nanocrystals likely form in the afterglow of the plasma where the gas temperature is sufficiently low. In fact the authors report that the nanocrystals have a log-normal size distribution [67], consistent with high agglomeration rates, which may indicate a low charge state of the nanoparticles, due to the quickly decreasing plasma density in the afterglow plasma.

Sankaran and Giapis [68] used a microdischarge at atmospheric pressure to dissociate the silane precursor and generate silicon particles (Figure 3.8d). The synthesis at atmospheric pressure is advantageous as it simplifies the further treatment of the nanocrystals. For instance, in the study by Sankaran

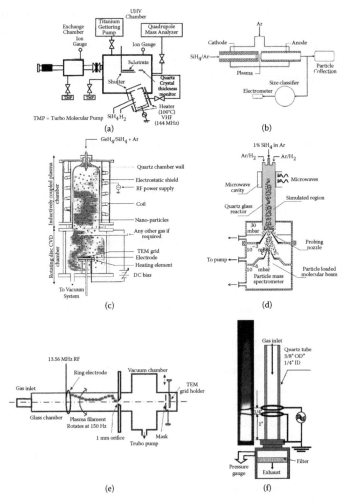

FIGURE 3.8

Different plasma reactors for the synthesis of nanocrystals: (a) ultra-high frequency discharge with pulsed gas injection. (From Ifuku, T., M. Otobe, A. Itoh, and S. Oda. 1997. Fabrication of nanocrystalline silicon with small spread of particle size by pulsed gas plasma. *Japan. J. Appl. Phys.* 36: 4031–4034. With permission.) Continuous flow reactors: (b) microdischarge plasma (From Sankaran, R. M., D. Holunga, R. C. Flagan, and K. P. Giapis. 2005. Synthesis of blue luminescent Si nanoparticles using atmospheric-pressure microdischarges. *Nano Lett.* 5: 531–535. With permission.), (c) ICP plasma reactor (From Gorla, C. R., S. Liang, G. S. Tompa, W. E. Mayo, and Y. Lu. 1997. Silicon and germanium nanoparticle formation in an inductively coupled plasma reactor. *J. Vac. Sci. Technol. A*, 15: 860. With permission.), (d) microwave plasma (From Giesen, B., H. Wiggers, A. Kowalik, and P. Roth. 2005. Formation of Si-nanoparticles in a microwave reactor: Comparison between experiments and modelling. *J. Nanoparticle Res.* 7: 29–41. With permission.), (e) filamented capacitive discharge (From Bapat, A., C. Anderson, C. R. Perrey et al. 2004. Plasma synthesis of single-crystal silicon nanoparticles for novel electronic device applications. *Plasma Phys. Controlled Fusion* 46: B97–B109. With permission.), (f) uniform capacitive plasma (From Mangolini, L., E. Thimsen, and U. Kortshagen. 2005. High-yield plasma synthesis of luminescent silicon nanocrystals. *Nano Lett.* 5: 655–659. With permission.)

et al. [68], particles were *in situ* classified by size using a differential mobility analyzer.

Different schemes using capacitive coupling of RF power were reported. In the first scheme [27,69] (Figure 3.8e), a capacitive plasma was formed by applying radiofrequency power at 13.56 MHz to a ring electrode that surrounded the reactor quartz tube with ~5 cm diameter. A grounded metal flange about 10 cm downstream of the ring electrode served as counterelectrode. The gas flow rate in the reactor was adjusted such that the pressure in the discharge tube was ~1.5 Torr (200 Pa) and the gas residence time in the plasma region was ~5 s. Using this approach, the authors could synthesize the almost perfect silicon nanocubes with an average size of ~35 nm, as well as faceted and spherical germanium nanocrystals [70]. In this scheme, the plasma often formed a rotating filament that was attributed to a thermal instability.

In Mangolini et al. [28], a flow-through capacitive plasma reactor for the synthesis of sub-10 nm semiconductor nanocrystals was introduced (Figure 3.8f). The pressure in this reactor is similar to the one in Bapat et al. [27], but higher flow rates are used to drag the particles out of the plasma zone and keep the residence times in the plasma in the range of 2 to 6 ms. Depending on the residence time, the diameter of the silicon nanocrystals studied in Mangolini et al. [28] could be adjusted between 3 and 6 nm by changing the residence time between 2 and 6 ms with little dependence on the precursor density. Size distributions were found to be rather narrow with standard deviations of 10% to 15% of the average particle size. In subsequent studies, the authors demonstrated that this approach could be extended to p- and n-doped silicon crystals (discussed below), as well as germanium [71], gallium nitride [72], indium phosphide [73], and silicon-germanium alloy [74].

3.3.2 Nanocrystal Doping

The electrical properties of semiconductors can be adjusted over wide ranges by introducing small amounts of impurity dopants [75], such as phosphorous or boron in the case of silicon. The controlled doping of semiconductor nanocrystals is also expected to have important consequences on the development of nanocrystal-based devices. For the case of II-VI and IV-VI semiconductor nanocrystals, doping was found to be a significant challenge [76,77], requiring a deep understanding of the kinetic effects involved in the inclusion of dopants in the particles during the growth process [76].

For group IV nanocrystals, some experience had been gained with approaches other than plasma synthesis. Baldwin et al. [78] reported a liquid phase approach capable of phosphorous doping of silicon crystals. However, the authors provided little details on the effects of dopants on the nanocrystals but focused more on nanocrystal films. Fujii and coworkers demonstrated the doping of Si nanocrystals embedded in silicon oxide with phosphorous (P) [79], boron (B) [80], and indium (In) [81]. The impurities were introduced by cosputtering of the dopants, Si and SiO_2. The resulting doped silicon-rich SiO$_x$ films were then annealed for extended periods of time at ~1100°C, leading to the formation of Si nanocrystals. The authors could show that the dopants were successfully incorporated into the Si nanocrystals and were electrically active [82,83].

Work on doping of free-standing nanocrystals with plasmas is in the beginning stages. Wiggers and coworkers reported producing doped silicon nanocrystals with their plasma approach [84,85] but focused on films based on nanocrystals and did not investigate the state of the dopants in the nanocrystals. Pi and Kortshagen [86] reported work on P and B doping of silicon crystals in the capacitively coupled plasma reactor shown in Figure 3.8f by introducing phosphine and di-borane dopant precursors together with the silane gas flow. The authors found a number of interesting results with respect to the incorporation of dopants into the silicon nanocrystals. For instance, they determined that P gets incorporated into the nanocrystals with about 100% efficiency (i.e., all P contained in the phosphine precursor is incorporated into the nanocrystals). Boron on the other hand was found to have a doping efficiency of only ~10%. Pi and Kortshagen also studied the influence of doping on the photoluminescence (PL) intensity of the silicon nanocrystals and found that B strongly quenches the PL even at low concentrations, while P doping at low concentrations enhances PL and leads to a quenching at higher concentration. These results were surprisingly consistent with those found in the studies by Fujii and coworkers [79,80], even though the synthesis and doping conditions used by these two groups were drastically different. Fujii and coworkers had attributed the PL enhancement at low P concentrations with the termination of silicon dangling bonds by P atoms. Finally, Pi and Kortshagen were able to determine the position of the dopant atoms in the plasma-produced silicon nanocrystals by removing the outer shell of the doped silicon nanocrystals. This was achieved by letting the particles oxidize and then removing the oxide shell with hydrofluoric acid. By comparing the dopant concentrations before and after etching, the authors determined that P mainly resides at the particle surface while B is incorporated into the particle core. These first studies have started to build an understanding of the doping of plasma-produced silicon nanocrystals, but there is still significant need for further studies to understand the impact of doping at both the nanocrystal and the nanoparticle film levels.

3.3.3 Plasma Surface Treatment and Functionalization of Nanocrystals

Controlling the surface of nanostructured materials is a difficult but crucial task. Optical and electronic properties strongly depend on the surface configuration, and surface functionalization also determines the processability of the nanoparticles. For nanocrystals grown in the liquid phase, coordinating ligands are often used to avoid particle agglomeration [5]. In plasma reactors, nanocrystals are created without ligands, which presents both a challenge and an opportunity. Plasma-produced nanocrystals will likely agglomerate during collection, and special steps need to be taken to deagglomerate the particles and allow for further processing. On the other hand, the fact that particles are created with bare surfaces enables one to experiment with various schemes to further modify the surface.

Imparting solubility of nanocrystal in specific solvents is of great importance. Solubility is achieved when the steric hindrance of organic molecules attached to the nanocrystal surfaces helps to overcome the van der Waals force between particles. Colloidal dispersions of semiconductor nanocrystals are useful for a range of applications from solution-processed devices to semiconductor films produced with printing or coating techniques. Several authors have demonstrated that colloidal suspensions can be produced from freestanding silicon nanocrystals [87–89].

The surface treatment with organic molecules such as 1-alkenes in order to produce stable nanocrystal colloids also has strong impact on the optical properties of the nanoparticles. In Jurbergs et al. [90] and Mangolini et al. [91], Mangolini, Kortshagen, and coworkers demonstrated that exceptionally efficient PL intensities can be achieved through surface functionalization with molecules such as 1-dodecene. Silicon nanocrystals covered with a native oxide layer have low PL quantum yields and are insoluble in most solvents, while silicon nanocrystals functionalized with 1-dodecene under careful avoidance of oxygen exhibited quantum yields as high as 60% to 70% [90,91]. Surface alkylation not only stabilizes the particle in solution, but it also removes dangling bonds, which act as nonradiative recombination centers and quench photoluminescence [92]. For the studies above, nanocrystals were generated in the gas phase with a nonthermal plasma and then transferred into the liquid phase for functionalization. This is a complex process that needs to be executed under oxygen-free and moisture-free conditions. The process is also time consuming, because functionalization in the liquid phase requires reaction times of several minutes up to hours. Mangolini and Kortshagen recently introduced an all-gas-phase approach to the surface functionalization of silicon nanocrystals [93]. In this scheme, silicon nanocrystals were synthesized using the plasma reactor described in Figure 3.8f and then immediately injected into a second, capacitively coupled plasma reactor, as shown in Figure 3.9. A vaporized organic monomer was added to the second stage of the reactor, where the plasma provides the activation energy for the surface grafting. This was concluded based on the fact that the reaction was not successful when the nanocrystals were simply exposed to the precursor vapor while not applying the second plasma discharge. The residence time of nanocrystals in the second plasma is approximately 1 second, but the same surface coverage with organic molecules could be achieved as in a many-hour-long liquid phase reaction. Moreover, different from the liquid phase reaction, in the two-stage plasma approach the functionalization is performed while the particles are nonagglomerated and still in flight in the second plasma, allowing for the functionalization with shorter ligands. Organic monomers with different functional groups were tested, showing

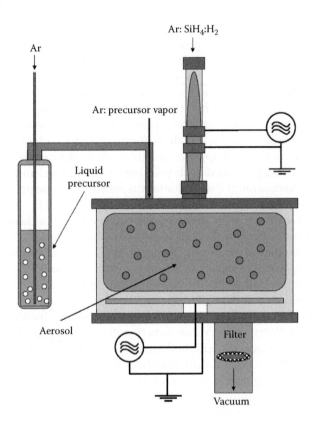

FIGURE 3.9
(See color insert.) Two-stage plasma for the synthesis and plasma-aided surface grafting of organic surfactant molecules. (Reproduced from Mangolini, L., and U. Kortshagen. 2007. Plasma-assisted synthesis of silicon nanocrystal inks. *Adv. Mater.* 19: 2513–2519. With permission.)

that the reactivity of the organic molecules with the nanocrystal surface was still largely controlled by the reactivity of a functional end group.

Another application of the two-stage plasma approach was presented in Pi et al. [94] and Liptak et al. [95]. In these studies, silicon nanocrystals were etched in the second plasma to reduce their size. Nanocrystals with diameters smaller than 3 nm are difficult to produce in nonthermal plasmas, because during the initial particle nucleation the cluster density often exceeds the positive ion density and agglomeration is fast [26]. With a two-stage plasma process the nanocrystal size can be reduced by performing an *in-flight* etch step in the second stage of the reactor. In Pi et al. [94,96], the authors applied an *in-flight* tetrafluorocarbon (CF_4) etch step in a second-stage plasma to the silicon nanocrystals synthesized in the first stage. This treatment enabled the authors to achieve silicon nanocrystals that luminesced across the entire range of the optical spectrum as shown earlier in Figure 3.1. Without etch treatment, nanocrystals remained larger and luminescence was limited to the orange red–near infrared range of the spectrum. A further benefit of the CF_4-treatment was that nanocrystals were covered in a CF-polymeric surface layer. In Liptak et al. [95], a similar etch treatment with SF_6, was reported, which left the nanocrystals with a fluorine terminated surface. This surface functionalization did not prevent the oxidation of silicon crystals, but it enabled comparatively high PL quantum yields for oxidized silicon crystals. The precise effect of fluorine on the optical properties of the silicon nanocrystals still remains to be understood.

3.4 Development of Devices Based on Plasma-Produced Nanocrystals

A wide array of devices that include semiconductor nanocrystals have been conceived, fabricated, and tested for many different applications. One of the first applications of semiconductor nanostructures was for electronic applications [97,98]. Single nanoparticle–based devices such as transistors have been seen as the ultimate in miniaturization. Oda's group in Nishiguchi and Oda [99] and Fu et al. [100] reported the fabrication of a single particle–based electronic device based on plasma-produced nanocrystals. The authors randomly deposited silicon nanocrystals and then made contact to them by depositing a polycrystalline silicon layer over them. The authors succeeded in fabricating a single particle transistor using a wrap-around gate, and reported electrical behavior that was consistent with single-electron transport through the nanocrystal. Single nanocrystal transistors were also demonstrated by Ding et al. [101]. The devices were based on the cubic silicon nanocrystals with an average size of 35 nm a wrap-around gate. Successful transistor behavior was demonstrated using I-V measurements with varying gate-source voltages. Electrical measurements showed the turn-off characteristics of a p-channel device.

Oda and coworkers successfully fabricated flash memory devices based on plasma-produced silicon nanocrystals [102,103]. In Hinds et al. [103], the authors first defined a field effect transistor channel by electron lithography and subsequently deposited silicon nanocrystals onto the channel directly from a plasma. The deposited silicon crystals were then embedded in a gate oxide, as shown in Figure 3.10. Clear signatures of charge trapping in (writing) and detrapping from (erasing) the silicon nanocrystals were found.

Although the field of single particle–based electronic devices shows great promise, in the last few years the interest has shifted toward the demonstration of devices that are based on ensembles of many particles. In Holman et al. [104], the authors demonstrated the first field effect transistor based on films of plasma produced silicon or germanium nanocrystals. Plasma produced silicon

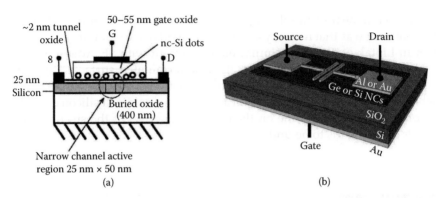

FIGURE 3.10
(a) A silicon-nanocrystal flash memory device. (From Hinds, B. J., T. Yamanaka, and S. Oda. 2001. Emission lifetime of polarizable charge stored in nano-crystalline Si based single-electron memory. *J. Appl. Phys.* 90: 6402–6408. With permission.) (b) A field effect transistor device using solution processed silicon or germanium nanocrystal films in the channel region. (From Holman, Z. C., C. Y. Liu, and U. R. Kortshagen. 2010. Germanium and silicon nanocrystal thin-film field-effect transistors from solution. *Nano Lett.* 10: 2661–2666. With permission.)

and germanium nanocrystals were dispersed in 1,2-dichlorobenzene, and 20 to 30 nm thick nanocrystal films were produced by spin-casting the dispersed nanocrystals onto prefabricated transistor substrates, see Figure 3.10b. Source and drain contacts were vacuum deposited onto the nanocrystal films. Both silicon and germanium films showed clear gating. The doping of the films was found to sensitively depend on the surface chemistry.

Nanostructured materials are also intensively investigated for thermoelectric devices. Thermoelectric devices convert thermal energy into electrical energy and are expected to help recover a part of the waste heat discharged from all thermal cycles. The effectiveness of thermoelectric materials is proportional to a dimensionless figure of merit

$$ZT = \frac{S^2 \sigma T}{k},$$

with S the Seebeck coefficient, σ the electrical conductivity, k the thermal conductivity of the material, and T the absolute temperature [105]. Many materials have ZT values around 1 [106], however, ZT larger ~4 is desired for efficient implementation of thermoelectric technology. To date, ZT exceeding unity have been achieved mainly form materials of the family (Bi_xSb_{1-x}) (Se_yTe_{1-y}) [105]. Nanograined materials have shown significant promise, because quantum-confinement effects may enhance the Seebeck coefficient S, while the reduction of the grain size is expected to inhibit phonon transport, thus reducing the thermal conductivity [107]. Poudel et al. [108] produced 20 nm, p-type BiSbTe via ball milling and then sintered them into a film via hot-pressing. The nanometer-sized structure resulted in a 40% improvement of the ZT figure of merit with respect to the bulk material, likely because of the reduction in thermal conductivity through enhanced phonon-scattering.

Plasma synthesis of nanoparticles is promising for silicon germanium alloys (SiGe), which are interesting for high temperature (>600°C) thermoelectric applications [106]. Joshi et al. [109] prepared ~20 nm particles of p-type doped $Si_{0.8}Ge_{0.2}$ through ball-milling and subsequently sintered the particles using hot-pressing. This nanostructured material showed ZT values as high a 0.95, which is a 50% improvement over the previous record based on p-type $Si_{0.8}Ge_{0.2}$ with grain sizes in the micrometer range [110]. Wiggers and coworkers recently reported a series of studies on the properties of films produced by laser annealing of plasma-produced silicon nanocrystals. Laser intensities exceeding 60 mJ/cm^2 resulted in partial melting and recrystallization of the silicon nanocrystal films and in a significant improvement of the electrical conductivity [66,84]. While ZT was not reported in these studies, the authors found a Seebeck coefficient of ~300 µV/K.

The application of semiconductor nanocrystals for photovoltaic energy conversion has attracted significant interest over the last few years. Various considerations, such as the effect of the emission of CO_2 on the global climate, the U.S. strategic need for energy independence, and the predicted increase in the cost of fossil fuels, motivate a renewed interest in solar electricity. At present electricity produced by direct solar-to-electricity conversion is still more expensive than electricity produced by burning coal, the cheapest form of electrical power currently available [111]. Reducing the cost of photovoltaic power generation is required to enable its widespread utilization. This can be achieved by developing new and cheaper manufacturing techniques and by increasing the power conversion efficiency.

The main parameter controlling the conversion efficiency of solar cells is the optical band gap of the absorbing material. The semiconductor layer absorbs photons with energies higher than the band gap, generating electron-hole pairs with the same energy of the absorbed photon. The charge carriers quickly relax to the band gap energy by interactions with the lattice, leading to a loss of useful energy and a loss of efficiency [112]. By analyzing these losses Shockley and Queisser [113] calculated a theoretical maximum efficiency of ~31% for a solar cell made from a single absorber under solar irradiation. Naturally, the reports of multi-exciton generation (MEG) in small semiconductor nanocrystals have generated great excitement, because the generation of multiple excited electron-hole pairs for each photon absorption event would enable the fabrication of solar cells that overcome the Shockley-Queisser limit [114]. MEG has been demonstrated in various nanocrystal materials [115–117], including plasma-produced silicon nanocrystals [118]. These reports used optical spectroscopy to probe the presence of multiple excitons in the nanocrystal volume after absorption of a single photon with energies higher than an integer multiple of the band gap energy. The first demonstration of electrical collection of charges generated via the MEG mechanism was given in Sambur et al. [119], but the device used for these experiments consisted of a single monolayer of lead sulfide quantum dots absorbed onto a TiO_2, a structure with poor optical absorption and low conversion efficiency. Moreover, the report in Sambur et al. [119] generated question with respect to the efficiency of the MEG process, and with respect to the energy needed to observe its onset. Such controversy was also ignited by the study reported in Nair and Bawendi [120], which failed to detect MEG in II-VI nanocrystals using optical spectroscopy. It is clear that MEG is not a well-understood process.

Semiconductor nanocrystals may also enable manufacturing processes that could significantly lower the cost of solar panels. Nanocrystals with the appropriate surface functionalization can be formulated into printable "inks" that can then be applied onto the substrate using nonvacuum, scalable roll-to-roll techniques. Talapin and Murray [121] first demonstrated that films cast from colloidal PbSe nanocrystals can be highly conductive. Similar approaches have been reported in the patent literature [122] for plasma-produced silicon nanocrystals. These approaches use thermal annealing [123] to sinter printed layers of plasma-produced semiconductor nanocrystals. The reduction in the melting point with respect of the bulk for small nanocrystals likely aids the sintering process.

A variety of other designs are promising for the production of low-cost, efficient solar cells. For instance, hybrid organic-inorganic solar cells combine the capability of producing low-cost organic solar cells with the good electrical transport properties of inorganic solar cells. In Liu et al. [124] the authors demonstrate the first hybrid organic-inorganic solar cell based on an organic hole conductor, poly-3-hexylthiophene (P3HT), and plasma-produced silicon nanocrystals, see Figure 3.11. This study built on previous studies of hybrid

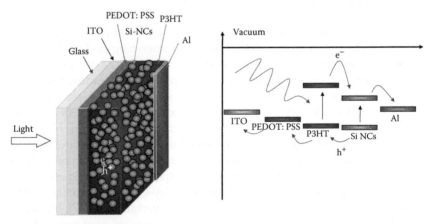

FIGURE 3.11
A hybrid solar cell based on a network of silicon nanocrystals and poly-3-hexylthiophene (P3HT). The silicon crystals act as electron transport medium, P3HT as a hole transport medium. The schematic energy level diagram demonstrates the transport of electrons and holes to their respective electrodes. (Reproduced from Liu, C.-Y., Z. C. Holman, and U. R. Kortshagen. 2009. Hybrid solar cells from P3HT and silicon nanocrystals. *Nano Lett.* 9: 449–452. With permission.)

solar cells [14,125] based on CdSe nanostructures. In Liu et al. [124], electrons were conducted in a percolating network of plasma-produced silicon nanocrystals while P3HT acted as the hole conductors. The conversion efficiency was about 1.2%, which is not sufficient for commercialization at this point, but it is a promising first step that leaves large room for improvement. Moreover, this device was fabricated using a single vacuum step (the plasma synthesis of nanocrystals) and was assembled without any vacuum or high-temperature step, raising the hope that such devices can be manufactured at very low cost.

Another interesting application of semiconductor nanocrystals is in light-emitting devices. Quantum confined nanocrystals exhibit size-tunable optical luminescence with narrow line widths, and their solution-processability makes them promising for low-cost display applications. Colvin et al. [11] presented the first light-emitting device based on CdSe nanocrystals with an external quantum yield of <0.1%. Today the highest reported external efficiencies of nanocrystal light-emitting devices are around 2% [126,127].

Most of the reports on nanocrystal-based LED have been based on II-VI quantum dots, which have excellent optical properties but have the obvious drawback of being composed of toxic heavy metals. Group IV quantum confined nanocrystals on the other hand are generally believed to be nontoxic, and have photoluminescence efficiency that equal those of II-VI quantum dots [91]. However, little work has been done to explore whether efficient light emission from electrically stimulated group IV quantum dots can be obtained. Ligman et al. [128] presented a device in which native oxide capped silicon nanocrystals were dispersed within a conductive polymer poly(9-vinylcarbazole) (PVK) [128]. The silicon nanocrystal-PVK hybrid devices showed simultaneous emission from PVK and silicon nanocrystals, however, with overall low external quantum efficiencies. More recently Kortshagen and coworkers demonstrated an improved hybrid organic-inorganic LED based on silicon quantum dots with a 0.6% external quantum yield [129]. In these devices, organic electron and hole injection layers were used to inject carriers into the emissive nanocrystal layer. In recent efforts, external quantum efficiencies of these devices could be improved to 8.6%, exceeding the reports of other nanocrystal systems by almost a factor of three.

3.5 Conclusions

Nonthermal plasmas offer unique advantages for nanocrystal synthesis when compared to other gas phase processes. They provide a highly reactive environment, which enables high production rates. They generate particles with a unipolar charge distribution when the particle density is lower than the ion density, which slows agglomeration and leads to the production of nanocrystals with small sizes and narrow size distributions. The nanoparticle charge also decreases diffusion losses of nanocrystals to the reactor wall, increasing the production rate. Exothermic reactions at the nanoparticle surface increase particle temperatures to several hundreds of Kelvin beyond the background gas temperature. This mechanism depends on particle size and is likely responsible for the formation of high-quality nanocrystals even for materials with high crystallization temperatures.

A variety of nonthermal plasma reactors for the synthesis of free-standing nanocrystals have been described. Beyond the synthesis of nanocrystals, plasmas can also be used to controllably dope the crystals and to perform *in-flight* surface chemistry. Surface functionalization with various ligands and surfactants is critical to terminate surface defects, stabilize the particles against oxidation in air, and impart solubility in various solvents.

Finally, we discussed how the properties of plasma-produced nanocrystals are of great interest for many applications, from the fabrication of nanoelectronic devices, to thermoelectric materials, to photovoltaic devices, to light-emitting devices. Promising examples of devices based on semiconductor nanocrystals have been discussed, proving that nonthermal plasmas are capable tools for the controlled, large-scale production of semiconductor nanocrystals that are hard to produce with liquid-phase synthesis or other gas phase approaches.

Acknowledgments

This work was supported in part by the National Science Foundation under grants CBET-0500332, CBET-0756326, and the MRSEC program of the National Science Foundation under grant DMR-0819885.

References

1. Selwyn, G. S., J. Singh, and R. S. Bennett, In situ laser diagnostic studies of plasma-generated particulate contamination. *Journal of Vacuum Science & Technology A*, 1989. 7(4): 2758–2765.
2. Butte, R., R. Meaudre, M. Meaudre, S. Vignoli, C. Longeaud, J. P. Kleider, and P. R. I. Cabarrocas, Some electronic and metastability properties of a new nanostructured material: hydrogenated polymorphous silicon. *Philosophical Magazine B-Physics of Condensed Matter Statistical Mechanics Electronic Optical and Magnetic Properties*, 1999. 79(7): 1079–1095.

3. Staebler, D. L. and C. R. Wronski, Reversible conductivity changes in discharge-produced amorphous Si. *Applied Physics Letters*, 1977. 31(4): 292–294.
4. Kortan, A. R., R. Hull, R. L. Opila, M. G. Bawendi, M. L. Steigerwald, P. J. Carroll, and L. E. Brus, Nucleation and growth of CdSe on ZnS quantum crystallite seeds, and vice versa, in inverse micelle media. I, 1990. 112: 1327–1332.
5. Murray, C. B., D. J. Norris, and M. G. Bawendi, Synthesis and characterization of nearly monodisperse CdE (E=S, Se, Te) semiconductor nanocrystallites. I, 1993. 115: 8706–8715.
6. Brus, L., Quantum crystallites and nonlinear optics. Applied physics A: Materials science & processing, 1991. 53(6): 465–474.
7. Alivisatos, A. P., Semiconductor Clusters, Nanocrystals, and Quantum Dots. I, 1996. 271(5251): 933–937.
8. Kovalev, D., H. Heckler, G. Polisski, and F. Koch, Optical properties of Si nanocrystals. *Physica Status Solidi (b)*, 1999. 215(2): 871–932.
9. Goldstein, A. N., C. M. Echer, and A. P. Alivisatos, Melting in semiconductor nanocrystals. *Science*, 1992. 256(5062): 1425–1427.
10. Gerberich, W. W., W. M. Mook, C. R. Perrey, C. B. Carter, M. I. Baskes, R. Mukherjee, A. Gidwani, J. Heberlein, P. H. McMurry, and S. L. Girshick, Superhard silicon nanospheres. *Journal of the Mechanics and Physics of Solids*, 2003. 51(6): 979–992.
11. Colvin, V. L., M. C. Schlamp, and A. P. Alivisatos, Light-emitting diodes made from cadmium selenide nanocrystals and a semiconducting polymer. *Nature*, 1994. 370(6488): 354–357.
12. Coe, S., W.-K. Woo, M. Bawendi, and V. Bulovic, Electroluminescence from single monolayers of nanocrystals in molecular organic devices. *Nature*, 2002. 420(6917): 800–803.
13. Huynh, W. U., X. Peng, and A. P. Alivisatos, CdSe nanocrystal rods/poly(3-hexylthiophene) composite photovoltaic devices. *Advanced Materials*, 1999. 11(11): 923–927.
14. Huynh, W. U., J. J. Dittmer, and A. P. Alivisatos, Hybrid nanorod-polymer solar cells. *Science*, 2002. 295(5564): 2425–2427.
15. Leschkies, K. S., R. Divakar, J. Basu, E. Enache-Pommer, J. E. Boercker, C. B. Carter, U. Kortshagen, D. J. Norris, and E. S. Aydil, Photosensitization of ZnO nanowires with CdSe quantum dots for photovoltaic devices. *Nano Letters*, 2007.
16. Dubertret, B., Skourides, D. J. Norris, V. Noireaux, A. H. Brivanlou, and A. Libchaber, In vivo imaging of quantum dots encapsulated in phospholipid micelles. *Science*, 2002. 298(5599): 1759–1762.
17. Hirasawa, M., T. Orii, and T. Seto, Size-dependent crystallization of Si nanoparticles. *Applied Physics Letters*, 2006. 88: 093119/1–093119/3.
18. Littau, K. A., P. J. Szajowski, A. J. Muller, A. R. Kortan, and L. E. Brus, A luminescent silicon nanocrystal colloid via a high-temperature aerosol reaction. *J. Phys. Chem.*, 1993. 97: 1224–1230.
19. Ostraat, M. L., J. W. De Blauwe, M. L. Green, L. D. Bell, M. L. Brongersma, J. Casperson, R. C. Flagan, and H. A. Atwater, Synthesis and characterization of aerosol silicon nanocrystal nonvolatile floating-gate memory devices. *Applied Physics Letters*, 2001. 79(3): 433–435.
20. Batson, P. E. and J. R. Heath, Electron energy loss spectroscopy of single silicon nanocrystals: The conduction band. *Physical Review Letters*, 1993. 71(6): 911.

21. Ehbrecht, M. and F. Huisken, Gas-phase characterization of silicon nanoclusters produced by laser pyrolysis of silane. *Phys. Rev. B*, 1999. 59(4): 2975–2985.
22. Li, X., Y. He, S. S. Talukdar, and M. T. Swihart, Process for preparing macroscopic quantities of brightly photoluminescent silicon nanoparticles with emission spanning the visible spectrum. *Langmuir*, 2003. 19: 8490–8496.
23. Cui, C. and J. Goree, Fluctuations of the charge on a dust grain in a plasma. *IEEE Transactions on Plasma Science*, 1994. 22(2): 151–158.
24. Matsoukas, T. and M. Russell, Particle charging in low-pressure plasmas. *Journal of Applied Physics*, 1995. 77(9): 4285–4292.
25. Schweigert, V. A. and I. V. Schweigert, Coagulation in low-temperature plasmas. *J. Phys. D: Appl. Phys.*, 1996. 29: 655.
26. Kortshagen, U. and U. Bhandarkar, Modeling of particulate coagulation in low pressure plasmas. *Phys. Rev. E*, 1999. 60(1): 887–898.
27. Bapat, A., C. Anderson, C. R. Perrey, C. B. Carter, S. Campbell, and U. Kortshagen, Plasma synthesis of single crystal silicon nanoparticles for novel electronic device application. *Plasma Physics and Controlled Fusion*, 2004. 46: B97.
28. Mangolini, L., E. Thimsen, and U. Kortshagen, High-yield plasma synthesis of luminescent silicon nanocrystals. *Nano Letters*, 2005. 5(4): 655–659.
29. Watanabe, Y., Formation and behaviour of nano/micro-particles in low pressure plasmas. *Journal of Physics D-Applied Physics*, 2006. 39(19): R329–R361.
30. Swihart, M. T., Electron affinities of selected hydrogenated silicon clusters (SixHy, x = 1–7, y = 0–15) from density functional theory calculations. *Journal of Physical Chemistry A*, 2000. 104(25): 6083–6087.
31. Howling, A. A., J.-L. Dorier, and C. Hollenstein, Negative ion mass spectra and particulate formation in radio frequency silane plasma deposition experiments, in *Appl. Phys. Lett.* 1993. 1341.
32. Howling, A. A., L. Sansonnens, J. L. Dorier, and C. Hollenstein, Time-resolved measurements of highly polymerized negative ions in radio frequency silane plasma deposition experiments. *Journal of Applied Physics*, 1994. 75(3): 1340–1353.
33. Howling, A. A., C. Courteille, J. L. Dorier, L. Sansonnens, and C. Hollenstein, From molecules to particles in silane plasmas. *Pure and Applied Chemistry*, 1996. 68(5): 1017–1022.
34. Perrin, J., C. Böhm, R. Etemadi, and A. Lloret, Possible routes for cluster growth and particle formation in RF silane discharges. *Plasma Sources Sci. Technol.*, 1994. 3: 252.
35. Fridman, A. A., L. Boufendi, T. Hbid, B. V. Potapkin, and A. Bouchoule, Dusty plasma formation: physics and critical phenomena. Theoretical approach. *Journal of Applied Physics*, 1995. 79(3): 1303–1314.
36. Bouchoule, A. and L. Boufendi, Particulate formation and dusty plasma behaviour in argon-silane RF discharge. *Plasma Sources Sci. Technol.*, 1993. 2: 204.
37. Gallagher, A., Model of particle growth in silane discharges. *Physical Review E*, 2000. 62(2): 2690.
38. Gallagher, A., A. A. Howling, and C. Hollenstein, Anion reactions in silane plasma. *Journal of Applied Physics*, 2002. 91(9): 5571–5580.
39. Kortshagen, U. R., U. V. Bhandarkar, M. T. Swihart, and S. L. Girshick, Generation and growth of nanoparticles in low-pressure plasmas. *Pure and Applied Chemistry*, 1999. 71(10): 1871–1877.

40. Bhandarkar, U., U. Kortshagen, and S. L. Girshick, Numerical study of the effect of gas temperature on the time for onset of particle nucleation in argon-silane low-pressure plasmas. *J. Phys. D: Appl. Phys.*, 2003. 36: 1399–1408.

41. Bhandarkar, U., M. T. Swihart, S. L. Girshick, and U. Kortshagen, Modelling of silicon hydride clustering in a low-pressure silane plasma. *J. Phys. D: Appl. Phys.*, 2000. 33: 2731–2746.

42. Watanabe, Y., M. Shiratani, T. Fukuzawa, H. Kawasaki, Y. Ueda, S. Singh, and H. Ohkura, Contribution of short lifetime radicals to the growth of particles in SiH4 high frequency discharges and the effects of particles on deposited films. *Journal of Vacuum Science & Technology A*, 1996. 14(3): 995–1001.

43. Kawasaki, H., H. Ohkura, T. Fukuzawa, M. Shiratani, Y. Watanabe, Y. Yamamoto, S. Suganuma, M. Hori, and T. Goto, Roles of SiH3 and SiH2 radicals in particle growth in rf silane plasmas. *Japanese Journal of Applied Physics Part 1-Regular Papers Short Notes & Review Papers*, 1997. 36(7B): 4985–4988.

44. Bernstein, I. B. and I. N. Rabinowitz, Theory of electrostatic probes in a low-density plasma. *Physics of Fluids*, 1959. 2(2): 112–121.

45. Allen, J. E., B. M. Annaratone, and U. de Angelis, On the orbital motion limited theory for a small body at floating potential in a Maxwellian plasma. *Journal of Plasma Physics*, 2000. 63: 299–309.

46. Matsoukas, T., M. Russell, and M. Smith, Stochastic charge fluctuations in dusty plasmas. *Journal of Vacuum Science & Technology a-Vacuum Surfaces and Films*, 1996. 14(2): 624–630.

47. Ratynskaia, S., S. Khrapak, A. Zobnin, M. H. Thoma, M. Kretschmer, A. Usachev, V. Yaroshenko, R. A. Quinn, G. E. Morfill, O. Petrov, and V. Fortov, Experimental determination of dust-particle charge in a discharge plasma at elevated pressures. *Physical Review Letters*, 2004. 93(8): 085001.

48. Khrapak, S. A., S. V. Ratynskaia, A. V. Zobnin, A. D. Usachev, V. V. Yaroshenko, M. H. Thoma, M. Kretschmer, H. Hofner, G. E. Morfill, O. F. Petrov, and V. E. Fortov, Particle charge in the bulk of gas discharges. *Physical Review E*, 2005. 72(1): 016406.

49. Zobnin, A., A. Nefedov, V. Sinel'shchikov, and V. Fortov, On the charge of dust particles in a low-pressure gas discharge plasma. *Journal of Experimental and Theoretical Physics*, 2000. 91(3): 483–487.

50. Lampe, M., V. Gavrishchaka, G. Ganguli, and G. Joyce, Effect of trapped ions on shielding of a charged spherical object in a plasma. *Physical Review Letters*, 2001. 86(23): 5278.

51. Lampe, M., R. Goswami, Z. Sternovsky, S. Robertson, V. Gavrishchaka, G. Ganguli, and G. Joyce, Trapped ion effect on shielding, current flow, and charging of a small object in a plasma. *Journal of Applied Physics*, 2003. 10(5): 1500–1513.

52. Gatti, M. and U. Kortshagen, Analytical model of particle charging in plasmas over a wide range of collisionality. *Physical Review E*, 2008. 78: 046402 1–6.

53. Cabarrocas, P. R.i., Gay, and A. Hadjadj, Experimental evidence for nanoparticle deposition in continuous argon-silane plasmas: Effects of silicon nanoparticles on film properties. *J. Vac. Sci. Technol. A*, 1996. 14(2): 655–659.

54. Bapat, A., C. R. Perrey, S. A. Campbell, C. B. Carter, and U. Kortshagen, Synthesis of highly oriented, single-crystal silicon nanoparticles in a low-pressure, inductively coupled plasma. *J. Appl. Phys.*, 2003. 94(3): 1969–1974.

55. Daugherty, J. E. and D. B. Graves, Particulate temperature in radio frequency glow discharges. *Journal of Vacuum Science & Technology A*, 1993. 11(4): 1126–1131.
56. Mangolini, L. and U. Kortshagen, Selective nanoparticle heating: another form of nonequilibrium in dusty plasmas. *Physical Review E*, 2009. 79: 026405 1–8.
57. Frenzel, U., U. Hammer, H. Westje, and D. Kreisle, Radiative cooling of free metal clusters. *Zeitschrift für Physik D Atoms, Molecules and Clusters*, 1997. 40(1): 108–110.
58. Koleske, D. D., S. M. Gates, and B. Jackson, Atomic H abstraction of surface H on Si: an Eley-Rideal mechanism? *J. Chem. Phys.*, 1994. 101(4): 3301–3309.
59. Sriraman, S., S. Agarwal, E. S. Aydil, and D. Maroudas, Mechanism of hydrogen-induced crystallization of amorphous silicon. *Nature*, 2002. 418: 62–65.
60. Otobe, M., T. Kanai, T. Ifuku, H. Yajima, and S. Oda, Nanocrystalline silicon formation in a SiH$_4$ plasma cell. *Journal of Non Crystalline Solids*, 1996. 198–200: 875–878.
61. Ifuku, T., M. Otobe, A. Itoh, and S. Oda, Fabrication of nanocrystalline silicon with small spread of particle size by pulsed gas plasma. *Japan. J. Appl. Phys.*, 1997. 36: 4031–4034.
62. Roca i Cabarrocas, P., T. Nguyen-Tran, Y. Djeridane, A. Abramov, E. Johnson, and G. Patrairche, Synthesis of silicon nanocrystals in silane plasmas for nanoelectronics and large area electronic devices. *J. Phys. D: Appl. Phys.*, 2007. 40: 2258–2266.
63. Viera, G., M. Mikikian, and E. Bertran, Atomic structure of the nanocrystalline Si particles appearing in nanostructured Si thin films produced in low-temperature radiofrequency plasmas. *Journal of Applied Physics*, 2002. 92(8): 4684–4694.
64. Gorla, C. R., S. Liang, G. S. Tompa, W. E. Mayo, and Y. Lu, Silicon and germanium nanoparticle formation in an inductively coupled plasma reactor. *J. Vac. Sci. Technol. A*, 1997. 15: 860.
65. Knipping, J., H. Wiggers, B. Rellinghaus, Roth, D. Konjhodzic, and C. Meier, Synthesis of high purity silicon nanoparticles in a low pressure microwave reactor. *Journal of Nanoscience and Nanotechnology*, 2004. 4(8): 1039–1044.
66. Lechner, R., H. Wiggers, A. Ebbers, J. Steiger, M. S. Brandt, and M. Stutzmann, Thermoelectric effect in laser annealed printed nanocrystalline silicon layers. *Physica Status Solidi-Rapid Research Letters*, 2007. 1(6): 262–264.
67. Giesen, B., H. Wiggers, A. Kowalik, and Roth, Formation of Si-nanoparticles in a microwave reactor: comparison between experiments and modelling. *Journal of Nanoparticle Research*, 2005. 7: 29–41.
68. Sankaran, R. M., D. Holunga, R. C. Flagan, and K. P. Giapis, Synthesis of blue luminescent Si nanoparticles using atmospheric-pressure microdischarges. *Nano Letters*, 2005. 5(3): 537–541.
69. Bapat, A., M. Gatti, Y.-P. Ding, S. A. Campbell, and U. Kortshagen, A plasma process for the synthesis of cubic-shaped silicon nanocrystals for nanoelectronic devices. *J. Phys. D: Appl. Phys.*, 2007. 40: 2247–2257.
70. Cernetti, P., R. Gresback, S. A. Campbell, and U. Kortshagen, Nonthermal plasma synthesis of faceted germanium nanocrystals. *Chemical Vapor Deposition*, 2007. 13(6–7): 345–350.
71. Gresback, R., Z. Holman, and U. Kortshagen, Nonthermal plasma synthesis of size-controlled, monodisperse, freestanding germanium nanocrystals. *Applied Physics Letters*, 2007. 91(9): 093119.

72. Anthony, R., E. Thimsen, J. Johnson, S. Campbell, and U. Kortshagen. A nonthermal plasma reactor for the synthesis of gallium nitride nanocrystals. In *Materials Research Society* 2006.

73. Gresback, R., R. Hue, W. L. Gladfelter, and U. R. Kortshagen, Combined plasma gas-phase synthesis and colloidal processing of InP/ZnS core/shell nanocrystals. *Nanoscale Research Letters*, 2011. 6.

74. Pi, X. D. and U. Kortshagen, Nonthermal plasma synthesized freestanding silicon-germanium alloy nanocrystals. *Nanotechnology*, 2009. 20: 295602 1–6.

75. Sze, S.M., *Physics of semiconductor devices*. 1981, New York: John Wiley & Sons.

76. Erwin, S. C., L. Zu, M. I. Haftel, A. L. Efros, T. A. Kennedy, and D. J. Norris, Doping semiconductor nanocrystals. *Nature*, 2005. 436: 91–94.

77. Norris, D. J., A. L. Efros, and S. C. Erwin, Doped nanocrystals. *Science*, 2008. 319(5871): 1776–1779.

78. Baldwin, R. K., J. Zou, K. A. Pettigrew, G. J. Yeagle, R. D. Britt, and S. M. Kauzlarich, The preparation of a phosphorus doped silicon film from phosphorus containing silicon nanoparticles. *Chemical Communications*, 2006(6): 658–660.

79. Fujii, M., A. Mimura, S. Hayashi, and K. Yamamoto, Photoluminescence from Si nanocrystals dispersed in phosphosilicate glass thin films: Improvement of photoluminescence efficiency. *Applied Physics Letters*, 1999. 75(2): 184–186.

80. Fujii, M., S. Hayashi, and K. Yamamoto, Photoluminescence from B-doped Si nanocrystals. *Journal of Applied Physics*, 1998. 83(12): 7953–7957.

81. Matsumoto, K., M. Fujii, and S. Hayashi, Photoluminescence from Si nanocrystals embedded in in doped SiO2. *Japanese Journal of Applied Physics Part 2-Letters & Express Letters*, 2006. 45(12–16): L450–L452.

82. Mimura, A., M. Fujii, S. Hayashi, D. Kovalev, and F. Koch, Photoluminescence and free-electron absorption in heavily phosphorus-doped Si nanocrystals. *Physical Review B*, 2000. 62(19): 12625–12627.

83. Fujii, M., Y. Yamaguchi, Y. Takase, K. Ninomiya, and S. Hayashi, Photoluminescence from impurity codoped and compensated Si nanocrystals. *Applied Physics Letters*, 2005. 87(21).

84. Lechner, R., A. R. Stegner, R. N. Pereira, R. Dietmueller, M. S. Brandt, A. Ebbers, M. Trocha, H. Wiggers, and M. Stutzmann, Electronic properties of doped silicon nanocrystal films. *Journal of Applied Physics*, 2008. 104(5).

85. Stegner, A. R., R. N. Pereira, K. Klein, R. Lechner, R. Dietmueller, M. S. Brandt, M. Stutzmann, and H. Wiggers, Electronic transport in phosphorus-doped silicon nanocrystal networks. *Physical Review Letters*, 2008. 100(2): 026803.

86. Pi, X. D., R. Gresback, R. W. Liptak, S. A. Campbell, and U. Kortshagen, Doping efficiency, dopant location, and oxidation of Si nanocrystals. *Applied Physics Letters*, 2008. 92: 123102 1–3.

87. Li, X., Y. He, and M. T. Swihart, Surface functionalization of silicon nanoparticles produced by laser-driven pyrolysis of silane followed by HF-HNO$_3$ etching. *Langmuir*, 2004. 20: 4720–4727.

88. Hua, F., M. T. Swihart, and E. Ruckenstein, Efficient surface grafting of luminescent silicon quantum dots by photoinitiated hydrosilylation. *Langmuir*, 2005. 21(13): 6054–6062.

89. Nelles, J., D. Sendor, A. Ebbers, F. M. Petrat, H. Wiggers, C. Schulz, and U. Simon, Functionalization of silicon nanoparticles via hydrosilylation with 1-alkenes. *Colloid Polymer Science*, 2007. 285: 729–736.

90. Mangolini, L., D. Jurbergs, E. Rogojina, and U. Kortshagen, High efficiency photoluminescence from silicon nanocrystals prepared by plasma synthesis and organic surface passivation. *Physica Status Solidi (c)*, 2006. 3(11): 3975–3978.

91. Jurbergs, D., E. Rogojina, L. Mangolini, and U. Kortshagen, Silicon nanocrystals with ensemble quantum yields exceeding 60%. *Applied Physics Letters*, 2006. 88: 233116 1–3.

92. Godefroo, S., M. Hayne, M. Jivanescu, A. Stesmans, M. Zacharias, O. I. Lebedev, G. Van Tendeloo, and V. V. Moshchalkov, Classification and control of the origin of photoluminescence from Si nanocrystals. *Nature Nanotechnology*, 2008. 3(3): 174–178.

93. Mangolini, L. and U. Kortshagen, Plasma-assisted synthesis of silicon nanocrystal inks. *Advanced Materials*, 2007. 19: 2513–2519.

94. Pi, X. D., R. W. Liptak, J. D. Nowak, N. Pwells, C. B. Carter, S. A. Campbell, and U. Kortshagen, Air-stable full-visible-spectrum emission from silicon nanocrystals synthesized by an all-gas-phase plasma approach. *Nanotechnology*, 2008. 19(24).

95. Liptak, R. W., B. Devetter, J. H. Thomas, U. Kortshagen, and S. A. Campbell, SF(6) plasma etching of silicon nanocrystals. *Nanotechnology*, 2009. 20(3).

96. Pi, X. D., R. W. Liptak, S. A. Campbell, and U. Kortshagen, In-flight dry etching of plasma-synthesized silicon nanocrystals. *Applied Physics Letters*, 2007. 91(8).

97. Cui, Y. and C.M. Lieber, Functional nanoscale electronic devices assembled using silicon nanowire building blocks. *Science*, 2001. 291(5505): 851–853.

98. Huang, Y., X. F. Duan, Y. Cui, L. J. Lauhon, K. H. Kim, and C. M. Lieber, Logic gates and computation from assembled nanowire building blocks. *Science*, 2001. 294(5545): 1313–1317.

99. Nishiguchi, K. and S. Oda, Electron transport in a single silicon quantum structure using a vertical silicon probe. *Journal of Applied Physics*, 2000. 88(7): 4186–4190.

100. Fu, Y., M. Willander, A. Dutta, and S. Oda, Carrier conduction in a Si-nanocrystal-based single-electron transistor-II. Effect of drain bias. *Superlattices and Microstructures*, 2000. 28(3): 189–198.

101. Ding, Y., Y. Dong, A. Bapat, J. D. Nowak, C. B. Carter, U. R. Kortshagen, and S. E. Campbell, Single nanoparticle semiconductor devices. *IEEE Transactions on Electron Devices*, 2006. 53(10): 2525–2530.

102. Banerjee, S., S. Huang, T. Yamanaka, and S. Oda, Evidence of storing and erasing of electrons in a nanocrystalline-Si based memory device at 77 K. *J. Vac. Sci. Technol. B*, 2002. 20(3): 1135–1138.

103. Hinds, B. J., T. Yamanaka, and S. Oda, Emission lifetime of polarizable charge stored in nano-crystalline Si based single-electron memory. *Journal of Applied Physics*, 2001. 90(12): 6402–6408.

104. Holman, Z. C., C.-Y. Liu, and U. R. Kortshagen, Germanium and silicon nanocrystal thin-film field-effect transistors from solution. *Nano Letters*, 2010. 10(7): 2661–2666.

105. Rowe, D. M., *Handbook of thermoelectric materials*. 1995, Boca Faton, FL: CRC Press.

106. Snyder, G. J. and E. S. Toberer, Complex thermoelectric materials. *Nature Materials*, 2008. 7(2): 105–114.

107. Dresselhaus, M. S., G. Chen, M. Y. Tang, R. G. Yang, H. Lee, D. Z. Wang, Z. F. Ren, J. P. Fleurial, and Gogna, New directions for low-dimensional thermoelectric materials. *Advanced Materials*, 2007. 19(8): 1043–1053.

108. Poudel, B., Q. Hao, Y. Ma, Y. C. Lan, A. Minnich, B. Yu, X. A. Yan, D. Z. Wang, A. Muto, D. Vashaee, X. Y. Chen, J. M. Liu, M. S. Dresselhaus, G. Chen, and Z. F. Ren, High-thermoelectric performance of nanostructured bismuth antimony telluride bulk alloys. *Science*, 2008. 320(5876): 634–638.

109. Joshi, G., H. Lee, Y. C. Lan, X. W. Wang, G. H. Zhu, D. Z. Wang, R. W. Gould, D.C. Cuff, M.Y. Tang, M.S. Dresselhaus, G. Chen, and Z.F. Ren, Enhanced thermoelectric figure-of-merit in nanostructured p-type silicon germanium bulk alloys. *Nano Letters*, 2008. 8(12): 4670–4674.

110. Vining, C. B., W. Laskow, J. O. Hanson, R. R. Vanderbeck, and P. D. Gorsuch, Thermoelectric properties of pressure-sintered SI0.8GE0.2 thermoelectric alloys. *Journal of Applied Physics*, 1991. 69(8): 4333–4340.

111. Lewis, N. S., G. Crabtree, A. J. Nozik, M. R. Wasielewski, and Alivisatos, Basic research needs for solar energy utilization, O.o.B.E.S. US Department of Energy, Editor. 2005.

112. Green, M. A., Third generation photovoltaics: Ultra-high conversion efficiency at low cost. *Progress in Photovoltaics*, 2001. 9(2): 123–135.

113. Shockley, W. and H. J. Queisser, Detailed balance limit of efficiency of p-n junction solar cells. *Journal of Applied Physics*, 1961. 32(3): 510–519.

114. Hanna, M. C. and A. J. Nozik, Solar conversion efficiency of photovoltaic and photoelectrolysis cells with carrier multiplication absorbers. *Journal of Applied Physics*, 2006. 100(7).

115. Schaller, R. D., M. Sykora, J. M. Pietryga, and V. I. Klimov, Seven excitons at a cost of one: redefining the limits for conversion efficiency of photons into charge carriers. *Nano Letters*, 2006. 6(3): 424–429.

116. Schaller, R. D. and V. I. Klimov, High efficiency carrier multiplication in PbSe nanocrystals: implications for solar energy conversion. *Physical Review Letters*, 2004. 92(18): 186601 1–4.

117. Ellingson, R. J., M. C. Beard, J. C. Johnson, U. Yu, O. I. Micic, A. J. Nozik, A. Shabaev, and A. L. Efros, Highly efficient multiple exciton generation in colloidal PbSe and PbS quantum dots. *Nano Letters*, 2005. 5(5): 865–871.

118. Beard, M. C., K. P. Knutsen, U. Yu, J. M. Luther, Q. Song, W. K. Metzger, R. J. Ellingson, and A. J. Nozik, Multiple exciton generation in colloidal silicon nanocrystals. *Nano Letters*, 2007. 7(8): 2506–2512.

119. Sambur, J. B., T. Novet, and B. A. Parkinson, Multiple Exciton Collection in a Sensitized Photovoltaic System. *Science*, 2010. 330(6000): 63–66.

120. Nair, G. and M. G. Bawendi, Carrier multiplication yields of CdSe and CdTe nanocrystals by transient photoluminescence spectroscopy. *Physical Review B*, 2007. 76: 081304 1–4.

121. Talapin, D. V. and C. B. Murray, PbSe nanocrystal solids for n- and p-channel thin film field-effect transistors. *Science*, 2005. 310: 86–89.

122. Lemmi, F., E. V. Rogojina, Yu, D. Jurbergs, H. Antoniadis, and M. Kelman. Methods for creating a densified group IV semiconductor nanoparticle thin film. 2008. Innovalight, Inc.

123. Bet, S. and A. Kar, Laser forming of silicon films using nanoparticle precursor. *Journal of Electronic Materials*, 2006. 35(5): 993–1004.

124. Liu, C.-Y., Z. C. Holman, and U. R. Kortshagen, Hybrid solar cells from P3HT and silicon nanocrystals. *Nano Letters*, 2009. 9(1): 449–452.
125. Gur, I., N. A. Fromer, C.-P. Chen, A. G. Kanaras, and A. P. Alivisatos, Hybrid solar cells with prescribed nanoscale morphologies based on hyperbranched semiconductor nanocrystals. *Nano Letters*, 2007. 7(2): 409-414.
126. Coe-Sullivan, S., Large-area ordered quantum-dot monolayers via phase separation during spin-casting. *Advanced Functional Materials*, 2005. 15(7): 1117–1124.
127. Niu, Y. H., A. M. Munro, Y. J. Cheng, Y. Q. Tian, M. S. Liu, J. L. Zhao, J. A. Bardecker, I. Jen-La Plante, D. S. Ginger, and A. K. Y. Jen, Improved performance light-emitting diodes quantum dot layer. *Advanced Materials*, 2007. 19(20): 3371.
128. Ligman, R. K., L. Mangolini, U. Kortshagen, and S. A. Campbell, Electroluminescence from surface oxidized silicon nanoparticles dispersed within a polymer matrix. *Applied Physics Letters*, 2007. 90: 061116 1-3.
129. Cheng, K.-Y., R. J. Anthony, U. Kortshagen, and R. J. Holmes, Hybrid silicon nanocrystals-organic light-emitting devices for infrared electroluminescence. *Nano Letters*, 2010.
130. Cheng, K.-Y., R. Anthony, U. R. Kortshagen, and R. J. Holmes, High-Efficiency Silicon Nanocrystal Light-Emitting Devices. *Nano Letters*, 2011.

4

Microscale Plasmas for Metal and Metal Oxide Nanoparticle Synthesis

Davide Mariotti
R. Mohan Sankaran

CONTENTS

4.1 Introduction

Microscale plasmas are a special class of electrical discharges formed in geometries where at least one dimension is reduced to submillimeter length scales.[1] As a result of their unique pD (where p is pressure and D is the smallest physical dimension) scaling, *microplasmas* are characterized by high-pressure stability,[2] non-equilibrium thermodynamics,[3,4] non-Maxwellian electron energy distribution functions (EEDFs),[5] and high electron densities.[4] Overall, these properties have attracted a great deal of interest from the plasma community for a wide range of applications including nanoparticle synthesis.[6]

Low-pressure, large-scale plasmas have been successfully implemented in integrated circuit (IC) manufacturing to define micro- and nanoscale patterns, but these top-down approaches cannot be used to produce materials below approximately 10 nm in size because of limitations associated with lithography.[7] Microplasmas allow nanometer-sized particles to be grown from the "bottom-up" by homogeneously nucleating particles from atomic level constituents. Particle nucleation and growth are controlled by the microreactor geometry, enabling the formation of narrow size distributions of nanoparticles *in a single step*. Compared to wet chemical methods, microplasma-based approaches to nanoparticle synthesis are clean (e.g., no surfactants), scalable, and more amenable to device fabrication. In this chapter, we discuss the novel features of microplasma sources and the recent progress that has been made in the area of nanoparticle synthesis.

4.2 Physics of Microscale Plasmas

Confining plasmas to small dimensions results in new physical behavior. Two general parameters that vary as the size of a plasma is decreased are the *surface-to-volume ratio* and the *electrode spacing*. The surface-to-volume ratio, which increases as the size of the plasma decreases, alters the overall energy balance and leads to plasma stability (or instability) within different regions of the operating parameter space. Decreasing the electrode spacing changes the electric field distribution and, thus, impacts the charge distribution and overall plasma neutrality. Overall, these effects have strong implications on the energy distribution of the different species (electrons, ions, radicals, neutrals) and on the physical structure of the plasma, either simultaneously or independently. There is now significant evidence in the literature that these transitions take place when the dimensions of the plasma are reduced to the micron scale.[8–11] Traditional properties and scaling laws for plasmas do not hold at these dimensions, justifying the existence of a different regime of operation, hereafter referred to as the microplasma regime (MPR).

A number of interesting experimental and theoretical results related to fundamental characterization of microplasmas have been reported in recent years, underpinned by basic energy balances and extrapolations from traditional plasma parameters. One important aspect that must be considered for materials applications is how to control the energy flow that will determine the transition from a non-equilibrium (i.e., low temperature) to equilibrium (i.e., thermal) plasma. Non-equilibrium plasmas are more desirable for materials applications because of the possibility of opening additional chemical pathways. In addition, low-temperature processes allow temperature-sensitive materials to be used. Simple energy considerations can show that reducing the plasma volume affects the overall energy flow, as determined by zero-dimensional (0D) energy balance equations. In general, confinement causes a simultaneous increase in electron temperature and electron density.[12] Electrons gain energy as a result of stronger electric fields and exchange energy through collisions with other species (i.e., neutrals, ions, etc.).[9] The energy balance is then determined by the electron density and the effective rate of energy exchange through inelastic collisions (e.g., vibrational-rotational energy exchange). The collision frequency for heating of neutral atoms or molecules, υ_{heat}, can be expressed as the following:

$$\upsilon_{heat} = n_e K_e(T_e) \tag{4.1}$$

where n_e is the electron density, K_e is the collision rate, and T_e is an effective electron temperature (see below for more details on the definition of effective electron temperature). During each collision, an average energy (ε) is exchanged and the power density (P_{heat}/V) transferred as thermal energy can be written as

$$\frac{P_{heat}}{V} = \varepsilon(T_e)n_g n_e K_e(T_e) \tag{4.2}$$

where n_g is the gas density. In addition, we must consider the power loss (P_{cool}) per unit volume[13]:

$$\frac{P_{cool}}{V} = n_g C_{pl}(T_g - T_0)\nu_{hr} \tag{4.3}$$

where T_g is the temperature of the neutral gas atoms or molecules and T_0 is the wall temperature, ν_{hr} is the frequency of heat removal, and C_{pl} is the heat capacity of the plasma. Assuming that heat conduction is the dominant cooling mechanism, Equation (4.3) can be expressed as a function of parameters that are easier to measure[13]:

$$\frac{P_{cool}}{V} \propto n_g \frac{T_g^{3/2}}{pD^2} \tag{4.4}$$

where, as before, D is a parameter that refers to the size of a microplasma in one dimension, and p is the pressure.[14–17] For a plasma at steady state (i.e., no gas flow), the power transferred to the gas must be equal to the power lost through conduction; therefore, we can relate Equation (4.2) to Equation (4.4) and obtain the following relationship:

$$D^2 \propto \frac{T_g^{3/2}}{p\varepsilon n_e K_e} \quad \text{or} \quad p \propto \frac{T_g^{3/2}}{D^2 \varepsilon n_e K_e} \tag{4.5}$$

The terms K_e and ε are generally a function of the EEDF. A complete analysis that takes into account the actual EEDF would be quite complicated considering that for microplasmas, the EEDF is non-Maxwellian (similarly to other high-pressure plasmas), and an extended high energy tail exists.[5,18,19] Nonetheless, in the range of electron energies considered here, both K_e and ε can be shown to increase with the average electron energy or the effective electron temperature. The effective electron temperature, T_e, is useful when dealing with non-Maxwellian electron energy distributions[20,25] and as a measure of the EEDF, we could now generally say that K_e and ε increase monotonically with T_e, at a slower rate for lower effective electron temperatures and more rapidly as inelastic collisions become more significant at higher effective electron temperatures. Hereafter, the word *effective* will be omitted with the implicit understanding that microplasmas are characterized by an unknown energy distribution for the electrons.[9,20]

The relationships found in Equation (4.5) can be used for a basic analysis of the effect of confinement on plasmas. In Figure 4.1, the pressure (p) is plotted as a function of the dimension of a plasma (D) where the thermal equilibrium limit (i.e., $T_e = T_g$) is indicated by the solid curve; the non-equilibrium regime ($T_e > T_g$) is the area indicated in gray. Near the thermal equilibrium curve, defined as the range of $T_g = 0.1T_e$ to $T_e = T_g^{13}$, the plasma is likely to undergo thermal instabilities. We first focus on the factors that define non-equilibrium conditions versus thermal conditions with respect to the plasma size.

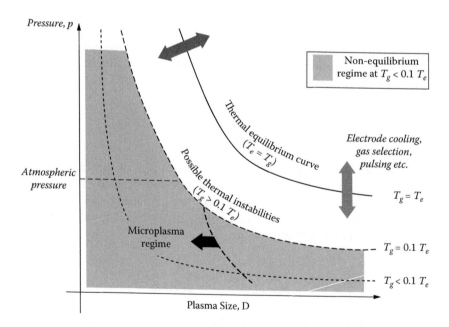

FIGURE 4.1
Different thermal regimes of plasma operation as a function of operating pressure and size (D) that is defined as the minimum dimension in any one direction. The transition from non-equilibrium to thermal equilibrium is indicated by the solid and dashed lines. (Adapted with permission from D. Mariotti and R. M. Sankaran, *J. Phys. D* 43, 323001 (2010), copyright 2010 Institute of Physics.)

We emphasize that Figure 4.1 is representative of a given input power, electrode geometry, gas composition, gas flow rate, and so forth. Thus, by modifying the configuration, operating conditions, or the plasma system, the thermal curve is shifted, and it is possible to achieve non-equilibrium within a range of volume-pressure combinations. In fact, a number of approaches have been used to maintain non-equilibrium conditions (see red double arrows in Figure 4.1). Generally, these approaches impact either the frequency of heat removal term or the K_e and ε terms. For example, if the input power is increased, the thermal curve in Figure 4.1 would shift down and to the left, resulting in a smaller non-equilibrium region, as expected for plasmas operated at higher powers. On the other hand, when electrodes are water cooled, the frequency of heat removal, ν_{hr}, is enhanced (see Equation 4.3), and the pressure at which thermal equilibrium is reached (see Equation 4.5) is shifted to a higher value. Similarly, other techniques that are used to cool a plasma, such as gas flow, will cause the thermal curve to shift up and to the right. Thus, external cooling permits a non-equilibrium plasma to be maintained at atmospheric pressure and large volume.

The gas composition also affects the range of volume-pressure conditions where non-equilibrium characteristics can be sustained. Comparison of discharges formed in He and Ar is illustrative of the role of gas composition on these properties. Helium has been commonly used as the supply gas to produce non-equilibrium, large volume plasmas at atmospheric pressure.[21] Although the ionization energy for He is relatively high (24.6 eV), the minimum energy required for electron-induced excitation of He gas atoms (>21.2 eV) is similar. As a result, electrons transfer their energy effectively via ionization processes, allowing He plasmas to be ignited and sustained at relatively low power. As previously mentioned, lower powers shift the thermal equilibrium curve in Figure 4.1 up and to the right, resulting in a larger window for non-equilibrium operation at atmospheric pressure, in agreement with previous experiments in He. In comparison, Ar is characterized by a larger difference between the lowest electron-induced excitation (>11.5 eV), and the ionization energy (15.8 eV) and excitation of Ar metastables can play a significant role in energy consumption. Consequently, a larger fraction of electrons are involved in inelastic collisions that do not contribute to ionization, and higher power is required to sustain an Ar plasma. Therefore, the thermal equilibrium curve for Ar plasmas is shifted down and to the left and non-equilibrium operation at atmospheric pressure requires that the discharge volume be reduced. Other gas compositions, including molecular gases that may be of greater importance for materials applications, will be characterized by an even smaller non-equilibrium region because the energy of the electrons will be reduced by additional collisional pathways.

Let us now consider plasmas that are not confined to submillimeter dimensions (i.e., bulk plasmas). These large-scale, low-pressure plasmas with a characteristic electron temperature much higher than the gas temperature are identified by a point in the pD space that lies within the gray area of Figure 4.1, to the right of the MPR. If now the pressure is increased, the point moves up and, eventually, a state of thermal equilibrium is reached. This is verified by Equation (4.5) where the consequence of increasing pressure is either plasma extinction due to a decreasing electron temperature (which decreases K_e and ε) or decreasing electron density, or thermal equilibrium, whereby the electron temperature and gas temperature approach one another.[10,22] It follows that in order to maintain non-equilibrium at atmospheric pressure, the volume of the plasma must be reduced. (Note that it has been observed experimentally that non-equilibrium characteristics are readily achieved at atmospheric pressure for submillimeter plasmas.[10,12]) Overall, this analysis illustrates that one of the most notable and interesting properties of microplasmas is the potential to separately control two system temperatures (electron and gas temperature) simply by reducing the size of the plasma. Combined with the parameters previously discussed (e.g., electrode cooling, convective gas flow, gas composition, etc.), this provides a set of additional control "knobs" to tune the non-equilibrium environment. Thus, microplasmas offer a wide range of process conditions for the synthesis of nanomaterials, with the possibility of exploring new operating regimes (e.g., pressures above atmospheric).[23]

The simplified 0D analysis above provides significant insight into the effect of plasma confinement on the energy balance and local thermodynamic equilibrium. However, plasma confinement has other more complicated implications that are not captured by this analysis. The Debye length is one of the traditional parameters used to characterize plasmas and is given by $\lambda_D = (\varepsilon_0 kT_e/(n_e q))^{1/2}$, where ε_0 is the electric constant, k is the Boltzmann constant, and q is the elementary charge. If a plasma is confined to a cavity comparable in size to the Debye length, shielding of the charge by the plasma and quasineutrality may no longer be preserved.[24] For plasmas with an electron temperature of 3 eV and an electron density of 10^{16} m^{-3}, a typical Debye length is approximately 10^{-4} m. Because microplasmas have been found to contain larger electron densities ($\sim 10^{18}$ m^{-3}) and slightly higher electron temperatures,[5,12,18,25,26] a departure from Debye shielding would probably occur at dimensions of 10^{-4} to 10^{-5} m. For sustainment to occur at these conditions, a plasma cannot be governed by the same mechanisms as in the case where the Debye length is orders of magnitude smaller than the plasma volume. Therefore, a regime of plasma operation may exist where the conventional definition of Debye length is no longer applicable. When a transition to this new regime occurs, the electron density distribution is incompatible with long-range steady-state plasma neutrality, as established by the Debye length, and the plasma can be sustained and achieved only dynamically.[27] Overall, the instantaneous charge imbalance can be used to define the MPR, and the classical definition of a plasma is challenged.[9,24]

4.3 Schemes for Nanoparticle Synthesis

Two general approaches schematically depicted in Figure 4.2 have been developed to synthesize nanoparticles in a microplasma. In the first scheme, vapor precursors such as molecular gases (e.g., SiH_4), liquid vapors (e.g., C_2H_5OH), or solid vapors (e.g., ferrocene) are introduced and dissociated in a microplasma to form reactive radicals that polymerize and homogeneously nucleate nanoparticles (Figure 4.2a). Alternatively, solid metal wires (e.g., Au, Mo) have been evaporated in a microplasma to form metal vapors that condense to nucleate particles (Figure 4.2b). The two approaches are discussed in detail in the following sections.

4.3.1 Gas-Phase Nucleation from Vapor Precursors

Gas-phase nucleation of nanoparticles in a microplasma from vapor precursors is a natural extension of previous studies with low-pressure plasmas[28]

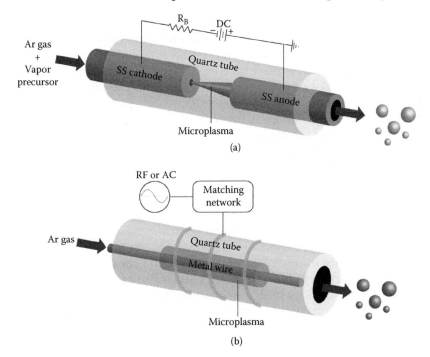

FIGURE 4.2
(See color insert.) Possible configurations for microplasma synthesis of metal nanoparticles. (a) Vapor precursors such as organometallic compounds are dissociated in a direct current (DC) microplasma to form radicals that can homogeneously nucleate nanoparticles. (b) Solid metal wires are evaporated and sputtered by a radio frequency (RF) or alternating current (AC) microplasma to form vapors that condense and nucleate nanoparticles.

and larger-scale atmospheric plasma jets.[29] There are many different precursors available, generally referred to as chemical vapor deposition (CVD) or metal-organic chemical vapor deposition (MOCVD) precursors, that can be non-thermally dissociated in a microplasma to grow different types of materials (i.e., metals, semiconductors, and oxides).

The first demonstration of gas-phase nucleation in a microplasma was carbon nanomaterials by Shimizu et al.[30] In that work, the source configuration consisted of a quartz capillary tube with a tungsten wire inserted inside as an electrode. A second electrode was placed as a coil on the outside of the capillary tube, and the microplasma jet was operated with ultra-high frequency (UHF) power. Experiments were performed with gas mixtures of CH_4 (0.5%) and Ar flowing through the microplasma into air and products deposited directly onto metal substrates. Different powers were found to induce morphological changes in the deposited film with the most promising result being the observation of spherical nano-onions about 30 nm in diameter. Carbon nanomaterials such as nanorods have also been synthesized in pulsed microplasmas formed in pure CH_4.[31]

Later studies have used a different electrode geometry and power coupling to nucleate and grow nanomaterials from vapor precursors. Sankaran et al. used a stainless-steel capillary tube and mesh as the cathode and anode, respectively, to form a DC microjet and synthesize Si[32], Fe,[33] and Ni[33] nanoparticles (Figure 4.2a). The particles were generated in the gas phase and detected by an aerosol-size classification system that was directly coupled to the continuous-flow, atmospheric-pressure microplasma as an online diagnostic tool.[34] The *in situ* measurements of the as-grown particles, along with careful control of the precursor vapor concentration, were essential to optimizing reactor conditions for the preparation of narrow distributions of ultrasmall nanoparticles less than 5 nm in diameter.[35] Figure 4.3 shows steady-state aerosol results for Ni nanoparticles synthesized in a microplasma from sublimed vapors of nickelocene in Ar. At 2 ppm, the geometric mean diameter and standard deviation, obtained by a log-normal fit, were found to be 2.2 nm and 1.13, respectively. Compared to low-pressure plasma experiments,[36] the particles synthesized by this route are much smaller because of the shorter residence times.[32] Atmospheric-pressure operation also reduces the minimum concentration of precursor required to nucleate particles and minimizes safety concerns related to handling of dangerous gases or other chemicals.[37] As the precursor concentration was raised to 2.6 and 5.2 ppm, the mean diameters shifted to 3.1 and 4.7 nm, respectively, while the standard deviations increased slightly to 1.16 and 1.22, respectively. These results are explained by a qualitative argument: at higher precursor concentrations, the density of radical species in the microplasma increases, driving particle growth and leading to larger mean diameters.

Aerosol measurements were confirmed by collecting the nanoparticles and performing transmission electron microscopy (TEM) analysis (Figures 4.4a through 4.4c). High-resolution images reveal that the particles grown by this

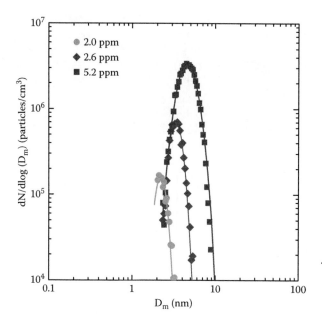

FIGURE 4.3

Aerosol measurements of as-grown Ni nanoparticles prepared in a direct current (DC) microplasma at indicated vapor concentrations of nickelocene in Ar. (Adapted with permission from W. -H. Chiang and R. M. Sankaran, *J. Phys. Chem. C.* 112, 17920 (2008), copyright 2009 American Chemical Society.)

method are spherical and crystalline (see insets of Figures 4.4a through 4.4c). The lattice spacing of 0.20 nm observed for one particle is comparable to the (111) crystalline plane of bulk Ni (Figure 4.4b inset). Histograms obtained by sizing and counting particles in the corresponding TEM images (Figures 4.4d through 4.4f) were found to be in agreement with aerosol results. These studies reflect the promise of microplasma-based synthesis routes for continuous and tunable preparation of high-purity nanometer-sized (i.e., 1 to 5 nm) particles at atmospheric pressure.

A distinct advantage of nucleating nanoparticles in a microplasma from a vapor source is the ability to mix different precursors to synthesize bimetallic or alloyed nanoparticles.[38] This has been demonstrated by the formation of Ni_xFe_{1-x} nanoparticles from vapors of nickelocene and ferrocene, where x refers to the atomic concentration of Ni in the as-grown bimetallic nanoparticles. The ratio of Ni to Fe was tuned by varying the relative vapor concentrations of the respective metallocenes while the mean size of the bimetallic nanoparticles was controlled by the total precursor vapor concentration in Ar. The actual amount of Ni and Fe in the nanoparticles was confirmed by *ex situ* materials analysis. Figure 4.5a shows a TEM micrograph of bimetallic nanoparticles synthesized from a vapor mixture of 27:73 nickelocene:ferrocene. The particles are spherical, crystalline, and

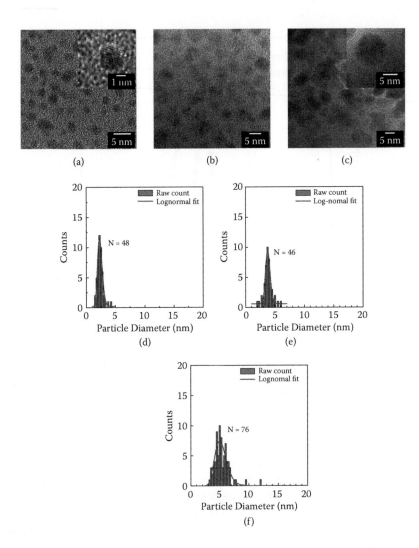

FIGURE 4.4
Transmission electron microscopy (TEM) analysis of as-grown Ni nanoparticles prepared in a direct current (DC) microplasma at various nickelocene vapor concentrations in Ar of (a) 2, (b) 2.6, and (c) 5.2 ppm (inset: HRETM image of a Ni nanoparticle exhibiting a lattice spacing of 0.20 nm). Histograms of Ni nanoparticles obtained from corresponding TEM images for nickelocene vapor concentrations in Ar of (d) 2, (e) 2.6, and (f) 5.2 ppm. (Adapted with permission from W. -H. Chiang and R. M. Sankaran, *J. Phys. Chem. C.* 112, 17920 (2008), copyright 2009 American Chemical Society.)

unagglomerated, similar to as-grown monometallic Ni nanoparticles. High-resolution TEM (HRTEM) of a representative particle shows a lattice spacing of 0.21 nm, slightly larger than the (111) crystalline plane for Ni nanoparticles (Figure 4.5a inset). The chemical composition of individual nanoparticles was assessed by energy-dispersive spectroscopy (EDX) (Figure 4.5b).

The ratio of the intensities of lines corresponding to Ni Kα and Fe Kα radiation for three samples grown with varying relative amounts of metallocene vapor confirms that the particle composition is precisely tuned by controlling reactor conditions. To evaluate the crystalline or alloy structure of the bimetallic nanoparticles, x-ray diffraction (XRD) was employed. The XRD

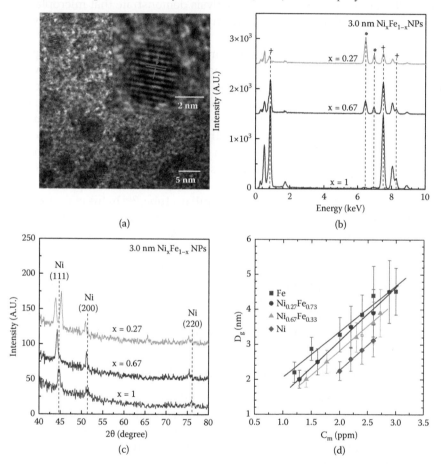

(a)

(b)

(c)

(d)

FIGURE 4.5

(a) Transmission electron microscopy (TEM) of as-grown $Ni_{0.27}Fe_{0.73}$ nanoparticles prepared in a direct current (DC) microplasma at a vapor concentration of 1.97 ppm ferrocene and 0.50 ppm nickelocene in Ar (inset: high-resolution transmission electron microscopy [HRTEM] image of a $Ni_{0.27}Fe_{0.73}$ nanoparticle exhibiting a lattice spacing of 0.21 nm). (b) Energy-dispersive spectroscopy (EDX) spectra of Ni_xFe_{1-x} (x = Ni at%) nanoparticles with Ni(+) and Fe(*) peaks indicated. (c) X-ray diffraction (XRD) characterization of Ni_xFe_{1-x} (x = Ni at%) nanoparticles with indices for Ni face-centered cubic (fcc) peaks indicated. (d) Geometric mean diameter of as-grown Ni_xFe_{1-x} nanoparticles versus metallocene vapor concentration in Ar for direct current (DC) microplasma synthesis. Error bars correspond to the geometric standard deviation of the particle size distributions. (Adapted with permission from D. Mariotti and R. M. Sankaran, *J. Phys. D* 43, 323001 (2010), copyright 2010 Institute of Physics.)

patterns of the same three samples show that when Fe is incorporated into a Ni nanoparticle, the Ni face-centered cubic (fcc) structure expands slightly (Figure 4.5c). At high atomic fractions of Fe (>50%), the bimetallic nanoparticles also exhibit body-centered cubic (bcc) crystal structure, as indicated by the appearance of new diffraction peaks (see Figure 4.5c). Overall, the aerosol results combined with materials analysis demonstrate that microplasma synthesis is capable of independently tuning the size and composition of nanoparticle alloys (Figure 4.5d).

Although DC coupling is convenient for igniting and sustaining microplasmas because it precludes the need for matching networks, there are several drawbacks. As in the case of low-pressure DC glow discharges, the power consumption is higher than if high-frequency (i.e., AC, RF, or higher) is used, leading to power losses and thermal effects (i.e., ion heating). Perhaps more important for materials applications, the electrodes must be conductive and in contact with the reactive plasma, which could result in contamination of the as-grown material. To address these issues, Nozaki et al. developed a very high frequency (VHF) (144 MHz) powered microplasma with metal electrodes surrounding a quartz capillary tube.[39,40] In his experiments, $SiCl_4$ was used as a vapor precursor to produce Si nanocrystals that were directly deposited downstream of the microplasma onto glass substrates. By adjusting the H_2 concentration in the microplasma, the deposition rate and particle size were tuned to obtain size-dependent room temperature photoluminescence, a very promising result for optical applications. RF-powered atmospheric pressure microplasmas have also been implemented for carbon nanotube (CNT) growth.[41,42] Discharges were formed in CH_4/H_2 mixtures with a catalyst loaded on a substrate to nucleate and grow CNTs.

4.3.2 Evaporation/Sputtering of Solid Metal Wire Electrodes

An alternative approach for nanoparticle synthesis is to evaporate or sputter a sacrificial metal electrode via microplasma generation.[43] Typically, a fine metallic wire (0.03 to 0.3 mm diameter) is inserted inside a quartz or alumina microcapillary tube (0.1 to 0.8 mm internal diameter) with an external electrode coiled around the outside of the capillary tube[44-52] (Figure 4.2b). In a few cases, a substrate is used as the counterelectrode that facilitates material deposition.[53-55] Some added benefits of inserting a metal wire into a microplasma are a lowering of the breakdown voltage and operating power. This has been explained by thermionic emission of electrons, particularly for refractory metals,[57] and electric field enhancement as a result of the electrode geometry. Previous reports have almost exclusively focused on ultrahigh frequency (UHF) operation at 450 MHz, although lower frequencies are equally capable of sustaining such microplasmas. Possible advantages for UHF excitation are lower power consumption and the ability to sustain microplasmas at low temperature and low gas flows that approach static conditions. Pulsed power at 450 MHz has been also demonstrated for the synthesis of gold nanoparticles.[50]

Several metal-oxide (Cu,[49] W,[49] Mo,[44,47,54,55] Fe,[49] Zn[56]) and pure metal (Au,[50] Mo[45,51]) nanoparticles have been prepared from the respective metal wire precursors. A drawback of also using the wire as an electrode is that there is limited control over the wire position with respect to the capillary walls and, therefore, with respect to the external electrode. Variations in the radial or axial distance of the wire from the external electrode may affect the kinetics of the surface reactions and possibly the morphology or phase of the as-grown structures. Nonetheless, highly reproducible results have been obtained.[54,55] Another potential limitation is the consumption of the wire electrode over time. Complete deterioration of the wire can occur within minutes and cause the process to be shut down; continuous operation for longer times will require this issue to be addressed, for example, by incorporating a wire-feeding apparatus.

The temperature at the wire surface and of the gas (i.e., neutrals) can be controlled by varying the gas flow rate, input power, and the electrode geometry. Heating of the wire is generally attributed to inductive coupling,[49] but other forms of heating may also contribute including ohmic and heat evolved from exothermic reactions (e.g., oxidation). Gas temperatures have been determined by spectroscopic techniques and found to be in the range of 700 to 1800 K[12,25,56]; however, the lower bound was limited by the low intensity of the emission, and it may be possible to operate at still lower temperatures, down to room temperatures. In the case of a remote microplasma, temperature-sensitive substrates have been used even at high gas temperatures (e.g., epoxy glass[49] or paper[50]). In all cases, the microplasmas are characterized by high effective electron temperatures (>2 eV) and non-equilibrium.[12,25]

Tungsten- and Mo-oxide nanoparticle films have been formed with average particle diameters of 20 to more than 100 nm.[47,52] In both cases, the gas flow rate determined the size and crystalline phase of the metal-oxide nanoparticles. Higher gas flow rates were found to yield smaller particles. The crystalline phase was found to vary from lower to higher oxidation state with increasing gas flow rate. EDX and x-ray photoelectron spectroscopy (XPS) characterization of the wire after operation showed that oxidation occurred at the metal wire surface. Overall, these results suggest that at lower gas flow rates, the gas temperature is higher, leading to the evaporation of a higher flux of metal oxide moieties that then nucleate and form nanoparticles with larger mean diameter and higher oxygen content. At higher gas flow rates, molten droplets may leave the wire surface because of the lower gas temperature that prevents evaporation of the higher melting point species such as metal oxides. This analysis has been corroborated by optical emission analysis that clearly indicates the presence of atomic Mo/W in the gas phase at lower gas flow rates. It should be noted that only small changes to the gas flow rate (i.e., 5 to 30 sccm) are needed to dramatically change the mechanism for particle formation. Formation of ZnO films has also recently been demonstrated with high deposition rates.[56] The preparation of other metal-oxide materials such as CuO or Fe_2O_3 should also be possible.

Under certain operating conditions, anisotropic growth of metal oxide nanostructures has been observed. Figure 4.6a shows a TEM image of Mo-oxide nanorods grown by evaporating/sputtering a Mo wire in an Ar/O_2 microplasma jet. The nanorods are highly reproducible and uniform with lengths between 35 and 95 nm and widths between 6 and 27 nm. The oxygen concentration appears to be critical to controlling the shape of the nanostructures although further experiments are needed to clarify the growth mechanism. HRTEM reveals that the nanostructures are crystalline and exhibit a MoO_3 monoclinic phase (Figure 4.6b).

Pure metallic Au nanostructures have also been synthesized and deposited at room temperature by adding small amounts of H_2 (<4%) to the gas flow.[50] Pulse modulation of the input power was essential to avoid excessive gas heating by H_2.[57] The role of H_2 has not been clearly established, but it is likely that atomic hydrogen, formed in the plasma volume, reacts and reduces the oxide on the wire surface.

Solid precursors have been introduced into a microplasma by more conventional schemes (i.e., exposure of a substrate to a reactive plasma). Fe-coated Si substrates were used to produce Si nanocones, CNTs, and other nanostructures in the gas phase through a combination of etching (or sputtering or evaporation) and catalytic growth processes.[58–60] A drawback of this

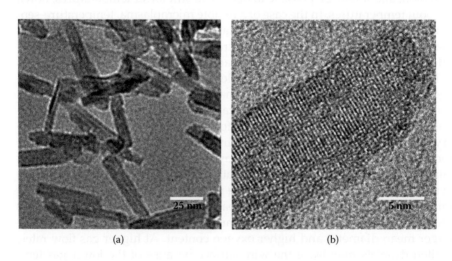

(a)   (b)

FIGURE 4.6
(a) Transmission electron microscopy (TEM) of molybdenum oxide nanorods synthesized by evaporating/sputtering a solid Mo wire in an atmospheric-pressure microplasma jet. (b) High-resolution transmission electron microscopy (HRTEM) image of a nanorod indicates that the nanostructures are crystalline with a lattice structure corresponding to MoO_3 monoclinic phase. The microplasma was operated with ultra-high frequency (UHF) power (10 W) in a gas mixture of Ar and O_2 (1%). (Courtesy of A. Chandra Bose.)

approach is that the formation and structure of as-grown nanomaterials are very sensitive to the distance of the substrate from the microplasma jet.

The mechanism for evaporation and sputtering of metal wire or films on substrates in a microplasma is not clear as there is no comparable process. Although wire flame spraying bears some similar features, microplasma-based systems are fundamentally different from these thermal processes because of the presence of energetic species (e.g., ions and electrons).[44,45,54] Because physical sputtering by ion bombardment is a function of the ion energy, we expect that sputtering may be reduced at the high pressures used in microplasmas due to collisional effects. However, the high surface temperature of the metallic wire could enhance sputtering of surface atoms at lower ion energies.[56] This picture has been complicated by experimental results indicating that Ar or He alone is usually insufficient to produce material.[45,51] In most cases, O_2, H_2, or CH_4 are needed which implies that reactive processes are important. There are a few examples of metal-oxide nanoparticles produced with only Ar and no O_2 gas.[49,54–56] The source of O_2 for oxide formation has been attributed to either the etching of the quartz capillary[49] or outside air leaking into the gas lines. Water vapor in the gas lines may also be responsible, as supported by the presence of OH radicals in the collected spectral emission.[47,56] Despite the uncontrolled source of oxygen, nanoparticles with reproducible size and composition have been synthesized. Curiously, Ar-O_2 mixtures produce similar results.[44,52] In either case, oxidation is believed to occur at the metal wire surface, and unlike thermal oxidation, oxygen radicals and ions should be involved in surface reactions leading to material formation.[61] More experiments are needed to reveal the separate roles of heating and other physical and chemical processes.

4.4 Potential for Scale-Up

Despite recent progress, widespread utilization of microplasmas for nano-particle synthesis will depend on the ability to scale-up current processes to industrial-level manufacturing or fabricate well-defined materials that are not capable of being produced by other methods. There are several issues that must be considered, including the quality of as-grown material in terms of homogeneity and structure, throughput (i.e., production rate), pro-cess cost, and process safety. The quality of as-grown material depends on the desired application—for example, in some cases, the nanoparticle size may be important and must be precisely controlled, whereas in others the composition is critical. Microplasmas are capable of producing a wide range of materials with well-defined structures and under various conditions, which could make them flexible for different applications. The throughput of material is currently a challenge because most previous studies are lab-scale and require only small amounts of material (e.g., nanograms); however, microplasmas can be operated in parallel to increase the production rate. The cost of microplasma-based processes is relatively low because of atmospheric-pressure operation. In addition, one can envision the development of small-scale portable apparatuses (i.e., bench top) that may be attractive for applications with lower manufacturing requirements.

Microplasma-based systems have been implemented and tested for continuous operation for more than 24 hours with no visible deterioration, making them suitable for large-scale production.[62] One of the key questions that remain unanswered is if microplasmas can provide the required throughput for industrial applications. A single DC microplasma reactor of the type shown in Figure 4.2a has been found to produce nanoparticles at a rate of about 10^{-10} g/min. This relatively low throughput can be easily increased by arranging microplasmas as a two-dimensional (2D) array. For example, an array consisting of 10,000 microplasmas (100 × 100) would still have a relatively small footprint no larger than 1 m², require only a power of approximately 100 kW, and achieve a production rate of 10^{-6} g/min. A drawback of this approach is that DC power is characterized by a high rate of charge losses. However, DC microplasmas are extremely simple to operate, and power supplies are commercially available at low cost. If the number of microplasmas in the array can be limited to less than ~200, other manufacturing needs could be met as well. For instance, a DC microplasma array would be suitable for niche applications such as nanoparticles for medical applications (e.g., drug delivery), where the quality of the product is more important than throughput.

The charge distribution and Debye length are relatively unaffected by electrode confinement in DC microplasmas. On the other hand, for microplasmas sustained with excitation frequencies above 1 MHz, the charge distribution will be significantly altered which provides an opportunity to tune the electron density and modify reaction rates. Thus, it may be possible to improve the material throughput in RF-driven microplasmas, as compared to DC microplasmas. This improvement is at the expense of somewhat more complex electronics (i.e., power supply, matching network, etc.) and a more carefully designed electrode configuration. Comparing microplasma systems that have been used to synthesize Si nanocrystals, it is apparent that DC power can produce high-quality nanocrystals with diameters below 2 nm and a standard deviation as low as 23%, but low production rates.[32] On the other hand, a VHF-powered microplasma has shown the potential for much higher manufacturing rates, with deposition rates reaching above 4 µm/min.[39] With a few assumptions, the material throughput of a single 144 MHz powered microplasma jet can be estimated to be approximately 10^{-6} g/min, which is four orders of magnitude higher than the DC microplasma. Again using the example of a 2D array of 10,000 microplasmas, the production rate is now 10^{-2} g/min. Unfortunately, the power required to run such a system would be above 100 kW, which is higher than that of a DC power system. In addition, the availability of RF power supplies with this power rating is more difficult; therefore, arrays of only about 300 to 400 microplasma reactors might be more feasible with commercially available systems, yielding production rates of ~10^{-4} g/min. Overall, this value compares favorably with other techniques: 10^{-5} g/min by low-pressure plasmas,[63,64] 10^{-4} g/min by low-pressure, continuous-flow plasmas,[36] 10^{-7} g/min by electrochemical etching,[65,66] and 10^{-7} g/min by laser processing in liquid.[67,68] Further improvements in production rates will require additional knowledge of microplasma processes.

Higher production rates may instead be achieved by using a solid precursor such as a metal wire.[47,50,52] A microplasma-based system operated at 450 MHz has been used to produce Mo- and W-oxide nanoparticles as a nanostructured thin film at a rate of 350 μm/s that is three orders of magnitude higher than the rate for Si nanocrystals by a VHF-powered microplasma.[52] If microplasmas are not used for large-scale manufacturing, they may still find applications on a smaller scale for fundamental research. Well-defined nanoparticles are essential to basic studies in catalysis, nanoelectronics, nanomedicine, or energy. An example is illustrated by a recent report that elucidated the role of a catalyst in CNT growth.[69] Precisely controlled bimetallic nanoparticles were prepared in a microplasma and used to establish a link between catalyst composition and the chirality of as-grown single-walled CNTs. This result gives insight into the mechanism for nanotube nucleation and may impact future nanotube-based electronic devices where high fractions of semiconducting or metallic nanotubes are needed.[70] More of these types of studies are expected in the near future as researchers continue to fabricate novel nanomaterials via microplasma synthesis.

4.5 Conclusions

The physical properties of microplasmas offer unique capabilities for the synthesis of a wide range of nanoparticle materials. Even though most of the recent reports have focused on small-scale synthesis, scale-up to manufacturing levels is also on the horizon where the features of microplasma-based systems are truly attractive (e.g., atmospheric operation, continuous, portable, etc.). Despite the potential that microplasmas hold for nanoparticle synthesis, some challenges remain. It is clear that a basic understanding of microplasmas, in terms of both plasma physics and reaction kinetics, is still lacking. Although there are some similarities with low-pressure plasmas, the high operating pressures and resulting modification to the energy exchange mechanisms cannot be neglected. Future efforts in modeling and simulation must include these effects. More importantly for applications, these studies must be connected to microplasma synthesis where chemically reactive gases, solid metal precursors, and liquids are used to produce desired materials. Thus, the scientific problems to be resolved are multidimensional and require a synergistic effort that links physics, diagnostics, and process.

References

1. See K. H. Becker, K. H. Schoenbach, and J. G. Eden, *J. Phys. D* 39, R55 (2006) and references therein.
2. S. -J. Park and J. G. Eden, *Appl. Phys. Lett.* 81, 4127 (2002).
3. P. Kurunczi, N. Abramzon, M. Figus, and K. Becker, *Acta Phys. Slovaca* 54, 115 (2004).
4. C. Penache, M. Miclea, A. Brauning-Demian, O. Hohn, S. Schossler, T. Jahnke, K. Niemax, and H. Schmidt-Bocking, *Plasma Sources Sci. Technol.* 11, 476 (2002).
5. F. Iza, J. K. Lee, and M. G. Kong, *Phys. Rev. Lett.* 99, 075004 (2007).
6. D. Mariotti and R. M. Sankaran, *J. Phys. D* 43, 323001 (2010).
7. See http://itrs.net: "The International Technology Roadmap for Semiconductors" (2010).
8. J. G. Eden and S. -J. Park, *Plasma Phys. Control. Fusion* 47, B83 (2005).
9. R. Foest, M. Schmidt, and K. Becker, *Int. J. Mass Spectrom.* 248, 86 (2006).
10. K. Tachibana, *IEEJ Trans.* 1, 145 (2006).
11. F. Iza, G. J. Kim, S. M. Lee, J. K. Lee, J. L. Walsh, Y. T. Zhang, and M. G. Kong, *Plasma Process. Polym.* 5, 322 (2008).
12. D. Mariotti, *Appl. Phys. Lett.* 92, 151505 (2008).
13. Y. P. Raizer, *Gas Discharge Physics*, 2nd ed. (Springer, Berlin, 1991).
14. D. Staack, B. Farouk, A. Gutsol, and A. Fridman, *Plasma Sources Sci. Technol.* 14, 700 (2005).
15. D. Staack, B. Farouk, A. Gutsol, and A. Fridman, *Plasma Sources Sci. Technol.* 17, 025013 (2008).
16. J. Benedikt, *J. Phys. D* 43, 043001 (2010).
17. P. J. Chantry, *J. Appl. Phys.* 62, 1141 (1987).
18. J. Choi, F. Iza, and J. K. Lee, *IEEE Trans. Plasma Sci.* 35, 1274 (2007).
19. C. G. Wilson, Y. B. Gianchandani, R. R. Arslanbekov, and V. Kolobov, *J. Appl. Phys.* 94, 2845 (2003).
20. D. Mariotti, Y. Shimizu, T. Sasaki, and N. Koshizaki, *Appl. Phys. Lett.* 89, 201502 (2006).
21. J. Park, I. Henins, H. W. Herrmann, and G. S. Selwyn, *J. Appl. Phys.* 89, 15 (2001).
22. W. Elenbaas, *The High Pressure Mercury Vapor Discharge.* (North-Holland, Amsterdam, 1951).
23. T. Tomai, H. Kikuchi, S. Nakahara, H. Yui, and K. Terashima, *Trans. Mater. Res. Soc. Jpn.* 33 355 (2008).
24. J. G. Eden and S. -J. Park, *Physics of Plasmas* 13, 057101 (2006).
25. D. Mariotti, Y. Shimizu, T. Sasaki, and N. Koshizaki, *J. Appl. Phys.* 101, 013307 (2007).
26. T. Farouk, B. Farouk, and A. Fridman, *IEEE Trans. Plasma Sci.* 38, 73 (2010).

27. A. Vogelsang, A. Ohl, H. Steffen, R. Foest, K. Schroder, K. -D. Weltmann, *Plasma Process. Polym.* 7, 16 (2010).
28. U. Kortshagen, *J. Phys. D.* 42, 113001 (2009).
29. N. Rao, S. Girschick, J. Heberlein, P. McMurry, S. Jones, D. Hansen, and B. Micheel, *Plasma Chem. Plasma Process.* 15, 581 (1995).
30. Y. Shimizu, T. Sasaki, T. Ito, K. Terashima, and N. Koshizaki, *J. Phys. D.* 26, 2940 (2003).
31. Q. Zou, M. Wang, Y. Li, and L. Zou, *J. Low Temp. Phys.* 157, 557 (2009).
32. R. M. Sankaran, D. Holunga, R. C. Flagan, and K. P. Giapis, *Nano Lett.* 5, 537 (2005).
33. W. -H. Chiang and R. M. Sankaran, *Appl. Phys. Lett.* 91, 121503 (2007).
34. R. P. Camata, H. A. Atwater, K. J. Valhalla, and R. C. Flagan, *Appl. Phys. Lett.* 68, 3162 (1996).
35. W. -H. Chiang and R. M. Sankaran, *J. Phys. Chem. C.* 112, 17920 (2008).
36. L. Mangolini, E. Thimsen, and U. Kortshagen, *Nano Lett.* 5, 655 (2005).
37. R. C. Flagan and M. M. Lunden, *Mat. Sci. Eng. A* A204, 113 (1995).
38. W-H. Chiang and R. M. Sankaran, *Adv. Mater.* 20, 4857 (2008).
39. T. Nozaki, K. Sasaki, T. Ogino, D. Asahi, and K. Okazaki, *Nanotechnology* 18, 235603 (2007).
40. T. Nozaki and K. Okazaki, *Pure Appl. Chem.* 78, 1157 (2006).
41. S. Kona, J. H. Kim, C. K. Harnett, and M. K. Sunkara, *IEEE Trans. Nanotech.* 8, 286 (2009).
42. H. Yoshiki, T. Okada, K. Hirai, and R. Hatakeyama, *Jpn. J. Appl. Phys.* 45, 9276 (2006).
43. Y. Shimizu, T. Sasaki, T. Ito, K. Terashima, and N. Koshizaki, *J. Phys. D* 36, 2940 (2003).
44. D. Mariotti, A. C. Bose, and K. Ostrikov, *IEEE Trans. Plasma Sci.* 37, 1027 (2009).
45. Y. Shimizu, K. Koga, T. Sasaki, and N. Koshizaki, *Cryst. Eng. Comm.* 11, 1940 (2009).
46. Y. Shimizu, A. C. Bose, T. Sasaki, D. Mariotti, K. Kirihara, T. Kodaira, K. Terashima, and N. Koshizaki, *Trans. Mater. Res. Soc. Jpn.* 31, 463 (2006).
47. A. C. Bose, Y. Shimizu, D. Mariotti, T. Sasaki, K. Terashima, and N. Koshizaki, *Nanotechnology* 17, 5976 (2006).
48. Y. Shimizu, T. Sasaki, C. Liang, A. C. Bose, T. Ito, K. Terashima, and N. Koshizaki, *Chem. Vap. Deposition* 11, 244 (2005).
49. Y. Shimizu, T. Sasaki, A. C. Bose, K. Terashima, and N. Koshizaki, *Surf. Coat. Technol.* 200, 4251 (2006).
50. Y. Shimizu, K. Kawaguchi, T. Sasaki, and N. Koshizaki, *Appl. Phys. Lett.* 94, 191504 (2009).
51. Y. Shimizu, K. Koga, T. Sasaki, D. Mariotti, K. Terashima, and N. Koshizaki, *Microprocess. and Nanotech.* 174 (2007).
52. Y. Shimizu, A. C. Bose, D. Mariotti, T. Sasaki, K. Kirihara, T. Suzuki, K. Terashima, and N. Koshizaki, *Jpn. J. Appl. Phys. Part 1* 45, 8228 (2006).
53. D. Mariotti, V. Svrcek, and D. -G. Kim, *Appl. Phys. Lett.* 91, 183111 (2007).
54. D. Mariotti and K. Ostrikov, *J. Phys. D* 42, 092002 (2009).
55. D. Mariotti, H. Lindstrom, A. C. Bose, and K. Ostrikov, *Nanotechnology* 19, 495302 (2008).
56. S. Stauss, Y. Imanishi, H. Miyazoe, and K. Terashima, *J. Phys. D* 43, 155203 (2010).

57. M. Wolter, I. Levchenko, H. Kersten, and K. Ostrikov, *Appl. Phys. Lett.* 96, 133105 (2010).
58. H. Shirai, T. Kobayashi, and Y. Hasegawa, *Appl. Phys. Lett.* 87, 143112 (2005).
59. Z. Yang, T. Kikuchi, Y. Hatou, T. Kobayashi, and H. Shirai, *Jpn. J. Appl. Phys. Part 1* 44, 4122 (2005).
60. Z. Yang, H. Shirai, T. Kobayashi, and Y. Hasegawa, *Thin Solid Films* 515, 4153 (2007).
61. U. Cvelbar, K. Ostrikov, and M. Mozetic, *Nanotechnology* 19, 405605 (2008).
62. R. M. Sankaran and K. P. Giapis, *J. Appl. Phys.* 92, 2406 (2002).
63. S. Nunomura, M. Kita, K. Koga, M. Shiratani, and Y. Watanabe, *J. Appl. Phys.* 99, 083302 (2006).
64. S. Nunomura, I. Yoshida, and M. Kondo, *Appl. Phys. Lett.* 94, 071502 (2009).
65. V. Svrcek, M. Kondo, K. Kalia, and D. Mariotti, *Chem. Phys. Lett.* 478, 224 (2009).
66. V. Svrcek, H. Fujiwara, and M. Kondo, *Acta Mater.* 57, 5986 (2009).
67. V. Svrcek, D. Mariotti, and M. Kondo, *Opt. Exp.* 17, 520 (2009).
68. V. Svrcek and M. Kondo, *Appl. Surf. Sci.* 255, 9643 (2009).
69. W. -H. Chiang and R. M. Sankaran, *Nat. Mater.* 8, 882 (2009).
70. W-H. Chiang, M. Sakr, X. P. A. Gao, and R. M. Sankaran, *ACS Nano* 3, 4023 (2009).

5

Large-Scale, Plasma-Assisted Growth of Nanowires

Uros Cvelbar
Mahendra K. Sunkara

CONTENTS

5.1 Introduction

With the development of new technologies, the demand for smaller-sized materials has rapidly grown. This search has eventually led to interest in nanosize materials, such as nanotubes (NTs), nanowires (NWs), quantum dots, nanopyramids, and nanopowders [1]. Several methods have now been developed to synthesize these novel materials. Today we can basically make most types of nanosize materials (e.g., semiconductors, oxides, sulfites, alloys, etc.) with different sizes, shapes, or even crystal structure. However, the synthesis processes have not been sufficiently controlled or suffer from poor reproducibility due to many parameters that can influence the synthesis reactions or a poor understanding of the growth process. Repeatability, of course, depends on the method and the synthesis. In general, material synthesis is easier to control in small volumes, systems, or reactors, for example, in a laboratory. However, materials must be produced on a large scale for widespread use. Therefore, the biggest challenge for nanomaterial synthesis is large-scale and large-quantity production. In this chapter, we describe our efforts to make and scale up the production of inorganic nanowires using plasma-based approaches.

5.2 Plasma-Enhanced Vapor-Liquid-Solid Growth of Nanowires

In 1964, Wagner and Ellis demonstrated vapor-liquid-solid (VLS) growth of one-dimensional silicon (Si) crystals (i.e., Si nanowires) [2]. In their experiments, a silicon substrate covered with Au particles was exposed to gaseous $SiCl_4/H_2$ at 950°C. The VLS mechanism, illustrated in Figure 5.1, involves a catalyst metal cluster forming a liquid (i.e., molten) alloy with the gas precursor (i.e., Au and Si) and precipitation of a solid phase from the catalyst as a one-dimensional solid material. The detailed mechanism consists of four steps: (1) diffusion in the vapor phase, (2) chemical reaction at the vapor-liquid interface, (3) diffusion in the liquid phase, and (4) crystallization at the solid-liquid interface [3–6]. The rate-limiting step could be either chemical reaction

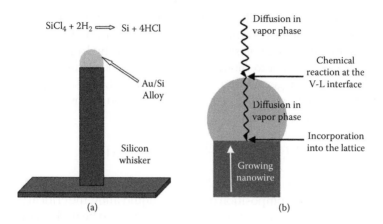

FIGURE 5.1

(a) Vapor-liquid-solid (VLS) growth mechanism for silicon whiskers using gold as the catalyst. Silicon is provided in the gas phase through $SiCl_4$ and flows over gold clusters supported on a silicon substrate. (b) Various chemical and transport processes taking place during VLS growth of nanowires: (1) vapor phase diffusion; (2) chemical reaction at the vapor-liquid interface; (3) liquid phase diffusion; and (4) incorporation into the lattice at the nanowire-liquid interface.

at the V-L interface or crystallization. The role of the molten metal cluster at the whisker tip is to (1) reduce the activation energy for the chemical reaction taking place at the vapor-liquid interface and (2) enhance the sticking coefficient through dissolution [3]. The competition between dissolution from the molten alloy cluster and deposition on the planar substrates determines whether one-dimensional growth or polycrystalline deposition occurs. From this picture, it is clear that the growth of one-dimensional structures usually relies on catalytic metals such as iron (Fe), nickel (Ni), gold (Au), and cobalt (Co), and process temperatures above the eutectic temperatures for the alloy of interest [7].

The use of plasma activation during VLS growth can help reduce the growth temperature needed for substrate heating. Gas phase radicals such as atomic hydrogen and others can recombine on the catalyst particle surfaces at faster rates than heating alone, enhancing the growth process. However, the main implication of plasma activation is that one can utilize any type of molten metal cluster to catalyze the growth of one-dimensional structures. This is because radical species such as atomic hydrogen can enhance the dissociation and dissolution of gaseous species into a given metal under molten conditions. Also, one can accomplish such dissociation kinetics at low enough temperatures that the metal clusters do not undergo evaporation.

5.2.1 Plasma-Enhanced Vapor-Liquid-Solid (VLS) Growth of Silicon and Germanium Nanowires Using Noncatalytic Metals

Low-melting metals such as gallium (Ga), indium (In), aluminum (Al), and bismuth (Bi) are generally considered to be noncatalytic metals for dehydrogenation reactions. These metal clusters cannot initiate one-dimensional growth of silicon or other related materials in a thermal process. However, in the presence of plasmas involving atomic hydrogen, these metal clusters have been shown to catalyze the growth of one-dimensional structures including silicon, germanium, and their alloys [1,8–10]. The presence of atomic hydrogen enables silane or silyl decomposition on molten Ga surface through the following reaction:

$$\mathrm{SiH}_x(g) + x\,\mathrm{H}(g) \xrightarrow{\mathrm{Ga}} \mathrm{Ga-Si}(l) + x\,\mathrm{H}_2(g) \tag{5.1}$$

Here, we discuss results for Ga metal clusters in more detail [1,11,12]. Because of the low equilibrium composition of Si in the liquid alloy, the critical nucleus size for silicon crystallization from a Ga-Si melt tends to be in the nanometer range, according to classical nucleation theory for solute precipitation from dissolved solutions. At a temperature of 400°C, using typical values for surface free energy and molar volume, the critical nucleus diameter is estimated to be around 2 nm with a modest dissolved Si concentration of 1 at%. In comparison, other systems involving transition metals or noble metals as liquid media exhibit equilibrium compositions in excess of 20 to 30 at% and thus the critical nuclei sizes exceed the equilibrium sizes of droplets, which are around 0.2 microns. The low miscibility and high surface tension in the Ga-Si system influence the nuclei to form a concave droplet and ensure that further growth of these nuclei is one-dimensional. The diameter of the nanowires depends upon the substrate temperature and the corresponding supersaturation of Si in Ga-Si liquid alloy that then controls the crystallization rate and the resulting diameters. At temperatures around 400°C or lower, a high density of silicon nanowires is formed from micron-sized Ga droplets as shown in Figure 5.2a. Plasma-enhanced chemical vapor deposition (PECVD) involving 2% silane in hydrogen in the presence of chlorine resulted in Ga droplet tip-led nanowire growth as shown in Figure 5.2b. Similar results are possible with other metals including In, Sn, Al, and Bi for solutes such as Si, Ge, and their alloys. In all of these experiments, a microwave plasma with 400 to 800 W power was used. *In situ* optical absorption spectroscopy showed the presence of various silyl species including SiH radicals. We suggest that the atomic hydrogen mediates the Si dissolution on Ga surface with the following reaction:

$$\mathrm{SiH} + \mathrm{H} \rightarrow \mathrm{Si} + \mathrm{H}_2 \text{ (on Ga surface)}$$

The resulting Si atoms on the Ga surface diffuse into the bulk and on the surface resulting in an enhanced sticking coefficient as compared to the underlying substrate. PECVD on Ga droplets resulted in defect-free single-crystalline silicon nanowires with no or minimal oxide shell. The native oxide shell is significantly thinner than that previously reported for silicon nanowires

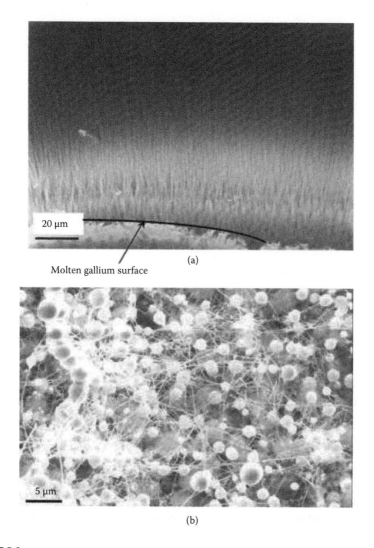

FIGURE 5.2
(a) Bulk nucleation and growth of Si nanowires from molten Ga surface when silane was sup-plied through the gas phase in microwave plasma; and (b) Si nanowires grown from smaller Ga droplets with silane, hydrogen, and chlorine in microwave plasma. (Reproduced with permission from Sharma, S., and M. K. Sunkara, Direct synthesis of single-crystalline silicon nanowires using molten gallium and silane plasma. *Nanotechnology*, 2004. 15: 130–134.)

grown by traditional VLS and oxide-assisted techniques. Figure 5.3 shows a high-resolution transmission electron microscopy (HRTEM) image of a 60 nm thick nanowire. Ordered fringes confirm that the wires are highly crystalline and defect free with a lattice spacing close to that of bulk silicon. The growth direction was determined to be <100> from both HRTEM and the selected area (electron) diffraction (SAED) pattern as shown in the inset. A growth

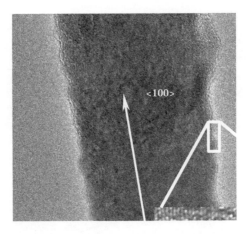

FIGURE 5.3
Low- and high-resolution transmission electron microscope images of Si nanowires synthesized using Ga droplets. The images show (100) growth direction for the resulting SiNWs. (Reproduced with permission from Sharma, S., and M. K. Sunkara, Direct synthesis of single-crystalline silicon nanowires using molten gallium and silane plasma. *Nanotechnology*, 2004. 15: 130–134.)

direction of <100> has not been previously reported for silicon nanowires grown using traditional VLS, where <111> and <211> are the most commonly observed growth directions [13]. However, <100> has been observed to be a possible growth direction for nanowires grown using supercritical fluid processing [14]. Silicon nanowires grown along a specific crystallographic directions are of particular interest for optoelectronic device applications where the emissive properties must be controlled [15]. Upon closer inspection of the HRTEM image of a 60 nm thick silicon nanowire in Figure 5.3, it can be seen that the nanowire surface is rough with an uneven amorphous surface layer between zero and 1.5 nanometers thick. The nonuniform amorphous layer could be attributed to the surface native oxide. The lack of a thick oxide shell is significant considering the fact that the nanowires had been exposed to room air for more than a month.

The observation of minimal to no oxide shell can be attributed to an inherent surface stabilization during nanowire growth. Such stabilization could occur by a partial hydrogen termination of the nanowire surface due to the presence of vicinal hydrogenated gallium species. Such stability to native oxide formation combined with a unique growth direction make these nanowires useful as interconnects in electronic devices. Additionally, a oxide-free and H-terminated silicon nanowires could be used to electrochemically reduce transition metal ions in solution to form nanoscale metal particles on the nanowire surface [16].

5.2.2 Plasma-Enhanced VLS Growth of Silicon Oxide and Nitride Nanowires

As schematically illustrated in Figure 5.4, PECVD with low-melting metal clusters enables nanowires of silicon and other single or alloyed materials to be synthesized [8]. Experiments involving nitridation are of particular interest via this approach. It is generally impossible to perform nitridation at low temperatures using nitrogen gas because of the nitrogen triple bond. Using microwave plasmas, we have shown that mixing silane and nitrogen diluted in hydrogen results in bulk nucleation and growth of silicon nitride (Si_xN_y) nanowires from gallium melt at reasonable temperatures less than 550°C. Nanowire growth was performed by exposing a quartz substrate covered with molten gallium film to a microwave plasma containing a mixture of 2% SiH_4/H_2, N_2, or O_2 for nitride and oxide nanowires, respectively. The experiments were conducted for durations ranging from 15 minutes to 3 hours and over a range of process conditions: 500 to 1100 W microwave power, and 10 to 50 Torr pressure, 0.5% to 50% N_2, O_2 in 2% SiH_4/H_2. The substrate temperature was determined using an infrared pyrometer to be approximately 550°C for microwave power of 750 W, 30 Torr pressure, and 40 sccm N_2 in 100 sccm 2% SiH_4/H_2.

The results show that the growth of silicon nitride nanowires is similar to that of silicon nanowires (i.e., multiple wires nucleated and grew from larger gallium droplets). Although low-magnification images show straight nanowires (Figure 5.5a), higher magnification images show the presence of some nanowires that are coiled with regular intervals similar to springs

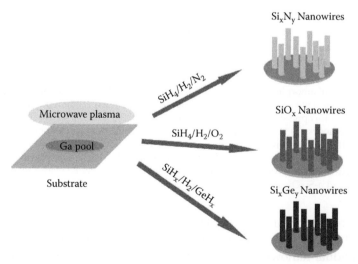

FIGURE 5.4

Synthesis of silicon, silicon oxide, silicon nitride, and Si-Ge alloy nanowires using plasma-enhanced chemical vapor deposition over molten Ga surface.

FIGURE 5.5
(a) High density of amorphous silicon nitride nanowires grown from a micron-sized gallium droplet exposed to a plasma containing silane and nitrogen. The experimental conditions were 700 W power, 30 Torr pressure, 50 sccm of 2% silane in hydrogen, and 100 sccm of nitrogen, and a duration of 3 hours. (b) Higher magnification of scanning electron microscope (SEM) image shown in (a) illustrating coiled morphology of the resulting nanowires; and (c) a SEM image illustrating the growth of high density of amorphous silicon dioxide nanowires from large Ga droplets. (Figures were reproduced with permission from Sunkara, M. K., et al., Bulk synthesis of a-SixNyH and a-SixOy straight and coiled nanowires. *Journal of Materials Chemistry*, 2004. 14(4): 590–594.)

(Figure 5.5b). The occurrences of the coiled morphology seem to be predominant over the entire sample and not limited to a particular location. In addition, individual nanowires growing from the gallium droplet converge to grow together as a bundle as indicated in Figure 5.5c.

The observation of high density and high growth rates (hundred microns per hour) for silicon nitride nanowires suggest that the dissolution of Si-N and Si-NH compounds is kinetically fast and highly selective to molten gallium. The equilibrium gas phase calculations did not indicate the presence of SiN or SiNH compounds in the gas phase at temperatures lower than 600°C. At this time, the dissolution mechanism is not understood. We believe that the chemisorption of atomic hydrogen or NH_x species increases the dissolution of Si-NH complexes into molten gallium.

Experiments using silicon and oxygen in a hydrogen plasma resulted in SiO_x nanowires. Interestingly, the results for silicon oxide nanowires are similar to that of silicon nitride nanowires—that is, high density (on the order of $10^{12} cm^{-2}$) of wires nucleate and grow from large gallium pools. The nanowires were typically less than 70 nm in diameter and several tens of microns long.

5.2.3 Plasma-Enhanced VLS Growth of Si and Ge Alloy Nanowires

Vapor phase precursors of Si and Ge can be combined to nucleate and grow SiGe alloy nanowires. In our experiments, a well-mixed powder of Si and Ge was placed next to a quartz substrate coated with a thin film of gallium and exposed to microwave plasma with H_2, N_2 gases flowing at 15 sccm and 100 sccm, respectively. The experiments were carried out at different microwave powers ranging from 500 to 700 W and a pressure of 30 Torr. The hydrogen/chlorine plasma reacted with the well-mixed Si and Ge powders to generate various Si and Ge species in the vapor phase. The optical emission spectrum in Figure 5.6a indicates that the dominant reactive species are SiH^* (silyl radicals), SiH_2Cl_2 (dichlorosilane), and $GeH_{4-n}Cl_n$ ($n = 0, 1, 2, 3, 4$) (chlorogermanes). The absence of a GeH^* peak at 265 nm could be due to excess hydrogen present in our experiments as in the case of previous reports [17]. Integrating the intensity of the peaks shows that the fraction of Ge containing vapor phase species increases linearly with increasing microwave power as shown in Figure 5.6b.

Raman analysis of SiGe alloys can be used to identify three distinct peaks—that is, two peaks for the pure phases (Si and Ge) and one for the alloy (SiGe phase). The optical phonons in the alloy produce a local mode at a frequency of ~410 cm^{-1} [18]. The optical phonons in pure Ge and pure Si produce local modes at frequencies of ~300 cm^{-1} and ~ 520 cm^{-1}, respectively. Alloying of Si and Ge results in both peak broadening and peak shifting of the local modes of Si and Ge relative to that of the pure materials [19]. A typical Raman spectrum for our nanowire samples shown in Figure 5.7a contains the characteristic peaks for Ge at ~300 cm^{-1} and for the SiGe phase at ~400 cm^{-1}. For SiGe alloys with $X_{Ge} > 0.45$, the Raman intensity corresponding to Si at ~520 cm^{-1} is observed to be extremely low, similar to that reported in the literature for bulk alloys [20]. From the ratios of the Raman peak intensities, the Ge fractional composition (Ge_x) can be estimated for our synthesized alloy nanowires. In order to confirm the composition, the nanowires were also characterized by TEM using two methods: energy dispersive spectroscopy (EDS) and lattice

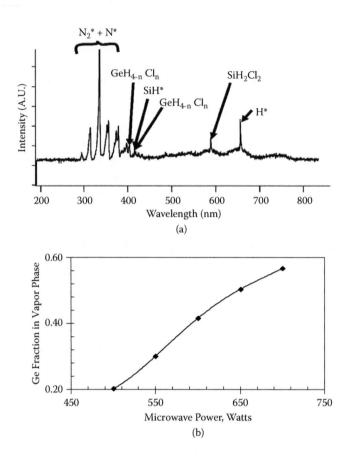

FIGURE 5.6
(a) A typical optical emission spectrum collected from microwave plasma during SiGe nano-wire growth. The synthesis conditions used were H_2 and N_2 flow rates at a ratio of 15:100, a pressure of 30 Torr, and microwave power of 500 W. (b) A plot showing the estimated Ge composition in the vapor phase as a function of microwave power. (Reproduced with permission from Meduri, P. et al., Controlled synthesis and Raman analysis of Ge-rich SixGe1-x alloy nanowires. *Journal of Nanoscience and Nanotechnology*, 2008. 8(6): 3153–3157.)

imaging. The lattice plane spacing determined using high-resolution TEM (HRTEM) can be related to the Si/Ge composition using Vegard's law [21]:

$$a_{GeSi} = x a_{Ge} + (1 - x) a_{Si}$$

where a_{Ge} and a_{Si} are the lattice constants of bulk Ge and bulk Si, respectively, and x is the fractional composition of Ge in the alloy. The compositional data obtained using Raman analysis were found to agree with that obtained by energy-dispersive X-ray spectroscopy (EDS) and HRTEM characterization. In Figure 5.7b, the Ge composition in the nanowires is plotted as a function of the Ge vapor concentration. As expected, the Ge fraction in the nanowires

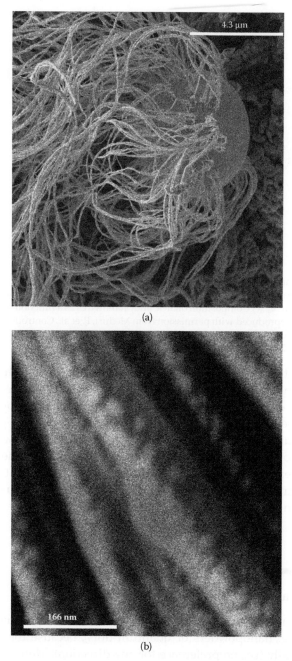

(a)

(b)

FIGURE 5.7
Synthesis of SiGe alloy nanowires at different microwave powers, 30 Torr pressure, and 15 sccm H_2 in 100 sccm N_2. (a) A high density of nanowires from a 2 micron-sized Ga droplet. (b) Raman spectra of resulting SiGe alloy nanowires indicating the characteristic peaks for SiGe alloy.
(continued)

(c)

FIGURE 5.7 (continued)
Synthesis of SiGe alloy nanowires at different microwave powers, 30 Torr pressure, and 15 sccm H_2 in 100 sccm N_2. (c) Composition of SiGe alloy nanowires as a function of Ge fraction in the vapor phase. (Reproduced with permission from Meduri, P. et al., Controlled synthesis and Raman analysis of Ge-rich SixGe1-x alloy nanowires. *Journal of Nanoscience and Nanotechnology,* 2008. 8(6): 3153–3157.)

increases with the Ge fraction in the vapor phase but is always higher in the nanowires than in the gas phase. In prior studies with bulk SiGe alloy deposition, Lovtsus et al. [24] showed that the Si fraction in the alloys increases with the synthesis temperature due to enhancement in the Si growth rate with temperature. In our case, the temperature also increased with increasing Ge vapor concentration, but in contrast, the Ge fraction in the nanowires increased instead of the Si fraction. This result suggests that in PECVD growth, the nanowire composition is controlled by the gas phase composition rather than the synthesis temperature.

5.3 Oxide Nanowires

The easiest shape of nanomaterials to synthesize in large quantities is a nanosphere (e.g., nanopowders or quantum dots), because these materials grow isotropically (i.e., no preference for any direction). More exotic shapes such as nanopyramids, nanoflowers, nanocones, and so forth, are difficult to synthesize because the growth direction must be carefully controlled. A shape that is of particular interest is a nanocylinder such as NWs, nanorods (NRs) or NTs, because of its potential for solar cells, batteries, and other photochemical and electrochemical applications. In this chapter, we focus on metal (e.g., iron and zinc) oxide NWs, which are desired for chemical sensors [23], optoelectronic devices [24], dye-solar cells [25,26], gas sensors [27,28], electrochromic devices [29], and so forth. Metal oxide NWs (e.g., iron oxide) have been synthesized by many routes: wet chemical treatments (sol-gel-mediated reactions [30–32], hydrothermal [33,34], or solvothermal processes [35], thermal and gas decomposition [36,37], chemical vapor deposition (CVD) [38], and plasma-assisted methods [39]. Wet chemical procedures that include sol-gel-mediated reactions, hydrothermal processes, or solvothermal processes and their variations, are widely spread for the synthesis of iron oxide or zinc oxide. However, these processes are typically difficult to control, yield rather small amounts of materials, and require a long time, as much as several hours (Figure 5.8). Additionally, there are problems with impurities such as reaction catalysts or surfactants, which must be separated from the synthesized NWs. Alternatively, thermal procedures have been developed, this includes flame synthesis [40], resistive heating [41], and thermal oxidation [42–44] or hot surface gas mixture decomposition (e.g., decomposition of gas mixture CO_2, SO_2, NO_2, H_2O, on 504 to 600°C heated Fe substrates [36] or decomposition of $Fe(CO)_5$ vapours on 300 to 400°C Si or glass substrates). These approaches can yield larger amounts of materials than wet chemical methods with minimal impurities. Moreover, the growth is well controlled, allowing the morphology to be tuned during growth. The disadvantages of thermal processes are that a mixture of different compositions (e.g., $Fe_{1-x}O$, α-Fe_2O_3, and Fe_3O_4) is often formed and the synthesis times are on the order of hours (Figure 5.8) [37,40]. There are a few recent reports of faster thermal growth to address this latter issue [40,41]. In comparison, CVD techniques for metal oxide NW synthesis generally need much less time. However, the growth of NWs requires templates or masks and even then the NWs are in

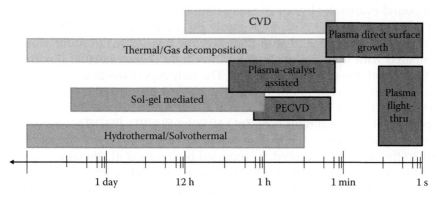

FIGURE 5.8
Different methods for large-scale synthesis of NWs and their characteristic time scale. (Reproduced. with permission from Cvelber, U. Towards large-scale plasma-assisted synthesis of nanowires. *Journal of Physics D – Applied Physics*, 2011. 44:174014)

most cases polycrystalline. In addition, CVD processes are harder to control, and it can be difficult to control the NW morphology. To address these issues, thermal and CVD processes have been combined to synthesize NWs, known as hot-wire or hot-filament CVD [45,46].

Large-scale, low-cost production of NWs would address a major impediment to widespread use of these novel materials in commercial applications. Specifically, there is need to synthesize these materials more rapidly. In the following sections, we discuss the recent development of plasma-assisted synthesis methods [47,48]. Generally, plasma-assisted methods can be divided into four categories. To describe the advantages and disadvantages of the different routes, we focus our discussion on iron oxide and zinc oxide NWs. The synthesis routes for iron oxide NWs (e.g., α-Fe_2O_3) can then be easily extended to other materials including CuO_2, V_2O_5, Nb_2O_5, or ZnO with SnO_2 NWs.

5.3.1 Fast Plasma Routes for Synthesis

5.3.1.1 Plasma-Enhanced Chemical Vapor Deposition

Plasma-enhanced chemical vapor deposition (PECVD) is a modified version of CVD that incorporates a plasma. PECVD has been used to deposit nanostructures including NWs from a gaseous precursor (vapor) to a solid product. A schematic of the typical process is presented in Figure 5.9a, where we can see that the precursor material, in our case metal M (Fe, Zn) or metal oxide, is evaporated or chemically sputtered from a target, interacts with the plasma and carrier gases, and then is deposited on a substrate surface. Here, it is worth mentioning that when we have a physical sputtering of target material into the plasma and deposition, the method is called plasma-enhanced physical vapor deposition (PEPVD). In chemical sputtering, the plasma ions react chemically with the target material and the so-formed compound evaporates by one of many complicated mechanisms.

PECVD allows ZnO NWs to be easily prepared but is more difficult for iron oxide NWs. This is due to the lower melting temperature for zinc as compared with iron. For this reason, there have been almost no reports of PECVD synthesis of iron oxide NWs. The only report worth mentioning is the synthesis of pyramid-line Fe_3O_4 NWs by Gao and coworkers [39], where they used CH_4 and N_2 to sputter hematite α-Fe_2O_3. In comparison, numerous reports exist for the synthesis of ZnO NWs by PECVD. PECVD is generally performed in a single step by self-assembly [49–51] or a two-step method [52]. In the single-step procedure reported by Iizuka *et al.*, the zinc is sputtered from a zinc target by an O_2/Ar plasma and reacts with oxygen on the substrate to form ZnO nanostructures [52,53]. The authors used a hollow-type magnetron (HTM) radio-frequency (RF) plasma source at a frequency of 13.56 MHz, where the sputtering rate of Zn was controlled by the biased voltage and the oxygen flow rate. In other cases, neutral oxygen atoms in the plasma have been found to play a key role in oxidation of sputtered Zn and

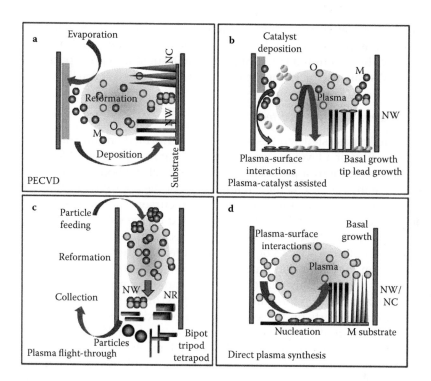

FIGURE 5.9

(See color insert.) Different plasma-based approaches for NW synthesis: (a) plasma-enhanced chemical vapor deposition (PECVD); (b) plasma-catalyst-assisted; (c) plasma flight-through; and (d) direct plasma synthesis routes. (Reproduced with permission from Cvelber, U. Towards large-scale plasma-assisted synthesis of nanowires. *Journal of Physics D – Applied Physics*, 2011. 44:174014.)

enable deposition of ZnO as NWs on the SiO_2 substrate [53]. Experiments reveal that the synthesized ZnO NWs are folded and bundled on the substrate [54]. The alignment of NWs on the substrate depends on the density of the sputtered Zn atoms that can be monitored by optical emission spectroscopy (OES) using the Zn/Ar spectral line intensity ratio [54]. The optimal line intensity ratio of Zn/Ar for NW synthesis has been reported to be 2/1. Zn has also been vaporized in an air ($N_2 + O_2$) plasma and deposited as ZnO NWs. This approach produces other shapes (e.g., hexagonal columns [55]) and can be performed with other carrier gases H_2 [56].

The two-step PECVD method for ZnO NW growth involves evaporation of solid zinc powder by a hollow-cathode discharge at temperatures in excess of 1000°C in a protective Ar atmosphere, followed by oxidation with atomic oxygen, and condensation from the gaseous phase on Si substrates [57]. In this case, the surface temperature, oxygen radical flux, quantity of zinc powder, and polarity are the most important factors for ZnO NW fabrication. The two-step method also allows the nucleation and the growth step to be effectively separated. For example, the oxygen content can be varied in a gas mixture with diethylzinc to create a precursor for zinc and oxygen on c-plane sapphire, Pt film on SiO_2/Si, and Si (100) substrates [52]. Special cases of ZnO NWs can also be formed (e.g., ZnO/a-C core-shell NWs) [58].

5.3.1.2 *Plasma-Catalyst Assisted*

Plasma-catalyst-assisted synthesis of NWs is performed on substrates coated with a thin film or a layer of unagglomerated catalyst particles. Metal oxide NWs grow with catalyst particles at their tips or at the bottom (Figure 5.9b) [59–61]. In many cases, a metal-containing gas or the appropriate alloyed catalyst particles are not readily available as a precursor. Therefore, solid metal or metal oxide nanoparticles must be deposited on the substrate by sputtering or evaporation. These particles are then used as nuclei for NW growth by reacting with plasma species or selectively attaching certain building units supplied from the plasma gas phase (Figure 5.9b). The size of the nuclei (catalyst particles) or the catalyst layer thickness can determine the thickness of the synthesized NWs on the sample surface [9,60,61].

There are no reports on the synthesis of iron oxide NWs by the plasma-catalyst-assisted route, whereas a few reports exist for ZnO NW growth. ZnO NWs have been grown via the VLS growth mechanism using gold (Au) nanoparticles as catalysts. There are drawbacks to VLS growth as it requires a very high temperature up to 925°C in order for the Zn vapor to dissolve in the Au catalyst. After saturation, Zn precipitates out of the droplet and is oxidized to form ZnO NWs [62].

5.3.1.3 *Plasma Flight-Through Method*

The plasma flight-through method describes various plasma processes that can reshape or synthesize new NWs from particles (e.g., solid–liquid–solid (SLS), solid–vapor–solid (SVS), and solid–liquid–vapor–solid (SLVS) growth mechanisms. Particles passing through a gaseous plasma, typically in a free fall, can either melt or vaporize and then interact with plasma species to form new types of NWs (Figure 5.9c) [63–65]. This method works with non-thermal or thermal plasmas operated typically at high frequencies and high gas pressures.

The plasma flight-through method is relatively new for NW synthesis. No reports exist for the synthesis of iron oxide NWs by this approach. The main problem could be the time iron or iron oxide particles spend inside the plasma that must be sufficiently long because iron and iron oxide have relatively high melting temperatures. In contrast, the synthesis of ZnO NWs by plasma-flight through has been reported in five papers by three different groups. Researchers from Taiwan used a DC plasma discharge operated at 70 kW and atmospheric pressure and fed Zn powders into the plasma jet with carrier gas. The powder was subsequently vaporized and oxidized [63,66]. They noted that the ratio between carrier gases significantly influenced the shape and length of NWs. In the case of N_2 carrier gas, the addition of Ar helped reshape the zinc powders from 10 μm spheres into elongated or NW-like structures, and less N_2 gas favored the formation of elongated rod/NW-like nanostructures [63]. Impurities in the feed, input power, and residence time were also found to affect the shape of the nanoparticles, in some cases resulting in tetrapot-like elongated structures. Increasing in the input power and residence were both found to produce ZnO NWs with higher aspect ratios [66].

Similar experiments have been carried out in a RF discharge operating at 4 MHz and 30 kW, with Ar and N_2 as the forming gas and sheath gas, respectively [67]. In order to get ZnO, oxygen gas was injected into the system together with the sheath gas. Different amounts of oxygen were found to change the length and diameter of the NWs. The optimal flow rate for NW synthesis was determined to be between 2.5 and 5 L/min. They concluded that the oxygen partial pressure and supersaturation of zinc vapor, controlled by the gas flow rates, play key roles in the growth of ZnO NWs or NRs [67] .

The highest aspect ratio and yield of ZnO NWs ever achieved by the plasma flight-through method was recently reported by Sunkara et al. They used a 2.45 GHz MW plasma jet reactor at atmospheric pressure and input power ranging from 300 W to 3 kW [47,48]. In their experiments, the sheath gas was air with a flow rate of 10 to 15 slpm. Reforming gases, H_2 (100 sccm) and O_2 (500 sccm), were also added to the sheath gas. This approach uses the same mechanism for particle reshaping as previously stated but was found to have a higher efficiency as 80% to 90% of all the powder input was reshaped into NWs in a single step. The low power consumption for the process combined with the high efficiency makes this a viable approach for industrial-scale production of NWs.

5.3.1.4 Direct Plasma Synthesis

Direct plasma synthesis is a process where NWs are grown by exposing a substrate to gaseous plasma. The method is simple and does not use any templates, catalysts, or vapor sources of the metal precursor. Metal foils or pieces are used as the metal source and exposed to gaseous plasma radicals that dissolve and, consequently, form nanostructures on the surface. The NWs are believed to grow by the SLS mechanism. This growth mechanism is typically used for low-melting metals or metal oxides [68]. In this case, the plasma helps liquefy the solid metal, solid metal oxide nuclei are created on the surface, and NWs nucleate and grow [68–70]. The selective growth at the molten metal, sometimes referred to as self-catalytic growth on molten phase, occurs through liquid-phase epitaxy. In the solid–solid (SS) phase, epitaxial NW growth occurs after a similar spontaneous nucleation when nuclei are created by the phase transformation of metal to metal oxide (Figure 5.9d). In this respect, direct plasma synthesis resembles plasma-catalyst-assisted growth, but the nuclei are created from the substrate, simplifying the overall process. Precise tailoring of NW shapes and lengths is possible by control-ling the plasma radical flux to the surface [71–73], the electrical conditions, and the surface temperature [74,75]. Because NWs are grown directly on metallic surfaces, additional processing steps are eliminated and that facili-tates applications [74–76].

Direct plasma synthesis was first reported in 2005 for the growth of Nb_2O_5 NWs on niobium foils [77]. The authors used a RF inductively cou-pled plasma (ICP) plasma, generated in O_2 gas at low gas pressures (50 to 200 Pa). Since then, the method has also been extended to other metal oxide NWs including hematite α-Fe_2O_3 [70,77,78]. The authors found that single-crystalline α-Fe_2O_3 NWs array growth can be controlled by the flux and dose of the neutral oxygen atoms and oxygen ions to the surface [79]. Because atoms and ions heat the surface through recombination processes, the released heat can be used to overcome the surface temperature of 570°C for Fe to undergo spontaneous nucleation to hematite phase α-Fe_2O_3 [78]. This can also be effectively achieved by controlling the electrical condi-tions on the iron surfaces exposed to the reactive oxygen plasma [74]. After nuclei are created, the surface temperature needs to be controlled, because an increase in temperature can result in different nanostructures (e.g., α-Fe_2O_3 nanobelt structures), due to increased mobility of atoms, or even absence of any nanostructures [78]. Therefore, the growth of NWs as opposed to other nanostructures is achieved in a narrow but well-defined temperature range. However, the temperature is not the most important growth parameter. The flux of neutral atoms and ions controls "supply and demand" and can tailor the shape, thickness, and length of NWs which emanate from nuclei [74,78,79].

Curiously, there are no reports of such growth for ZnO NWs. This is prob-ably due to the low heat of vaporization needed for the sublimation of zinc in vacuum that results in zinc evaporating from the surface before any stable ZnO nuclei can be created.

5.3.1.5 Mixed Plasma Routes

Mixed plasma routes refer to approaches that combine two or more of the previously described synthesis routes for NWs. An example is the synthesis of NWs with catalytic particles mixed together with other metallic particles in the plasma flight-through method. The catalytic particles represent the nucleation site from which vaporized metallic particles form NWs. Although there are no reports of zinc oxide or iron oxide NWs, other NW materials and NTs have been prepared by this approach [80–83]. PECVD and the plasma flight-through route where Zn powders are deposited into a MW plasma torch operated with O_2 gas and a swirling gas $N_2 + O_2$ (air) could be classified as a mixed plasma route because solid metal powders must first evaporate before reforming and depositing as ZnO NWs [84,85].

5.3.2 Parameters for Large-Scale Synthesis of NWs

5.3.2.1 Time

Time is one of the most important parameters for large-scale, economical nanomaterial synthesis. In Figure 5.10, the time required for synthesis of iron oxide NWs or NRs using various synthesis approaches is shown. Methods such as pulsed laser deposition (PLD) take 60 to 140 min to synthesize exotic ϵ-Fe_2O_3 NWs [86]. Improved process efficiency is found for thermal routes such as rapid flame synthesis, which can take a minimum of about 20 to 30 min [38], or resistive heating which only requires a couple of minutes [41]. Otherwise, thermal processes are generally longer, from 20 min to 12 hours or, in some cases, a couple of days [37,87–89]. The most time-efficient process reported to date is direct plasma synthesis of α-Fe_2O_3 NWs. This process takes only several minutes and produces large amounts of NWs (Figure 5.3). Plasma processes should be more efficient than thermal processes, because the plasma aids dissociation of molecules into radicals and other species that are very reactive with surfaces. The PECVD method has been found to be 10 times more time efficient than thermal CVD. Even larger discrepancies

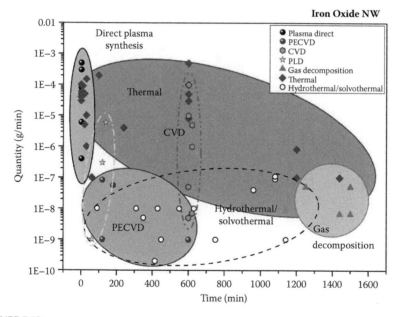

FIGURE 5.10

(See color insert.) Time required for the synthesis of iron oxide nanowires (NWs) versus quantity produced for plasma-assisted syntheses in comparison with other methods (e.g., thermal, gas decomposition, pulse laser deposition [PLD], chemical vapor deposition [CVD], and hydrothermal/solvothermal. (Marked areas denote reported results for specific method, but can always be extended toward smaller quantities.) (Reproduced with permission from Cvelber, U. Towards large-scale plasma-assisted synthesis of nanowires. *Journal of Physics D – Applied Physics*, 2011. 44:174014.)

are observed for wet chemical processes such as hydrothermal or solvo-thermal processes. These processes take a couple of hours or even days, and depend on many parameters and steps. Based on the reports summarized in Figure 5.10, the least time efficient process for NW synthesis is the gas decomposition method, which can take several days [90,91].

A similar plot is shown for zinc oxide NWs or NRs in Figure 5.11. The most time-efficient plasma synthesis process is the flight-through method, where Zn or ZnO particles are reformed while passing through the plasma. This process normally takes less than a second and has been reported to occur within a range of 1.7×10^{-4} to 5×10^{-3} min [47,48,63,66,67]. A combination of PECVD and plasma flight-through method is also very fast, where NWs are synthesized in only 1 s (1.7×10^{-2} min) [84,85]. The next fastest approach is direct plasma synthesis that takes several tens of seconds, followed by PECVD, which takes more than 10 min [53–58]. All plasma-assisted processes are found to be several orders of magnitude faster than thermal, wet chemical, and gas decomposition methods (not shown in Figure 5.11).

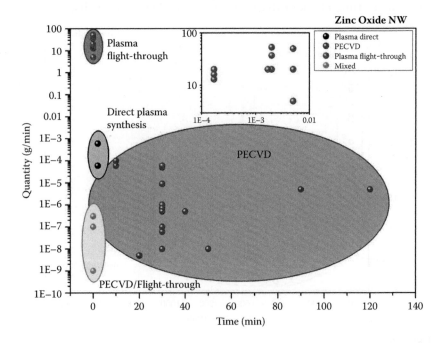

FIGURE 5.11
(See color insert.) Time required for the synthesis of ZnO nanowires (NWs) versus quantity produced for plasma-assisted syntheses in comparison with other methods (e.g., thermal, gas decomposition, PLD, CVD and hydrothermal/solvothermal. (Marked areas denote reported results for specific method, but can always be extended toward smaller quantities.) (Reproduced with permission from Cvelber, U. Towards large-scale plasma-assisted synthesis of nanowires. *Journal of Physics D – Applied Physics*, 2011. 44:174014.)

5.3.2.2 Quantity

Process time is important for large-scale synthesis, and the quantity of synthesized material (i.e., production rate) is also critical. Processes that are rapid but yield small amounts of material are generally not very useful. Figure 5.10 shows that plasma processing is superior to gas decomposition, hydrothermal, and solvothermal processes in terms of time scale. However, while high quantities of iron oxides have been reported for direct plasma synthesis (5×10^{-5} g min^{-1} of α-Fe_2O_3 NWs) [79,92], the PECVD method does not yield sufficiently high quantities of iron oxide NWs to be adequate for industrial use. The maximum quantity of pyramid-like Fe_3O_4 NWs reported for PECVD was 8.3×10^{-8} g min^{-1} [39]. PLD and CVD processes yield higher quantities, 6×10^{-6} g min^{-1} and 1×10^{-4} g min^{-1}, respectively [38,86]. The only process similar to direct plasma synthesis is thermal, where the highest quantity of synthesized α-Fe_2O_3 NWs reported was about 5×10^{-5} g min^{-1} [38,41,44].

Although iron oxide NWs have never been synthesized by the plasma flight-through method, the best results for ZnO NWs have been achieved by this method (Figure 5.11). Production rates of 20 to 50 g min^{-1} have been reported which would address any problems regarding scale up [47,48,63,66,67]. The production rate via plasma flight-through is followed by direct plasma synthesis on foils with a maximum of 6×10^{-4} g min^{-1}. Much smaller amounts of ZnO NWs have been synthesized by mixed methods [84,85] or by other PECVD methods [53–58]. The results for the other methods such as wet chemical procedures are far less and discouraging.

5.3.2.3 Quality

An additional requirement for industrial implementation of NWs is their quality, which encompasses their dimension (size), shape, crystallinity, crystalline orientation, surface density, electrical and mechanical properties, and purity. In many cases, even when process time and yield are appropriate for large-scale manufacturing, the quality of the synthesized material may not be adequate.

A critical aspect of NWs for applications is their shape and dimension defined by the NW length (l) and diameter (d). These two parameters are especially important for the plasma flight-through route, because most of the materials that pass through the plasma do not reform into NWs but remain as particles or form other tripod or tetrapod structures [66,67]. These alternative structures are in many cases not desirable and need to be separated from the NWs. At low temperatures or high feed rates, the density of plasma species may not be high enough to convert the fed materials. The result is a mixed product of nanoparticles and NWs that are typically stuck together and inseparable (Figure 5.12a). The quality of NWs can be related by the ratio of the length to the diameter, l/d. The optimal yield versus quality for ZnO NWs was reported for the plasma flight-through method with a 30 kW RF thermal torch operating at atmospheric

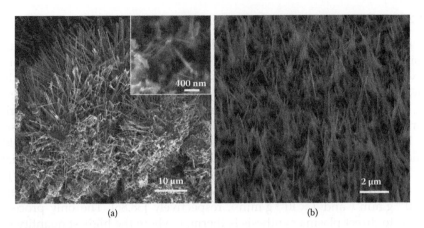

(a)   (b)

FIGURE 5.12
Scanning electron microscopy (SEM) images of (*a*) ZnO NWs synthesized from Zn powder by the plasma flight-through method in a 1.7 kW MW system and (*b*) α-Fe2O3 NWs synthesized by direct plasma synthesis on Fe substrate in a 1 kW RF ICP low-pressure plasma system. (Reproduced with permission from Cvelber, U. Towards large-scale plasma-assisted synthesis of nanowires. *Journal of Physics D – Applied Physics*, 2011. 44:174014.)

pressure, which gave 20 g min^{-1} and a l/d ratio of 14. Increasing the feed flow rate to 37 g min^{-1} and then 53 g min^{-1} was found to lower the l/d ratio to 8 and 2, respectively [93]. Improved conversion of NWs was achieved by Sunkara's group with a 1.7 kW MW plasma. Their process yielded 20 g min^{-1} with an l/d ratio of 50, and an overall conversion of 80% to 90% [48].

The purity of single NWs or NW arrays is higher with plasma processes than other approaches. Thermal routes normally produce different types of metal oxides. Wet chemical routes or gas decomposition methods can incorporate impurities from residuals, side products of synthesis, or foreign atoms. These impurities significantly affect the properties of the synthesized material. In thermal oxidation of a Fe foil, the thick basal layer contains multiple iron oxide phases (e.g., γ-Fe$_2$O$_3$ and Fe$_3$O$_4$, in addition to α-Fe$_2$O$_3$). In comparison, direct plasma synthesis only produces α-Fe$_2$O$_3$ NWs (Figure 5.12b). For applications as electrodes, thermally prepared samples will not be useful, because a thick iron oxide layer with multiple iron oxide phases will act as a trap for passing electrons, whereas, in the case, of plasma-synthesized NW arrays, electrons will easily pass through the layer without any significant losses. An additional advantage of plasma-synthesized NWs is the lattice superstructure, which in α-Fe$_2$O$_3$ NWs is characterized by periodic oxygen vacancy planes (Figure 5.13a) [33,78]. These planes may be created by thermally induced surface stress initiated during the fast epitaxial growth. These vacancies introduce p/n-type properties in the NWs without any chemical doping, and enable more efficient charge transport along the wires. In contrast, there are no such structures observed in the thermally prepared samples (Figure 5.13b), when the NWs grow slowly, in thermal equilibrium.

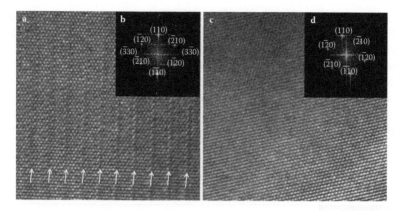

FIGURE 5.13

(*a*) High-resolution transmission electron microscopy (HRTEM) image of a typical (direct) plasma-grown α-Fe_2O_3 NW showing an edge-on view of several ordered oxygen vacancy planes (indicated by arrows). (b) Fast Fourier transform (FFT) of the images shown in (a) with indexing consistent with the [0 0 1] zone axis. An array of superlattice spots originating from the order oxygen vacancy planes is visible in addition to the α-Fe_2O_3 spots (indexed spots). The distance between the superlattice spots matches that of the 1/10 (3 –3 0) planes. (c) HRTEM image of a typical thermally grown α-Fe_2O_3 NW. Its FFT shown in (*d*) does not reveal any superlattice pattern. (Contributed by J Jasinski, UofL.) (Reproduced with permission from Cvelber, U. Towards large-scale plasma-assisted synthesis of nanowires. *Journal of Physics D – Applied Physics*, 2011. 44:174014.)

For many applications, single crystalline NWs are desirable. This is hard to achieve especially for PECVD and plasma-catalyst-assisted processes, where the long wires are composed of polycrystalline particles or catalyst atoms that are incorporated into the NW structure. Single-crystalline NWs are also often converted to polycrystalline ones when exposed to high temperatures during the synthesis process.

Last, the shape of NWs, not the *l/d* ratio, but the morphology, is important. Changing plasma parameters during growth allows the shape of NWs to be tailored in different ways. NWs can be thickened or narrowed during the growth or widened at the bottom, and so forth. The NW shape is modified by controlling the flux of neutral atoms or ions to the treated surface. If the incoming flux of ions and the ion energy are increased during synthesis, then the basal diameter of NWs widens, making them tip-like and appropriate for applications as AFM tips. However, in order to control the surface density of synthesized NWs, we need to control several plasma parameters simultaneously. Additionally, the plasma parameters must be controlled to optimize synthesis, because the flux of plasma species such as neutral atoms, ions, and metastables are intimately linked to growth dynamics. Unfortunately, there is still a lack of measurement techniques available to control these plasma species. In recent years, many groups have successfully developed new characterization tools to determine plasma parameters (e.g., catalytic probes [94–96] or laser induced fluorescence (LIF) [97–99] for measuring neutral atom densities, and Langmuir probes [100,101] to detect charged species). Furthermore, even optical emission spectroscopy is becoming more widely accessible and is used to monitor plasma processes [94,102,103]. All of these tools are necessary to measure and control plasma parameters in order to optimize the NW synthesis (i.e., control the processing time, quantity, and quality of NWs more efficiently). By optimizing the process, we can also use energy more efficiently. The energy efficiency of a process will play an important role in industrial scale up, where we will have to look at input energy per synthesized gram of NWs.

5.4 Large-Scale Production of Nanowires

Although catalyst-assisted vapor-liquid-solid schemes have been typically used in the past to synthesize NWs [62,104], other reaction schemes have been developed to synthesize metal oxide NWs. Many of the previously described synthetic methods refer to direct reaction between low melting point metals and O_2. In some cases, the reaction on micron-size molten metal clusters leads to high density NW growth from the large cluster [68]. In other cases, metal droplets are formed which initiate self-catalytic tip-led growth of NWs [105]. In all the above cases, the use of a substrate limits the ability to produce large quantities of NWs. High-throughput NW production can be achieved by gas-phase synthesis, wherein the reacted species are swept away from the reaction zone quickly and additional unreacted species are continuously introduced.

In the case of transition metals with high melting points, rapid oxidation of metal foils in highly dissociated O_2 plasmas can lead to metal oxide NW growth [77]. However, the plasma in this case is not expected to raise the temperature of the metal foils above their melting points. Hence, one-dimensional NW growth is believed to occur as a result of the low mobility of metal atoms. For low-melting metals, both thermal and plasma oxidation of substrate-supported metals occur at temperatures above the melting point [68]. A recent study showed that Zn powders could be oxidized to ZnO NWs using a high-power radio-frequency plasma or a DC thermal plasma [93,106].

Here we describe a large-scale approach for NW production at a rate of 5 kg/ day using a high-throughput plasma jet reactor that operates at atmospheric conditions. We applied this process to several metal powders including Zn, Al, Sn, and Ti, different particle sizes, and various plasma parameters.

5.4.1 Experimental Plasma Reactor for Bulk Production

Bulk production of NWs was performed with a newly designed reactor that efficiently generates microwave plasma inside a quartz tube. The reactor operates at pressures ranging from a few Torr to atmospheric pressures and powers ranging from 300 W to 3 kW. A schematic and a photograph of the reactor are presented in Figure 5.14. The reactor includes a sheath gas delivery chuck to protect the tube from heat generated by the high-power plasma discharge. This allows for prolonged operation of the plasma jet. The length of the plasma jet is 30 to 40 cm as shown in Figure 5.14b. Quartz tubes with diameters of 3.8 cm (1.5 inch) and 5 cm (2 inch) are used for the experiments. A metal rod with pointed ends is used to ignite the plasma. Metal powders are poured directly into the plasma cavity zone and then allowed to flow down by gravity with gas flow. Alternatively, the metal source is fed into the plasma jet by pressurized gas or a mechanical dispenser. The resulting powders are collected at the bottom either in a quartz flask or a filter bag.

Several experiments with different metal powder feeds were conducted in the microwave plasma jet reactor operated at powers ranging from 700 to 2000 W and 2.45 GHz frequency and 8 to 15 slpm of air as the sheath gas. Sn and Zn metal powders with sizes between 1 and 5 μm, Ti powders with sizes between 20 and 100 μm (65 μm average size), and Al powders with sizes between 3 and 4.5 μm were used. A gas mixture of 2 to 4 slpm of Ar and 100 to 500 sccm of O_2 (the oxidative gas) were used at atmospheric pressures. The resulting metal oxide powders were characterized using a FEI Nova600

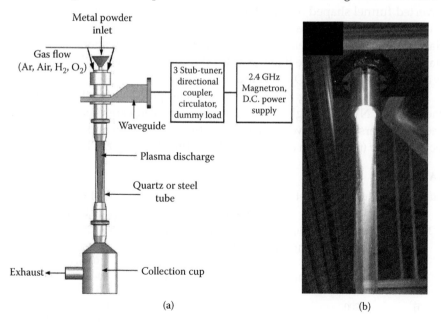

(a)　　　　　　　　　　　　　　　　(b)

FIGURE 5.14

(a) The microwave plasma jet reactor showing all the essential components. (b) Photograph of the high-density plasma jet discharge at 2 kW power produced in a quartz tube of 3.8 cm diameter with a plasma flame length of about 30 to 40 cm. (Reproduced with permission from Kumar, V. et al., Gas-phase, bulk production of metal oxide nanowires and nanoparticles using a microwave plasma jet reactor. *Journal of Physical Chemistry C*, 2008. 112(46): 17750–17754.)

FE-SEM, a Renishaw in-via micro-Raman/photoluminescence spectroscope and Bruker D8 Advance model x-ray diffractometer (XRD).

5.4.2 Technology Assessment for Large-Scale Production

5.4.2.1 Different NW Material Systems

The process conditions and resulting morphologies of NWs are summarized in Table 5.1. The resulting NW powders were collected from the quartz cup, filter papers placed in the exhaust, and the quartz tube wall. The powder collected from filter paper had thinner NWs compared to other powders. The SEM images in Figures 5.15a through 5.15d show the morphologies of oxide NWs synthesized from Sn and Zn metal powders, respectively. The SnO_2 NWs tend to be straight but highly branched, whereas ZnO NWs have a flowery morphology. Nanostructures of ZnO in the shapes of tripod, brush, and comb (Figures 5.15e through 5.15g) were also observed. Figures 5.16a through 5.16d show SEM images of TiO_2 and Al_2O_3 NWs. TiO_2 NWs are more difficult to synthesize than NPs and very short 1D couplet morphology of NWs are obtained with the current reactor set-up. Al_2O_3 NWs tend to be inverted funnel shaped and protrude out from the bulk metal in a flowery pattern. Straight and isolated Al_2O_3 NWs have also been observed. The XRD data (shown in the supplementary document) and the corresponding Raman data (Figure 5.2h) indicate that the ZnO NWs are composed of the hexagonal wurtzite phase while SnO_2 NWs are rutile phase.

The experiments using Sn, Zn, and Al metal powders for the respective metal oxide NWs are highly reproducible. However, experiments using Ti metal powders did not always yield NWs. In the case of Sn and Zn, the product efficiency (the fraction of NWs to NPs) is about 80% to 90%. The remaining

TABLE 5.1

Summary of the Experimental Conditions Used for Nanowire (NW) Production

| Metal | Power, W | Flow Rates | | | Resulting Metal Oxide NW Characteristics |
		Air, slpm	H_2, sccm	O_2, sccm	
Sn	1200–1500	10–15	100	500	20–100 nm (dia); 10 μm (length)
Zn	1400–1700	10–15	100	500	50–100 nm (dia); 5 μm (length)
Ti	700	8–10	100	500	100–250 nm (dia); 1 μm (length)
Al	800	8–10	100	500	50–150 nm (dia); 3–5 μm in length

Source: From Kumar, V. et al., Gas-phase, bulk production of metal oxide nanowires and nanoparticles using a microwave plasma jet reactor. *Journal of Physical Chemistry C,* 2008. 112(46): 17750–17754. With permission.

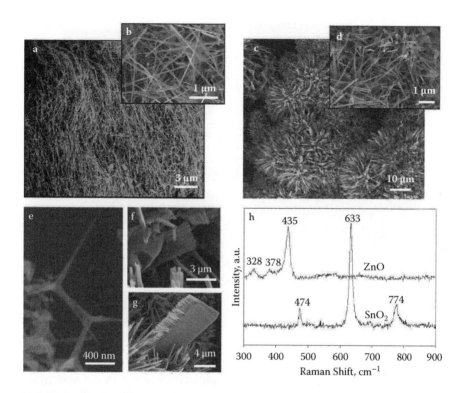

FIGURE 5.15
Low- and high-resolution scanning electron microscope (SEM) images of SnO_2 (a, b), and ZnO (c, d) NWs are shown. Nanostructures of ZnO such as tripod (e), nano brush (f), and nano-comb (g) are also observed. Raman spectra of as-synthesized SnO_2 and ZnO NWs are shown (h). (Reproduced with permission from Kumar, V. et al., Gas-phase, bulk production of metal oxide nanowires and nanoparticles using a microwave plasma jet reactor. *Journal of Physical Chemistry C*, 2008. 112(46): 17750–17754.)

10% to 20% is unreacted or partially oxidized metals and agglomerated metal oxide particles.

The above experiments were conducted using a metal powder feed of about 5 grams/minute which translates to a production capacity of 5 kilograms of metal oxide NWs per day when operated continuously. The reactor can be operated continuously with a recycle stream for unreacted metal particles and a continuous collection system for metal oxide NWs using filter bags. NW production will be limited by two factors: (1) the maximum solid loading in the entraining gas used for recycling and (2) the maximum amount of solid that can be treated by the plasma flame.

5.4.2.2 Postproduction Purification

The as-grown powders contain unreacted metal powder, metal oxide NPs, and metal oxide NWs. These can be purified using a simple gravity

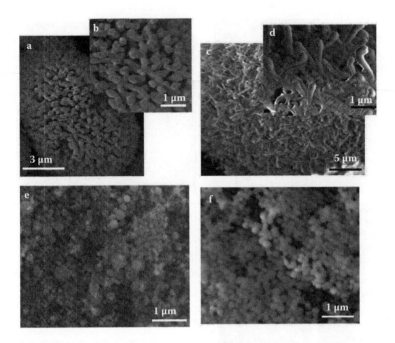

FIGURE 5.16
Low- and high-resolution scanning electron microscope (SEM) images of TiO_2 (a, b) and Al_2O_3 (c, d) nanowires are shown. As-synthesized nanoparticles (NPs) of TiO_2 (e) and Al_2O_3 (f) are also shown. (Reproduced with permission from Kumar, V. et al., Gas-phase, bulk production of metal oxide nanowires and nanoparticles using a microwave plasma jet reactor. *Journal of Physical Chemistry C*, 2008. 112(46): 17750–17754.)

sedimentation technique. As an illustration, the as-synthesized ZnO NW powders were dispersed in 1-methoxy 2-propanol, horn sonicated for a few minutes, and left for about 4 hours under gravity to settle. The top portion of the dispersion contained a high proportion (>95%) of NWs, whereas the bottom precipitate was mostly unreacted metals, partially oxidized metals, and agglomerated metal oxide particles. The results can be explained by agglomeration behavior of spherical particles versus NWs in solutions. In our previous study [107,108], we studied the dispersion behavior of NPs versus NWs using tungsten oxide and found that NPs tend to agglomerate into large spherical particles that then quickly settle due to gravity while NWs tend to be stable in dispersions. Thus, the NWs remain suspended in solution and can be easily separated from the NPs.

5.4.2.3 Effect of H_2, Water Vapor, and Metal Powder Diameter

Experiments were performed using H_2 at a flow rate of 100 to 500 sccm mixed with other gases (total gas flow rate of 8 to 12 slpm). In our earlier experiments with a low-pressure microwave plasma and thermal oxidation, the use of H_2/water vapor enhanced the growth of metal oxide NWs at low temperatures [68]. Experiments conducted using H_2 with Sn metal powders show improved quality, higher fraction of NWs, and improved size control (about 20 nm diameter) with lengths up to several microns (Figure 5.15a). Experiments were also conducted by introducing steam as a sheath gas instead of H_2. The effect of water vapor is similar to that of H_2—several active radicals, such as H and OH, are produced that can reduce metal oxides. In the presence of such radicals, the lateral growth of metal oxide nuclei is inhibited by maintaining a high surface energy [68]. Thus, the addition of H_2 or water vapor results in a reduction of the NW diameter. Experiments performed using powders with larger particle sizes (10 to 100 microns), however, showed a drastic decrease in the fraction of NWs produced. Therefore, the production of high-quality, bulk quantities of NWs appears to require H_2, water vapor, and small diameter metal powders.

5.4.2.4 Production of Metal Oxide NPs versus NWs

In the case of Ti and Al powders, experiments conducted with 1.3 kW, 10 slpm air, 2 slpm Ar, 100 and 500 sccm of H_2 and O_2 resulted in metal oxide NPs of the respective materials (Figures 5.16e and 5.16f). In comparison, experiments at lower powers such as 700 and 900 W resulted in NWs. At powers less than 900 W, the resulting Al_2O_3 NW powders clearly show bulk nucleation and growth of NWs from metal particle cores, as is seen in Figure 5.16. In the case of higher powers, both Ti and Al metal powders were converted into metal oxide vapors before condensing to form metal oxide NPs. We suggest that spherical NPs are produced at higher plasma power (with high reproducibility) due to homogeneous nucleation of metal oxide particles via condensation of metal oxide vapors, while NWs were produced at lower plasma power due to the nucleation and growth of metal oxide nuclei from molten metal particles. NP production is also readily observed in the case of Ti and Al. This is because of the high vapor pressures for the metal oxide species and the high exothermic heat associated with surface oxidation and radical recombination reactions. The Al_2O_3 NPs were more uniform in size compared to TiO_2 NPs because of the smaller metal powder feed size. The XRD spectra of TiO_2 NPs indicate that both anatase and rutile phases are present. The anatase fraction (f_A) was calculated to be about 80% using the following formula [109]: $f_A = (1 + 1.265 * I_R/I_A)^{-1}$, where I_R and I_A are the peak intensities of the rutile and anatase phases, respectively, in the XRD spectrum of the sample. The Raman spectra of the as-synthesized TiO_2 NPs correspond well with that of commercial powders (~95% anatase, from Alfa Aesar) with predominant peaks for anatase phase and a single peak for rutile phase. The XRD spectra of as synthesized Al_2O_3 NPs matches JCPDS data base # 016-0394 confirming the presence of δ-Al_2O_3 [110]. The δ phase transforms to α phase upon further heating. This phase does not give Raman bands due to reasons described elsewhere [111,112].

5.4.2.5 *Mechanism for Heating in Microwave Plasma*

Metal particles cannot absorb microwave radiation directly to heat up. In addition, the residence time for micron-size metal particles in the microwave plasma reactor is limited to a few seconds. However, the nucleation and growth of NWs indicate that the metal particles heat up to very high temperatures, enough for them to melt (at least at the surface). In order to understand the mechanism for heating in the microwave plasma reactor, we performed a simple order of magnitude analysis for heat transfer by conduction, convection, radiation, surface chemical reactions, and collisional heating. Our preliminary analysis suggests that convection and reactive heating are the only dominant modes. Collisional heating, which scales with $v/(\omega^2 + v^2)$, is ineffective in microwave plasmas because the collision to wave frequency ratio v/ω is much less than 1 [113,114]. Here, v is the electron-neutral collision frequency, and ω is the angular frequency of the electromagnetic field. For an electron to collide with a positively charged metal surface, the time it takes to reach the surface $(1/v)$ should be smaller than the time the surface takes to become negatively charged and repel the electron $(1/\omega$, due to changing polarity of microwave-induced electric field).

In order to melt a Ti metal particle with a size of 65 μm, 0.9 mJ (milliJoule) of energy $(m * C_p * \Delta T + m * L$, m is mass of a single spherical particle of Ti, C_p is average specific heat capacity over the temperature range, ΔT is the temperature difference between the Ti melting point and ambient temperature and L is the latent heat of fusion) is required. Reactive heating can impart about 40 mJ of energy in a flight time of 0.25 sec through the plasma flame. Convective heating, assuming a gas temperature of 1000 K, can provide about 5 mJ to the metal particle. The reactive heating is estimated using the mass flux to the particle, with an average value for heat of reaction of 1000 kJ/mol and 20% contribution by radicals. The most important surface reactions involving O_2 and H_2 radical recombination as well as metal oxidation reactions are as follows:

$$H + H \rightarrow H_2 \qquad\qquad \Delta H = -444 \text{ kJ/mol}$$

$$O + O \rightarrow O_2 \qquad\qquad \Delta H = -505.3 \text{ kJ/mol}$$

$$Ti(s) + 2O \rightarrow TiO_2(s) \qquad\qquad \Delta H = -1445 \text{ kJ/mol}$$

$$Ti(s) + O_2 \rightarrow TiO_2(s) \qquad\qquad \Delta H = -939.6 \text{ kJ/mol}$$

Figure 5.17 shows a comparison of the energy required to melt and the energy supplied via reactive and convective heat transfer, as a function of particle size. The data show that the reactive energy is at least an order of magnitude higher than the convective energy. The analysis also shows that there is an upper size limit for metal particles (in hundreds of microns) beyond which the particles will not melt in the plasma by either of the heating

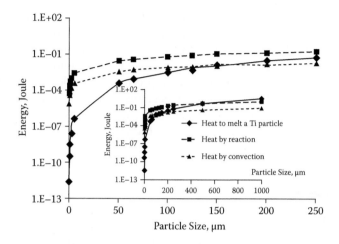

FIGURE 5.17
Comparison of energy required to melt a Ti metal of 65 μm size and reactive and convective energies imparted to the particle during its flight. Inset shows the comparison at larger particle sizes. Reactive energy is about 10 times higher than the convective energy. (Reproduced with permission from reference Kumar, V. et al., Gas-phase, bulk production of metal oxide nanowires and nanoparticles using a microwave plasma jet reactor. *Journal of Physical Chemistry C,* 2008. 112(46): 17750–17754.)

modes. Nevertheless, this simple analysis indicates that the reactive heating dominates the overall heating mechanism for metal particles in microwave plasmas unlike that mentioned in some of the modeling efforts published earlier [115,116].

In order to test our hypothesis that reactive heating is responsible for melting metal particles, comparative experiments were performed using N_2 plasma at 1.5 kW power and air plasma at 1.2 kW with gas flow rates of 12 slpm in both cases and the same particle size. The air plasma experiment showed large Ti metal particles getting converted to smaller Ti metal particles along with finer TiO_2 NPs. On the other hand, N_2 plasma experiments [8] did not show any significant alteration in the starting metal size. This clearly demonstrates that the metals in plasma are heated to a large extent reactively and not convectively. Convective heating, however, is more effective in heating small size particles. This was confirmed by pouring 100 nm size Zn particles into a N_2 plasma at 1.2 kW and 12 slpm flow rate. N_2 plasma treatment of about 10 μm size Al powders, on the other hand, did not show any change in particle size.

5.4.2.6 Experimental Observations of Nucleation and Growth of NWs

Experiments with Zn and Al metal powders often resulted in flowery morphologies, indicating a high density of nucleation. In many ways, these observations are similar to those observed when performing plasma oxidation of metal particles supported on substrates [68]. Experiments with Sn and Ti produced slightly different results, and the flowery type NW growth observed for Ti was not easily reproducible. In the case of Sn, the resulting NWs are very straight and long (several microns). Experiments with reactive vapor transport of Sn onto substrates [105] also produced very long, straight, and branched SnO_2 NWs. At this time, the mechanism for NW growth by direct oxidation of submicron size metal particles cannot be explained in simple terms and is the subject of our continued studies.

References

1. Sunkara, M. K., et al., Bulk synthesis of silicon nanowires using a low-temperature vapor-liquid-solid method. *Applied Physics Letters*, 2001. 79(10): 1546–1548.
2. Wagner, R. S., and W. C. Ellis, Vapor-liquid-solid mechanism of single crystal growth. *Applied Physics Letters*, 1964. 4(5).
3. Bootsma, G. A., and H. J. Gassen, A quantitative study on the growth of silicon whiskers from silane and germanium whiskers from germane. *Journal of Crystal Growth*, 1971. 10(3): 223–234.
4. Givargizov, E. I., Periodic instability in whisker growth. *Journal of Crystal Growth*, 1973. 20(3): 217–226.
5. Givargizov, E. I., Fundamental aspects of VLS growth. *Journal of Crystal Growth*, 1975. 31(December): 20–30.
6. Wang, H., and G. S. Fischman, Role of liquid droplet surface diffusion in the vapor-liquid-solid whisker growth mechanism. *Journal of Applied Physics*, 1994. 76(3): 1557–1567.
7. Morales, A. M., and C. M. Lieber, A laser ablation method for the synthesis of crystalline semiconductor nanowires. *Science*, 1998. 279(5348): 208–211.
8. Sunkara, M. K., et al., Bulk synthesis of a-SixNyH and a-SixOy straight and coiled nanowires. *Journal of Materials Chemistry*, 2004. 14(4): 590–594.
9. Sharma, S., and M. K. Sunkara, Direct synthesis of single-crystalline silicon nanowires using molten gallium and silane plasma. *Nanotechnology*, 2004. 15(1): 130–134.
10. Sharma, S. S., M. K. Sunkara, G. Lian, and Elizabeth C. Dickey, A non-traditional vapor-liquid-solid method for bulk synthesis of semiconductor nanowires. *Nanophase and Nanocomposite Materials*, 2002. 703: 123.
11. Meduri, P. et al., Composition controlled synthesis and Raman analysis of Ge-rich SixGe1-x nanowires. *Journal of Nanoscience and Nanotechnology*, 2008. 8(6): 3153–3157.
12. Sharma, S., and M. K. Sunkara, Direct synthesis of single-crystalline silicon nanowires using molten gallium and silane plasma. *Nanotechnology*, 2004. 15: 130–134.
13. Tan, T. Y., S. T. Lee, and U. Gosele, A model for growth directional features in silicon nanowires. *Applied Physics A—Materials Science and Processing*, 2002. 74(3): 423–432.

14. Holmes, J. D. et al., Control of thickness and orientation of solution-grown silicon nanowires. *Science*, 2000. 287(5457): 1471–1473.

15. Zianni, X., and A. G. Nassiopoulou, Directional dependence of the spontaneous emission of Si quantum wires. *Physical Review B*, 2002. 65(3).

16. Sun, X. H. et al., Surface reactivity of Si nanowires. *Journal of Applied Physics*, 2001. 89(11): 6396–6399.

17. deFreeze, M., V. L. Dalal, and J. Falter, *Materials Research Society Symposium Proceedings*, 2001. A.5.2.1: 664.

18. Feldman, D. W., M. Ashkin, and J. J. H. Parker, Raman scattering by local modes in germanium-rich silicon-germanium alloys. *Physical Review Letters*, 1966. 17(24): 1209–1212.

19. Ren, S. -F., W. Cheng, and P. Y. Yu, Microscopic investigation of phonon modes in SiGe alloy nanocrystals. *Physical Review B*, 2004. 69(23).

20. Weizman, M. et al., Phase segregation in laser crystallized polycrystalline SiGe thin films. *Thin Solid Films*, 2005. 487(1–2): 72–76.

21. Dismukes, J. P., L. Ekstrom, and R. J. Paff, Lattice parameter and density in germanium-silicon alloys. *Journal of Physical Chemistry* 1964. 68(10): 3021–3027.

22. Meduri, P. et al., Controlled synthesis and Raman analysis of Ge-rich SixGe1-x alloy nanowires. *Journal of Nanoscience and Nanotechnology*, 2008. 8(6): 3153–3157.

23. Comini, E., and G. Sberveglieri, Metal oxide nanowires as chemical sensors. *Materials Today*, 2010. 13(7–8): 28–36.

24. Fan, Z. Y. et al., Photoluminescence and polarized photodetection of single ZnO nanowires. *Applied Physics Letters*, 2004. 85(25): 6128–6130.

25. Law, M. et al., Nanowire dye-sensitized solar cells. *Nature Materials*, 2005. 4(6): 455–459.

26. Gubbala, S., et al., Band-edge engineered hybrid structures for dye sensitized solar cells based on SnO_2 nanowires. *Advanced Functional Materials*, 2008. 18: 2411–2418.

27. Arulsamy, A. D. et al., Reversible carrier-type transitions in gas-sensing oxides and nanostructures. *Chemphyschem*, 2010. 11(17): 3704–3712.

28. Deb, B. et al., Gas sensing behaviour of mat-like networked tungsten oxide nanowire thin films. *Nanotechnology*, 2007. 18(28).

29. Gubbala, S., J. Thangala, and M. K. Sunkara, Nanowire based electrochromic devices. *Solar Energy Materials and Solar Cells*, 2007. 91(9): 813–820.

30. Wang, H., T. Sekino, and K. Niihara, Magnetic mullite-iron composite nanoparticles prepared by solid solution reduction. *Chemistry Letters*, 2005. 34(3): 298–299.

31. Woo, K. et al., Sol-gel mediated synthesis of Fe_2O_3 nanorods. *Advanced Materials*, 2003. 15(20): 1761–1764.

32. Zhong, L. S., et al., Self-assembled 3D flowerlike iron oxide nanostructures and their application in water treatment. *Advanced Materials*, 2006. 18(18): 2426–2431.

33. Chen, D. L., and L. Gao, A facile route for high-throughput formation of single-crystal alpha-Fe_2O_3 nanodisks in aqueous solutions of Tween 80 and triblock copolymer. *Chemical Physics Letters*, 2004. 395(4–6): 316–320.

34. Pregelj, M. et al., Synthesis, structure, and magnetic properties of iron-oxide nanowires. *Journal of Materials Research*, 2006. 21(11): 2955–2962.

35. Jin, B., S. Ohkoshi, and K. Hashimoto, Giant coercive field of nanometer-sized iron oxide. *Advanced Materials,* 2004. 16(1): 48–51.
36. Fu, Y. Y. et al., Synthesis of large arrays of aligned alpha-Fe$_2$O$_3$ nanowires. *Chemical Physics Letters,* 2003. 379(3–4): 373–379.
37. Wang, R. M. et al., Bicrystalline hematite nanowires. *Journal of Physical Chemistry B,* 2005. 109(25): 12245–12249.
38. Chueh, Y. L. et al., Systematic study of the growth of aligned arrays of alpha-Fe$_2$O$_3$ and Fe$_3$O$_4$ nanowires by a vapor-solid process. *Advanced Functional Materials,* 2006. 16(17): 2243–2251.
39. Liu, F., J. Y. Lee, and W. J. Zhou, Multisegment PtRu nanorods: Electrocatalysts with adjustable bimetallic pair sites. *Advanced Functional Materials,* 2005. 15(9): 1459–1464.
40. Rao, P. M., and X. L. Zheng, Rapid catalyst-free flame synthesis of dense, aligned alpha-Fe$_2$O$_3$ nanoflake and CuO nanoneedle arrays. *Nano Letters,* 2009. 9(8): 3001–3006.
41. Rackauskas, S. et al., A novel method for metal oxide nanowire synthesis. *Nanotechnology,* 2009. 20(16).
42. Ghoshal, T. et al., Direct synthesis of ZnO nanowire arrays on Zn foil by a simple thermal evaporation process. *Nanotechnology,* 2008. 19(6).
43. Hiralal, P. et al., Growth and process conditions of aligned and patternable films of iron(III) oxide nanowires by thermal oxidation of iron. *Nanotechnology,* 2008. 19(45).
44. Nagato, K. et al., Direct synthesis of vertical alpha-Fe$_2$O$_3$ nanowires from sputtered Fe thin film. *Journal of Vacuum Science & Technology B,* 2010. 28(6): C6P11–C6P13.
45. Thangala, J. et al., Large-scale, hot-filament-assisted synthesis of tungsten oxide and related transition metal oxide nanowires. *Small,* 2007. 3(5): 890–896.
46. Vaddiraju, S., H. Chandrasekaran, and M. K. Sunkara, Vapor phase synthesis of tungsten nanowires. *Journal of the American Chemical Society,* 2003. 125(36): 10792–10793.
47. Kim, J. H., V. Kumar, B. Chernomordik, and M. K. Sunkara, Design of an efficient microwave plasma reactor for bulk production of inorganic nanowires. *Infomacije Midem,* 2008. 38: 237–243.
48. Kumar, V. et al., Gas-phase, bulk production of metal oxide nanowires and nanoparticles using a microwave plasma jet reactor. *Journal of Physical Chemistry C,* 2008. 112(46): 17750–17754.
49. Ostrikov, K., Colloquium: Reactive plasmas as a versatile nanofabrication tool. *Reviews of Modern Physics,* 2005. 77(2): 489–511.
50. Ostrikov, K. et al., Self-assembled low-dimensional nanomaterials via low-temperature plasma processing. *Thin Solid Films,* 2008. 516(19): 6609–6615.
51. Ostrikov, K., and A. B. Murphy, Plasma-aided nanofabrication: Where is the cutting edge? *Journal of Physics D—Applied Physics,* 2007. 40: 2223–2241.
52. Liu, X. et al., Growth mechanism and properties of ZnO nanorods synthesized by plasma-enhanced chemical vapor deposition. *J. Appl. Phys.,* 2003. 95: 3141.
53. Kumeta, K., H. Ono, and S. Iizuka, Formation of ZnO nanostructures in energy-controlled hollow-type magnetron RF plasma. *Thin Solid Films,* 2009. 518: 3522–3525.
54. Ono, H., and S. Iizuka, Growth of ZnO nanowires in hollow-type magnetron O-2/Ar RF plasma. *Thin Solid Films,* 2009. 518(3): 1016–1019.

55. Baxter, J. B., F. Wu, and E. S. Aydil, Growth mechanism and characterization of zinc oxide hexagonal columns. *Applied Physics Letters*, 2003. 83(18): 3797–3799.

56. Purohit, V. S. et al., ECR plasma assisted deposition of zinc nanowires. *Nuclear Instruments & Methods in Physics Research Section B—Beam Interactions with Materials and Atoms*, 2008. 266(23): 4980–4986.

57. Huo, C. et al., Different shapes of nano-ZnO crystals grown in-catalyst-free DC plasma. *Plasma Sources Science and Technology*, 2009. 11: 564.

58. Ra, H. W., D. H. Choi, S. H. Kim, and Y.-H. Im, Formation and characterization of ZnO/aC core-shell nanowires. *Journal of Physical Chemistry C*, 2009. 113: 3512–3516.

59. Levchenko, I. et al., Plasma-controlled metal catalyst saturation and the initial stage of carbon nanostructure array growth. *Journal of Applied Physics*, 2008. 104(7).

60. Mariotti, D., and K. Ostrikov, Tailoring microplasma nanofabrication: From nanostructures to nanoarchitectures. *Journal of Physics D—Applied Physics*, 2009. 42(9).

61. Meyyappan, M., and M. K. Sunkara, *Inorganic Nanowires: Applications, Properties and Characterization*, 2010. Boca Raton, FL: CRC Press.

62. Huang, M. H. et al., Catalytic growth of zinc oxide nanowires by vapor transport. *Advanced Materials*, 2000. 13(2): 113–116.

63. Ko, T. S. et al., ZnO nanopowders fabricated by dc thermal plasma synthesis. *Materials Science and Engineering B—Solid State Materials for Advanced Technology*, 2006. 134(1): 54–58.

64. Mariotti, D., A. C. Bose, and K. Ostrikov, Atmospheric-microplasma-assisted nanofabrication: Metal and metal-oxide nanostructures and nanoarchitectures. *IEEE Transactions on Plasma Science*, 2009. 37(6): 1027–1033.

65. Shimizu, Y., B.A., D. Mariotti, T. Sasaki, K. Kirihara, T. Suzuki, K. Terashima, and N. Koshizaki, Reactive evaporation of metal wire and microdeposition of metal oxide using atmospheric pressure reactive microplasma jet. *Japanese Journal of Applied Physics*, 2006. 45: 8228–8234.

66. Lin, H. F., S. C. Liao, and C. T. Hu, A new approach to synthesize ZnO tetrapod-like nanoparticles with DC thermal plasma technique. *Journal of Crystal Growth*, 2009. 311(5): 1378–1384.

67. Peng, H. et al., Plasma synthesis of large quantities of zinc oxide nanorods. *Journal of Physical Chemistry C*, 2007. 111(1): 194–200.

68. Sharma, S. and M.K. Sunkara, Direct synthesis of gallium oxide tubes, nanowires, and nanopaintbrushes. *Journal of the American Chemical Society*, 2002. 124(41): 12288–12293.

69. Cvelbar, U., K. Ostrikov, and M. Mozetic, Reactive oxygen plasma-enabled synthesis of nanostructured CdO: Tailoring nanostructures through plasma-surface interactions. *Nanotechnology*, 2008. 19(40).

70. Ostrikov, K. et al., From nucleation to nanowires: A single-step process in reactive plasmas. *Nanoscale*, 2010. 2(10): 2012–2027.

71. Cvelbar, U., and M. Mozetic, Behaviour of oxygen atoms near the surface of nanostructured Nb_2O_5. *Journal of Physics D—Applied Physics*, 2007. 40(8): 2300–2303.

72. Levchenko, I., and K. Ostrikov, Nanostructures of various dimensionalities from plasma and neutral fluxes. *Journal of Physics D—Applied Physics*, 2007. 40(8): 2308–2319.

73. Tam, E. et al., Ion-assisted functional monolayer coating of nanorod arrays in hydrogen plasmas. *Physics of Plasmas*, 2007. 14(3).
74. Cvelbar, U. et al., Control of morphology and nucleation density of iron oxide nanostructures by electric conditions on iron surfaces exposed to reactive oxygen plasmas. *Applied Physics Letters*, 2009. 94(21).
75. Drenik, A. et al., Catalytic probes with nanostructured surface for gas/discharge diagnostics: A study of a probe signal behaviour. *Journal of Physics D—Applied Physics*, 2008. 41(11).
76. Keem, K. et al., Photocurrent in ZnO nanowires grown from Au electrodes. *Applied Physics Letters*, 2004. 84(22): 4376–4378.
77. Mozetic, M. et al., A method for the rapid synthesis of large quantities of metal oxide nanowires at low temperatures. *Advanced Materials*, 2005. 17(17): 2138–2142.
78. Cvelbar, U. et al., Spontaneous growth of superstructure alpha-Fe_2O_3 nanowire and nanobelt arrays in reactive oxygen plasma. *Small*, 2008. 4(10): 1610–1614.
79. Cvelbar, U., and K. Ostrikov, Deterministic surface growth of single-crystalline iron oxide nanostructures in nonequilibrium plasma. *Crystal Growth & Design*, 2008. 8(12): 4347–4349.
80. Chiang, W. H., C. Richmonds, and R. M. Sankaran, Continuous-flow, atmospheric-pressure microplasmas: A versatile source for metal nanoparticle synthesis in the gas or liquid phase. *Plasma Sources Science & Technology*, 2010. 19(3).
81. Chiang, W. H. et al., Nanoengineering NixFe1-x catalysts for gas-phase, selective synthesis of semiconducting single-walled carbon nanotubes. *ACS Nano*, 2009. 3(12): 4023–4032.
82. Chiang, W. H., and R. M. Sankaran, Linking catalyst composition to chirality distributions of as-grown single-walled carbon nanotubes by tuning NixFe1-x nanoparticles. *Nature Materials*, 2009. 8(11): 882–886.
83. Mariotti, D., and R. M. Sankaran, Microplasmas for nanomaterials synthesis. *Journal of Physics D—Applied Physics*, 2010. 43(32).
84. Hong, Y. C. et al., ZnO nanocrystals synthesized by evaporation of Zn in microwave plasma torch in terms of mixture ratio of N-2 to O-2. *Physics of Plasmas*, 2006. 13(6).
85. Hong, Y. C., J. H. Kim, and H. S. Uhm, ZnO nanorods synthesized by self-catalytic method of metal in atmospheric microwave plasma torch flame. *Japanese Journal of Applied Physics Part 1—Regular Papers Brief Communications and Review Papers*, 2006. 45(7): 5940–5944.
86. Morber, J. R. et al., PLD-assisted VLS growth of aligned ferrite nanorods, nanowires, and nanobelts—Synthesis, and properties. *Journal of Physical Chemistry B*, 2006. 110(43): 21672–21679.
87. Han, Q. et al., Growth and properties of single-crystalline gamma-Fe_2O_3 nanowires. *Journal of Physical Chemistry C*, 2007. 111(13): 5034–5038.
88. Kim, C. H. et al., Magnetic anisotropy of vertically aligned alpha-Fe_2O_3 nanowire array. *Applied Physics Letters*, 2006. 89(22).
89. Lee, Y. C. et al., p-Type alpha-Fe_2O_3 nanowires and their n-type transition in a reductive ambient. *Small*, 2007. 3(8): 1356–1361.
90. Xu, L. et al., Formation, characterization, and magnetic properties of Fe_3O_4 nanowires encapsulated in carbon microtubes. *Journal of Physical Chemistry B*, 2004. 108: 10859–10862.

91. Yang, J. B. et al., Crystal and electronic structures of $LiNH_2$. *Applied Physics Letters*, 2006. 88(4).
92. Cui, Y., and C. M. Lieber, Functional nanoscale electronic devices assembled using silicon nanowire building blocks. *Science*, 2001. 291(5505): 851–853.
93. Peng, H. et al., Plasma synthesis of large quantities of zinc oxide nanorods. *Journal of Physical Chemistry C*, 2007. 111(1): 194–200.
94. Cvelbar, U. et al., Inductively coupled RF oxygen plasma characterization by optical emission spectroscopy. *Vacuum*, 2007. 82(2): 224–227.
95. Mozetic, M. et al., A method for the rapid synthesis of large quantities of metal oxide nanowires at low temperatures. *Advanced Materials*, 2005. 17(17): 2138–2142.
96. Mozetic, M. et al., Comparison of NO titration and fiber optics catalytic probes for determination of neutral oxygen atom concentration in plasmas and postglows. *Journal of Vacuum Science & Technology A*, 2003. 21(2): 369–374.
97. Gaboriau, F. et al., Comparison of TALIF and catalytic probes for the determination of nitrogen atom density in a nitrogen plasma afterglow. *Journal of Physics D-Applied Physics*, 2009. 42(5).
98. Gomez, S., P. G. Steen, and W. G. Graham, Atomic oxygen surface loss coefficient measurements in a capacitive/inductive radio-frequency plasma. *Applied Physics Letters*, 2002. 81(1): 19–21.
99. Mozetic, M., U. Cvelbar, Determination of the neutral oxygen atom density in a plasma reactor loaded with metal samples *Plasma Sources Science and Technology*, 2009. 18.
100. Hopkins, M. B., and W. G. Graham, Langmuir probe technique for plasma parameter measurement in a medium density discharge. *Review of Scientific Instruments*, 1986. 57: 2210–2214.
101. Schwabedissen, A., E. C. Benck, and J. R. Roberts, Langmuir probe measurements in an inductively coupled plasma source. *Physical Review E*, 1997. 55(3): 3450–3459.
102. Krstulovic, N. et al., Optical emission spectroscopy characterization of oxygen plasma during treatment of a PET foil. *Journal of Physics D—Applied Physics*, 2006. 39(17): 3799–3804.
103. Mafra, M. et al., Treatment of hexatriacontane by Ar-O-2 remote plasma: Formation of the active species. *Plasma Processes and Polymers*, 2009. 6: S198–S203.
104. Dick, K. A. et al., Synthesis of branched "nanotrees" by controlled seeding of multiple branching events. *Nature Materials*, 2004. 3(6): 380–384.
105. Rao, R. et al., Synthesis of low-melting metal oxide and sulfide nanowires and nanobelts. *Journal of Electronic Materials*, 2006. 35(5): 941–946.
106. Liao, S. -C. et al., DC thermal plasma synthesis and properties of zinc oxide nanorods. *Journal of Vacuum Science & Technology B: Microelectronics and Nanometer Structures*, 2006. 24(3): 1322–1326.
107. Thangala, J. et al., Large-scale, hot-filament-assisted synthesis of tungsten oxide and related transition metal oxide nanowires. *Small*, 2007. 3(5): 890–896.
108. Kozan, M. et al., In-situ characterization of dispersion stability of WO_3 nanoparticles and nanowires. *Journal of Nanoparticle Research*, 2008. 10: 599–612.
109. Spurr, R. A., and H. Myers, Quantitative analysis of anatase-rutile mixtures with an x-ray diffractometer. *Analytical Chemistry*, 1957. 29(5): 760–762.

110. Lippens, B. C., and J. H. d. Boer, Study of phase transformations during calcinations of aluminum hydroxides by selected area electron diffraction. *Acta Crystallography*, 1964. 17.
111. Aminzadeh, A., and H. Sarikhani-fard, Raman spectroscopic study of Ni/Al_2O_3 catalyst. *Spectrochimica Acta Part A: Molecular and Biomolecular Spectroscopy*, 1999. 55: 1421–1425.
112. Chen, Y. et al., NIR FT Raman spectroscopic studies of η-Al_2O_3 and Mo/η-Al_2O_3 catalysts. *Spectrochimica Acta Part A: Molecular and Biomolecular Spectroscopy*, 1995. 51(12): 2161–2169.
113. Conrads, H., and M. Schmidt, Plasma generation and plasma sources. *Plasma Sources Science and Technology*, 2000. (4): 441.
114. Ganachev, I. P., and H. Sugai, Production and control of planar microwave plasmas for materials processing. *Plasma Sources Science and Technology*, 2002. (3A): A178.
115. Wan, Y. P. et al., Model and powder particle heating, melting, resolidification, and evaporation in plasma spraying processes. *Journal of Heat Transfer*, 1999. 121(3): 691–699.
116. Fincke, J. R. et al., Modeling and experimental observation of evaporation from oxidizing molybdenum particles entrained in a thermal plasma jet. *International Journal of Heat and Mass Transfer*, 2002. 45(5): 1007–1015.

6

Cathodic Arc Discharge for Synthesis of Carbon Nanoparticles

Manish Chhowalla
H. Emrah Unalan

CONTENTS

6.1 Fullerene and Nanoparticle Formation in Carbon Cathodic Arc Deposition

It is well established that tetrahedral amorphous carbon (ta-C) with a diamond-like sp^3 bonding fraction of 85% can be deposited using a carbon cathodic vacuum arc [1]. The deposition of hard and highly elastic carbon films, which mostly consist of graphitic sp^2 bonding, has also been achieved using a graphite cathode with a localized high pressure of helium or nitrogen at the arc spot [2]. It has been shown through high-resolution electron microscopy and electron energy loss spectroscopy that covalent linking of curved graphitic planes via break up of fullerene-like bucky "onions" and nanotubes are the likely origin of the extremely unusual mechanical properties of these carbon films. The impingement of ionized carbon and

147

nanoparticles (C_{2n} molecules) simultaneously onto a substrate at high velocity is necessary to achieve the fullerene-like structure of these hard and elastic films. Therefore, understanding the formation mechanism of fullerenes and nanotubes from the cathodic arc process is important in controlling and understanding the properties of these films. Although the process for the production of fullerenes and nanotubes is well established [3,4], the mechanism for their formation is still unclear.

Behavior of nanometer-sized particles in silane plasmas has been extensively studied in order to reduce or eliminate them due to their detrimental effects on amorphous silicon microelectronic device fabrication. Plasmas containing such particles are referred to as dusty plasmas. The nanoparticles generated in our localized high-pressure carbon arc can be regarded as dust in the plasma, and principles of dusty plasmas can be used to understand their nucleation and growth kinetics. We examine the afterglow plasma of a carbon arc in a local high-pressure environment using Langmuir probe techniques to study the nucleation and initial growth of fullerenes and nanotubes [5]. Because nanometer-sized particles assume a sustainable negative charge, they can be detected in the afterglow plasma using a positively biased Langmuir probe. The electron saturation current decays sharply above 1 ms after the plasma has been switched off (T_{off}), while the heavier negatively charged particles can be detected for up to several milliseconds in the afterglow plasma. As summarized below, it is possible to obtain information on the mass and, therefore, the size of particles in the plasma from the ratio of positive and negative saturation currents in the afterglow. Using this method, we found that a minimal time of 0.5 s was needed after arc plasma initiation on a cold (room-temperature) graphite target for the formation of the C_{60} molecule.

The carbon arc discharge was formed on a graphite cathode by touching an earthed anode to it in the presence of high nitrogen pressure (~10 Torr). The high pressure near the discharge was created by injecting the gas through a 1 mm cavity close to the carbon arc spot [6]. The pressure near the substrate was held constant at 6 mTorr. A tungsten rod with a diameter of 0.8 mm and length of 7 mm immersed in the plasma 25 cm directly in front of the discharge was used as the Langmuir probe. The ratio of the afterglow saturation currents was monitored using an oscilloscope as a function of the discharge time, (T_{on}). In order to ensure that the target temperature remained constant during the experiments, the cathode was allowed to cool 15 to 20 minutes after each measurement. Traces were taken for carbon arc discharges in vacuum (10^{-6} Torr), and localized high pressure for comparison. All other conditions were maintained to be constant in both cases. Below, we provide an abridged version of the technique for the detection of nanometer-sized particles in the plasma. The ion saturation current for a Langmuir probe immersed in plasma with negative species is given by

$$I_s = eS\left[n_e(kT_e / 2\pi m)^{1/2} + (Q/e)n_-(kT_- / 2\pi M_-)^{1/2}\right] \qquad (6.1)$$

where e is the electronic charge, S the probe area, k the Boltzmann's constant, Q the charge of the ion, n_e, m_e, and T_e are the density, mass, and temperature of the electrons, respectively, and n_-, M_-, and T_- are the density, mass, and temperature of the negative ions. Due to their higher mobility, the electron current contribution to I_s is considered to be negligible ~1 ms after switching off the discharge.

The positive ion current contribution in the afterglow plasma can be found by

$$I_+ = eSn_+(kT_+ / 2\pi M_+)^{1/2} \tag{6.2}$$

where T_+ is the temperature of the positive ions. Preliminary measurements and calculations indicate that the gas pressure at the surface of the cathode is 10 Torr. At this pressure, the plasma density is dominated by the nitrogen species. The ionization of the nitrogen (either N_2^+ or N^+) is usually 1% to 2% in reactive cathodic arc evaporation [7]. The "frozen-in" [8] states of carbon above the cathode spot are expected to go through numerous thermalizing collisions. Therefore, the plasma near the substrate in our system more closely resembles a glow discharge rather than a cathodic vacuum arc plasma. Assuming that all the species detected by the probe are thermalized, then the ratio of T_+/T_- in the afterglow plasma estimated from the decay times of the saturation currents is approximately 1, indicating $T_+ \sim T_-$. For measurements taken after ~1 ms, combining Equations (6.1) and (6.2) yields the mass ratio M_-/M_+ given by

$$M_- / M_+ = (I_+ / I_s)^2 \tag{6.3}$$

The particle size can be estimated from the ratio of M_-/M_+. The negatively charged particles in plasmas with T_e close to T_i are maintained if the flux of ions and electrons is equal, and electron emission occurs if the electric field becomes higher than a threshold value. Assuming that the particle acts like a spherical Langmuir probe in the plasma, then the charge on it can be estimated from the orbital motion-limited theory for spherical probes, assuming that the particle size is significantly less than the Debye length. The charge ratio Q/e on a particle of <2 nm in this type of plasma with glow discharge like properties ($n_+ \sim n_-$) is estimated to be –1 [5]. Therefore, by measuring the saturation currents after turning off the discharge, it is possible to obtain information about subnanometer particle formation in the initial stages of a discharge. The particles are detected as negative ions in the afterglow of the discharge. The ratios of M_-/M_+ were calculated from the saturation currents 1 to 2 ms after turning off the discharge. The I_+ and I_- currents were measured at ±30 V, which was suitable for measuring the saturation currents. Several traces were taken to verify the reproducibility of the data. Typical traces taken from the afterglow plasma with the Langmuir probe biased at +30 V in vacuum and high pressure arc discharge are shown in Figures 6.1a and 6.1b, respectively. Figures 6.1a and 6.1b are inverted photographs of

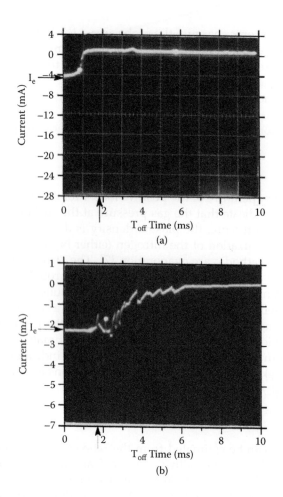

FIGURE 6.1

Oscilloscope traces of probe current versus time for vacuum (a) and localized high-pressure arc discharges (b). The electron saturation currents for both cases is marked with an arrow and labeled as I_e. The time at which the discharge was switched off is marked as T_{off} on the time axis. Note that the current decays smoothly and sharply in less than 0.5 ms for the vacuum case (a). The current trace in (b) is noisier, and a small signal can be detected several milliseconds after the discharge has been turned off. (Note: the minor offset above zero after $T_{off} > 2$ ms in (a) is due to poor registration between scale and oscillogram, the current decays to 0 A with no residual current in this case.)

oscilloscope traces. The current drawn initially represents the electron saturation current, I_e, which for the vacuum case (Figure 6.1a), sharply drops to zero after the discharge is switched off. In contrast, the saturation current for the high-pressure discharge (Figure 6.1b) is noisier than the vacuum case and decays less rapidly for $T_{off} = 1$ ms. However, the afterglow current does not decay to zero as in the vacuum case. A residual current I_s from slower negatively charged species in the afterglow plasma is detected several milliseconds after the discharge is shut off, as seen in Figure 6.1b.

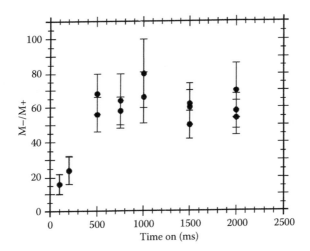

FIGURE 6.2
The calculated mass ratio (M_-/M_+) versus discharge time. The mass ratio rises to 56 ± 10 for a discharge time of 0.5 s or more above which it remains constant within the errors in the measurements. The error bars represent the noise in the ion and negative particle saturation currents in the afterglow plasma.

The M_-/M_+ mass ratio calculated from Equation (6.3) as a function of the discharge time (T_{on}) is plotted in Figure 6.2. The mass ratios were calculated using the saturation currents obtained from oscilloscope traces similar to Figure 6.1b immediately after the arc was extinguished. The mass ratio for $T_{on} < 0.5$ s is 24 ± 7 but increases to 56 ± 10 for $T_{on} > 0.5$ s, where it appears to saturate for larger T_{on} times. Mass spectroscopy was performed on the localized high-pressure discharge plasma in the region of Langmuir probe measurements in order to confirm the presence of C_{2n} molecules. Several mass spectra of the plasma during the discharge were collected with the spectrograph ionizer on. The mass spectrum of our localized high-pressure discharge verifies the presence of C_{60} and other C_{2n} molecules, as shown in Figure 6.3. The relative abundance of different C_{2n} species cannot be directly obtained from the mass spectrograph, as the spectrometer has lower sensitivity for the high masses [720 amu (C_{60}), 840 amu (C_{70})] close to the detection limit (1000 amu). No such C_{2n} molecular peaks are present in the mass spectrum for a discharge in vacuum, in agreement with the probe results. Therefore, assuming that the positive species is composed solely of C^+ species (12 amu), a mass ratio of 56 ± 10 corresponds to a C_{60} molecule. If the contribution to M_+ from some N_2^+ species of the gas jet near the cathode spot is considered, then a $C^+:N_2^+$ ratio of 95:5 will yield the 720 amu value of C_{60} for $M_-/M_+ = 56$. From Figure 6.2, it can be concluded that the threshold time for the formation of C_{2n} molecules with masses close to C_{60} is 0.5 s after the initiation of the arc discharge. The main source of error arises from fluctuations in the saturation currents, as seen in Figure 6.1b. The decay rates of the afterglow for a probe biased at –30 V in plasma generated in vacuum and localized high pressure were found to be similar, indicating that only a minor contribution to the positive ion saturation current comes from ionized C_{2n} clusters.

The growth process of nanoparticles starts from the formation of the C_2 molecule. The reaction time for the formation of the C_2 dimer by an excited carbon collision is of the order of 10 to 100 ns [9]. Gamaly and Ebbesen [9] estimated the time of growth for a nanotube of 1 mm in length to be 0.02 s. The time for the formation of the much smaller C_{60} molecules is expected to be much shorter, of the order of tens of nanoseconds. We attribute the rather large formation time of 0.5 s observed in our experiments to the time required for the bulk graphite electrode region in the vicinity of the microscopic arc spot to reach ~1200°C which is required for the formation of stable C_{60} [10]. The time scale for a large region (~mm) of the graphite cathode to reach this temperature is of the order of 1 s. Once the cathode has achieved a steady temperature, the formation of C_{60} along with larger C_{2n} molecules can be readily observed, as seen in Figure 6.3.

In this part of the chapter, we discussed the ability to analyze carbon nanoparticle formation in terms of a charged dusty plasma process. The detected carbon nanoparticles using a Langmuir probe technique are in agreement with the presence of fullerene-sized molecules in the plasma. The results indicate a minimal time of 0.5 s for the formation of a particle containing 56 carbon atoms, which is close to that of the stable C_{60} fullerene molecule.

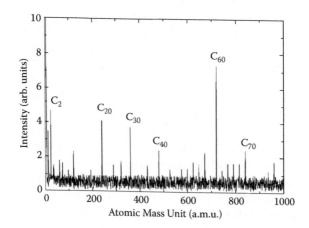

FIGURE 6.3
Atomic mass spectrum showing the presence of numerous carbon clusters ranging from C_2 to C_{80} confirming the presence of nanoparticles. Measurements in vacuum did not reveal any nanoparticles in the plasma. The mass spectrum was taken in the region where the Langmuir probe measurements were made. A spectrograph ionizer was used so that the atomic mass unit (amu) represents a ratio of mass to a single charge (m/q).

6.2 Properties of Carbon Onions Produced by an Arc Discharge in Water

Carbon nanomaterials have received a great deal of attention since the discovery of the C_{60} fullerene molecule [11] and the carbon nanotube [4]. C_{60} research in the early 1990s was enabled by the capability to produce large quantities (a few milligrams) of the material using the high-pressure arc discharge method [12]. The ability to readily generate high-purity carbon nanotubes has also led to a rapid expansion in exploration of its properties. Carbon nanotubes are being considered for a wide range of electronic and mechanical applications because of their extraordinary properties. However, for applications such as fuel cell electrodes and nanocomposite structural materials, large quantities (kilograms) of the material are desired. Presently, several industrial and governmental projects are underway to mass produce several kilograms of single and multiwalled carbon nanotubes in a cost-effective manner [13]. In addition to carbon nanotubes, spherical carbon onions are interesting, because they are expected to have superior lubrication properties. Nevertheless, onions can only be produced in minute quantities by electron beam irradiation of amorphous carbon using a transmission electron microscope at 700°C [14,15], annealing nanodiamonds at 1100 to 1500°C [16], implantation of 120 keV carbon ions in silver or copper [17], radio frequency plasma-enhanced chemical vapor deposition [18], and shock wave treatment of carbon soot [19]. An economical method to produce carbon onions in bulk quantities using an arc discharge in water was reported by our group [20].

The conventional methods used to fabricate carbon nanomaterials require vacuum systems to generate the plasma. Examples include arc discharge [12,21], laser ablation [22], and glow discharge [23]. These methods are not feasible for bulk production because of the high investment and running costs for the vacuum equipment and the low yield of the desired products. The vacuum processes also yield, in addition to the desired nanomaterials, unwanted contaminants (amorphous carbon and disordered nanoparticles) so that time-consuming and costly purification steps must be carried out. Therefore, a process that allows the generation of nanotubes or nano-onions with minimum contamination is desirable. Ishigami et al. [24] proposed a high yield method for multiwalled carbon nanotubes that does not require vacuum systems. In their method, an arc discharge was generated in liquid nitrogen between two carbon electrodes. Although their method is superior to conventional ones in terms of simplicity, the rapid evaporation of the liquid nitrogen poses a problem. After their report, an even more economical technique using water instead of liquid nitrogen was used to successfully produce nanotubes [25]. Working independently on a water arc, we found that high-concentration nano-onions can be obtained as floating powder on the water surface while the rest of the product emitted is found at the base of the water [11]. This indicates that nano-onions tend to naturally segregate from the other products, yielding a higher-purity product. In this chapter, we report further experimental details and several physical analyses of the floating powder and propose a model for the formation mechanism.

6.2.1 Experimental Details

In our method, a direct current (DC) arc discharge was generated in deionized water between two carbon electrodes. Our apparatus consisted of two graphite electrodes submerged in 2500 cm^3 of distilled water (resistance 51.4 MW with a gap of 1 mm) in a Pyrex® beaker. The arc discharge was initiated by contacting the 99.9% pure grounded anode (6 mm diameter) with a cathode (12 mm tip diameter) of similar purity submerged to a depth of 3 cm in distilled water. The discharge voltage and current were 16 to 17 V and 30 A, respectively. The arc discharge in water was found to be stable and could be run for several tens of minutes so long as a cathode–anode gap of ~1 mm was maintained. The discharge in our case can be characterized as an anodic arc as the smaller anode electrode is consumed. The anode consumption rate for these conditions was approximately 117.2 mg/min, whereas the cathode consumption was negligible.

A schematic of our apparatus is shown in Figure 6.4. The relevant components of the apparatus are labeled. A digital image of the arc discharge in water is also shown. The bright area between the electrodes indicates the arc plasma region. The plasma can also be visually observed to surround the anode, indicating the direction of plasma expansion. In addition, fine black powder emitted from the plasma ball region is readily visible. The evaporation rate of the water during arc discharge operation was measured

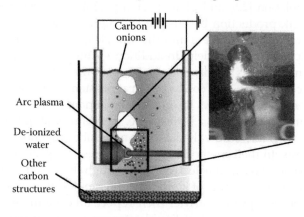

FIGURE 6.4
The apparatus used for arc discharge in water with a digital image of the discharge. (Reproduced with permission from Sano et al. [26].)

to be 99 cm^3/min, which is significantly less than that for liquid nitrogen (300 cm^3/min), making this process more cost effective and economical. Similar to the case of a conventional fullerene reactor, the carbon plasma is generated by thermal evaporation of the anode. Therefore, unlike cathodic arc plasmas, the carbon vapor is generated by thermionic rather than thermofield emission. The material collected from the water surface was weighed without purification to obtain the production rate and was found to be 3.6 mg/min. Under the same condition, the production rate of the sediment products was found to be 15.9 mg/min.

6.2.2 Results

The floating powder from the water surface was characterized by several microscopy techniques. High-resolution transmission electron microscopy (HRTEM) was performed on a JEOL 4000EX microscope operated at 400 kV. Unpurified material from the water surface was sprinkled onto holey carbon transmission electron microscopy (TEM) grids for investigation. A typical HRTEM image of the material is shown in Figure 6.5. Many nested onion-like particles with diameters of 30 to 35 nm can be seen in their agglomerated form. In addition, elongated nested particles similar to multiwalled nanotubes can also be observed. The elongated structure and the onion are shown in greater detail in the upper and lower panels of Figure 6.5, respectively. It should be noted that a more detailed study of the floating powder reported here shows two types of onion structure. The first is a well-crystallized onion structure with well-defined concentric shells as previously reported [21]. Here in Figure 6.5, we show not so well-crystallized onions along with elongated fullerene-like structures that are also present in the floating powder product of the water arc. It should be mentioned that observations using a lower-energy 200 kV TEM also showed the presence of onions, indicating that electron beam transformation is not the cause of the observed onions. Scanning electron microscopy (SEM) studies were performed on a Hitachi S800-FE microscope operated at 20 kV. To prepare the SEM sample, a silicon plate was directly dipped into the water to collect the raw material and dried in an oven at 100°C. Figures 6.6a and 6.6b show low- and high-magnification images taken using the SEM, respectively. It is readily seen that the spherical particles are agglomerated into clusters. The diameters of the particles estimated from Figure 6.6 ranged from 4 to 36 nm. Also, in the inset of Figure 6.6a, a "ball and bat" figure shows an elongated nanotube and a spherical onion.

The floating powder was also analyzed with a differential scanning calorimeter (DSC) (Seiko Instruments, DSC-2). The measurement was performed at a rate of 5 K/min from room temperature to 500°C in air. No significant endothermic and exothermic heat was detected in this temperature range except water evaporation. This indicates that the particles produced by arc discharge in water are free of volatile impurities.

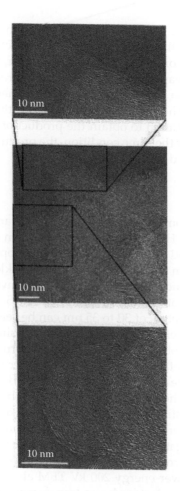

FIGURE 6.5
High-resolution transmission electron microscopy (HRTEM) images of onions and elongated nanoparticles produced by the arc discharge in water and collected as the floating powder on top of the water surface. (Reproduced with permission from Sano et al. [26]. Copyright 2002 American Institute of Physics.)

The specific surface area of the floating powder was determined by nitrogen gas adsorption based on the Brunauer–Emmett–Teller (BET) adsorption isotherm using an automated surface area analyzer (Coulter, OMNISORP100). The result showed an extremely large value of 984.3 m²/g, significantly larger than that reported for single- and multiwalled carbon nanotubes (SWNTs and MWNTs) [27–29]. We tabulated the surface area values for different carbon materials in Table 6.1. The floating powder produced by the water arc has a specific surface area that is approximately 3 to 5.5 times higher than that of other materials. The mean particle diameter can be derived from a simple correlation, assuming that the nanoparticles are uniformly spherical, according to

$$S = \frac{4\pi r_m^2}{4\rho\pi r_m^3 / 3} \tag{6.4}$$

where S, r_m, and ρ are the specific surface area, mean particle radius, and particle density, respectively. If 984.3 m²/g and a measured density of 1.64 g/cm³ are used for S and ρ, the mean diameter is determined to be 3.7 nm. This value compares well to the smallest diameter onion we have measured using microscopy. The diameter value determined from adsorption is calculated under an

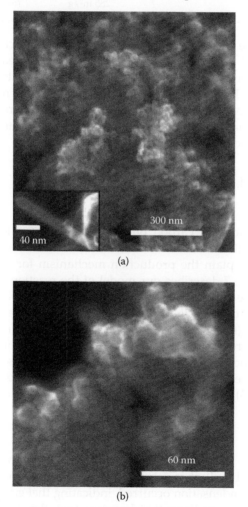

FIGURE 6.6
Scanning electron microscope (SEM) images of the floating powder produced by arc discharge in water. (a) Low-magnification image of the floating powder. The inset shows an elongated and a spherical (ball and bat) particle. (b) High-magnification SEM image of (a). (Reproduced with permission from Sano et al. [26]. Copyright 2002 American Institute of Physics.)

TABLE 6.1

Specific Surface Area Comparisons of Various Types of Carbon Materials

Sample	Specific Surface Area	Reference
Carbon onion and multiwalled carbon nanotubes (MWNTs)	984 m²/g	This work
Single-walled carbon nanotubes (SWNTs) and MWNTs	312 m²/g	Hernadi et al. [27]
MWNTs	178 m²/g	Inoue et al. [28]
Bundles of SWNTs	285 m²/g	Ye et al. [29]

assumption that the adsorption of nitrogen gas molecules occurs on the apparent surface area of the spherical particles. However, the real surface of the poorly crystallized onions is "rough" due to structural defects in the shells, as observed in the HRTEM image. This "surface roughness" increases the overall surface area of the particle. Because r_m is inversely proportional to S (Equation 6.4), adsorption enhancement can result in a smaller calculated mean particle diameter than the actual particle diameter.

6.2.3 Model for the Water Arc

6.2.3.1 Gas Bubble Formation

Two types of structures, carbon onions and elongated structures, were obtained in the water arc used in this study. Based on this result, we propose a model to explain the production mechanism for the two types of structures. Figure 6.7 describes our model of the reaction zone. There is a plasma zone between electrodes surrounded by a gas bubble due to vaporization of the surrounding liquid, as the arc temperature is estimated to be around 4000 K (the sublimation temperature of carbon). In fact, this gas bubble can be regarded as a microscale water-cooling reaction chamber that enables rapid quenching of the arc discharge. The main gas components are CO and H_2 produced by the reaction of C atomic vapor and H_2O at the gas–liquid interface [25] through:

$$C + H_2O \rightarrow CO + H_2$$

To measure the gas bubble formation rate, the vapor was trapped in a Pyrex dish placed above the discharge beaker. The trapped vapor was allowed to condense but no condensation occurred, indicating that gas bubbles did not include any water vapor. From the above reaction, the stoichiometric mole fraction of CO should be 50%. From the measured volumetric formation rate and this mole fraction value, we estimate the CO production rate to be 0.01 mol/min. As stated above, the corresponding production rates of the floating powder and the sediment products are 3.6 and 15.9 mg/min, respectively,

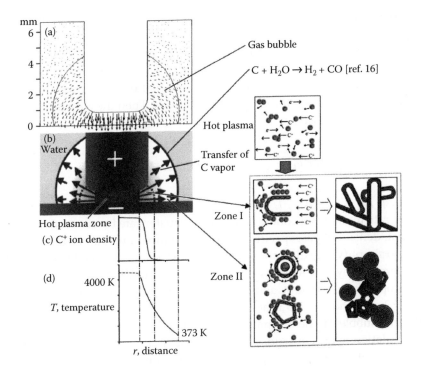

FIGURE 6.7
Proposed formation mechanism of onions in a water arc. (a) Relative electric field strength, shown by arrows, between a rod anode (17 V) and a flat cathode (ground) in a gas bubble surrounded by water. (b) Direction of thermal expansion from plasma to the water interface. (c) Qualitative ion density distribution. (d) Temperature gradient obtained from Equation (6.5). Assuming q_c, Q_R, and $dT/dt = 0$. The formation of elongated nanoparticles in zone (1) and onions in zone (2) are also shown schematically. (Reproduced with permission from Sano et al. [26]. Copyright 2002 American Institute of Physics.)

and the anode consumption rate is 117.2 mg/min. This indicates that approximately 83% of C was lost from the reaction system at a rate of 0.008 mol/min during arc discharge. This loss of carbon is attributed to the formation of CO. The discrepancy between the two values, 0.01 and 0.008 mol/min, may be caused by overestimation of the CO gas production rate due to other gas production by electrolysis. For higher accuracy in the mass balance, more elaborate gas analyses are necessary. To produce CO at the gas–liquid interface, C atomic vapor must be present at the interface. It is therefore reasonable to expect that C atomic vapor exists wholly in the bubble surrounding the discharge.

6.2.3.2 Nanoparticle Formation

The extremely sharp temperature gradient in the gas bubbles emanating from the hot plasma region to the gas–water interface is essential to cause rapid solidification of the vaporized carbon. The temperature at the hot plasma is estimated to be approximately 4000 K (the melting and boiling points of graphite are 3823 and 4203 K, respectively), while the temperature at the gas–water interface is the boiling point of water, 373 K. To estimate the approximate temperature gradient, we assume that the heat transfer occurs in a radial direction from the center of the plasma. Then the equation of heat balance can be expressed as

$$-\frac{d}{dr}(4\pi r^2 q_k) - \frac{d}{dr}(4\pi r^2 q_c) + 4\pi r^2 Q_R = 4\pi r^2 \rho C_p \frac{dT}{dt} \qquad (6.5)$$

where r, q_k, q_c, Q_R, ρ, C_p, T, and t are the distance from the arc center, heat transfer rate by thermal conductivity, heat transfer rate by convection, reaction heat, density of gas, specific heat of gas, temperature, and time, respectively. To simplify the calculation, q_c, Q_R, and dT/dt are assumed to be zero. If q_k is expressed by Fourier's law of thermal conductivity, $q_k = -kdT/dr$, where k is a proportional constant, and we assume boundary conditions of $T = 4000$ K at $r = 2$ mm and $T = 373$ K at $r = 5$ mm, the average temperature gradient in the gas bubble can be estimated to be 1209 K/mm. We note that this simple calculation is only provided to give a rough approximation of the temperature gradient. In reality, q_c, Q_R, and dT/dt cannot be neglected, and all parameters are highly space and time dependent. Now, the expansion rate of C vapor from the hot plasma zone to the cold region can be estimated by a simple approximation. If all the graphite consumed in the anode is assumed to be converted to C vapor, the volumetric C vapor expansion rate is calculated to be 5.3 10^{-5} m^3/s at 4000 K under ideal gas conditions (i.e., 1 atm pressure). If the hot plasma zone between electrodes with a 1 mm gap is assumed as a cylindrical zone with a 2 mm diameter, the expansion velocity can be estimated by dividing the volumetric expansion rate by the surface area of the assumed hot plasma zone, to be 4.2 m/s. This high-expansion velocity enables C vapor to transfer into the cold zone of the bubble readily. The cold zone can be categorized as two parts: (1) one part where the quenching of C vapor occurs within the ion current adjacent to the hot plasma zone and (2) another part without the ion current outside zone (1). Although we have not obtained the distribution of the ion current density in our system, we provide a map of the simulated electric field in a configuration that is close to our electrode shapes in Figure 6.7a. This simulation does not include the effect of the C vapor expansion. In zone (1), elongated structures such as nanotubes are expected to be produced because of their epitaxial growth in the C ion current. On the other hand, in zone 2, three-dimensional (3D) isotropic growth of nanoparticles is preferable because of the absence of an axis of symmetry. In this case, onions may be produced.

6.2.3.3 Flotation of Carbon Nanoparticles

Subsequent to the formation of onions, carbon particles coagulate into larger van der Waals crystals. We find that these large carbon clusters readily float to the top of the water surface. This floating powder remains separated at the surface of the water even after vigorous dispersion through ultrasonication. In order to investigate the mechanism responsible for floatation of the onion powder, we attempted to measure the actual density of the powder. To calculate the density, the true volume of the particles was estimated by soaking the particles in acetone to fill the void between particles. We measured a mean density of 1.64 g/cm^3. It is noteworthy that this value is higher than water although it is lower than graphite 2.25 g/cm^3. The density measured for our sample is comparable to that of well-known caged nanoparticles such as C_{60} and nanotubes with densities of 1.72 and 1.2 to 2 g/cm^3, respectively [30]. Hence, flotation of the particles cannot be ascribed to their weight but more likely to their hydrophobic surface. In fact, the particles can be dispersed well in organic solvents such as acetone, toluene, and *n*-hexane. In our arc-in-water system, onions naturally form agglomerates and float, separating themselves from other reaction by-products that settle to the bottom of the beaker.

In order to maximize the production rate of nanomaterials such as onions and nanotubes, several important aspects of the reaction should be optimized such as the current density, gas pressure, concentration of emitted species, and temperature gradient in the reaction zone. It must be emphasized that not only the production rate but also the selectivity of onions or nanotubes can be controlled because the formation depends on the directionality at the particle formation zone caused by the ion current (see Figure 6.7). One possible way to achieve this is to use a different electrode shape. If a narrow anode is used to decrease the relative area of the ion current zone compared to the isotropic quenching zone, the production of onions may increase.

In summary, we successfully produced carbon onions by an arc discharge in water. The arc discharge in water allows carbon onions to be synthesized in large quantities as opposed to more conventional vacuum processes. HRTEM and SEM analyses show onions with diameters ranging from 4 to 36 nm in the floating powder. The measured specific surface area of the particles was found to be extremely high, 984.3 m^2/g, indicating that the particles produced by this method are promising for gas storage. The mean particle diameter was calculated to be 3.7 nm from the specific surface area. The discrepancy between the particle size estimated by microscopic analysis and the specific surface area is ascribed to surface roughness from defective shells in the onions. Also it was found that the surface of the particles is hydrophobic because the particles float on water despite their density, 1.64 g/cm^3, being higher than that of water. To explain the production of onions and nanotubes, a model of the arc-in-water system in which there are two quenching zones was proposed. Based on this model, we propose that the physical characteristics of the nanoparticle products from the water-arc can be controlled to improve the quantity of production.

Acknowledgments

Authors acknowledge contributions of G. A. J. Amaratunga, C. J. Kiely, I. Alexandrou, R. A. Aharonov, R. Devenish, N. Sano, H. Wang, K. B. K. Teo and K. Iimura.

References

1. P. J. Fallon, V. S. Veerasamy, C. A. Davis, J. Robertson, G. A. J. Amaratunga, W. I. Milne, and J. Koskinen, *Phys. Rev. B* 48, 4777 (1993).
2. G. A. J. Amaratunga, M. Chhowalla, C. J. Kiely, I. Alexandrou, R. A. Aharonov, and R. Devenish, *Nature (London)* 383, 321 (1996).
3. W. Krätschmer, L. D. Lamb, K. Fostiropoulos, and D. R. Huffman, *Nature (London)* 347, 354 (1990).
4. S. Ijima, *Nature (London)* 354, 56 (1991).
5. T. Fukuzawa, M. Shiratani, and Y. Watanabe, *Appl. Phys. Lett.* 64, 3098 (1994).
6. A. Rogozin and R. F. Fontana, Presented at International Conference on Metallurgical Coatings and Thin Films 1996 (ICMCTF '96), San Diego, CA.
7. C. Bergman, *Surf. Coat. Technol.* 36, 243 (1988).
8. S. Anders and A. Anders, *J. Phys. D* 21, 213 (1988).
9. E. G. Gamaly and T. W. Ebbesen, *Phys. Rev. B* 52, 2083 (1995).
10. T. Guo, P. Nikolaev, A. G. Rinzler, D. T. Colbert, and R. E. Smalley, *Chem. Phys. Lett.* 243, 49 (1995).
11. H. W. Kroto, J. R. Heath, S. C. O'Brien, R. F. Curl, and R. E. Smally, *Nature (London)* 318, 162 (1985).
12. W. Kratschmer, L. D. Lamb, K. Fostiropoulos, and D. R. Huffman, *Nature (London)* 347, 354 (1990).
13. For example, The National Institute of Materials and Chemical Research (NIMCR) and Showa Denko KK, Japan, recently announced a project to develop a mass-production method to produce several hundred kilograms of nanotubes per day.
14. D. Ugarte, *Nature (London)* 359, 707 (1993).
15. F. Banhart, T. Fuller, Ph. Redlich, and P. M. Ajayan, *Chem. Phys. Lett.* 269, 349 (1997).
16. V. L. Kuznetsev, A. L. Chuvilin, Y. V. Butenko, I. Y. Mal'kov, and V. M. Tikov, *Chem. Phys. Lett.* 222, 343 (1994).
17. T. Cabioc'h, E. Thune, J. P. Riviere, S. Camelio, J. C. Girard, P. Guerin, M. Jaouen, L. Henrard, and P. Lambin, *J. Appl. Phys.* 91, 1560 (2002).

18. X. H. Chen, F. M. Deng, J. X. Wang, H. S. Yang, G. T. Wu, X. B. Zhang, J. C. Peng, and W. Z. Li, *Chem. Phys. Lett.* 336, 201 (2001).
19. K. Yamada, H. Unishige, and A. B. Sowaoka, *Naturwissenschaften* 78, 450 (1991).
20. N. Sano, H. Wang, M. Chhowalla, I. Alexandrou, and G. A. J. Amaratunga, *Nature (London)* 414, 506 (2001).
21. S. Ijima and T. Ichihashi, *Nature (London)* 363, 603 (1993).
22. A. Thess et al., *Science* 273, 483 (1996).
23. A. M. Cassell, J. A. Raymakers, J. Kong, and H. J. Dai, *J. Phys. Chem. B* 103, 6484 (1999).
24. M. Ishigami, J. Cumings, A. Zettl, and S. Chen, *Chem. Phys. Lett.* 319, 457 (2000).
25. Y. L. Hsin, K. C. Hwang, F. -R. Chen, and J. -J. Kai, *Adv. Mater.* 13, 830 (2001).
26. N. Sano, H. Wang, I. Alexandrou, M. Chhowalla, K. B. K. Teo, G. A. J. Amaratunga, K. Iimura, *J. Appl. Phys.* 92, 2783 (2002).
26. K. Hernadi, A. Fonseca, J. B. Nagy, A. Fudala, D. Bernaerts, and I. Kiricsi, *Appl. Catal., A* 228, 103 (2002).
27. S. Inoue, N. Ichikuni, T. Susuki, T. Uematsu, and K. Kaneko, *J. Phys. Chem. B* 102, 4689 (1998).
28. Y. Ye, C. C. Ahn, C. Witham, B. Fults, J. Liu, A. G. Rinzler, D. Colbert, K. A. Smith, and R. E. Smalley, *Appl. Phys. Lett.* 74, 2307 (1999).
29. R. Saito, G. Dresselhaus, and M. S. Dresselhaus, *Physical Properties of Carbon Nanotubes.* Imperial College Press, London (1999).

7

Atmospheric Plasmas for Carbon Nanotubes (CNTs)

Jae Beom Park
Se Jin Kyung
Geun Young Yeom

CONTENTS

7.1 Introduction

In the mid 1980s, Smalley and coworkers at Rice University (Houston, Texas) discovered fullerenes [1], geometric cage-like structures of C60 carbon atoms composed of hexagonal and pentagonal faces similar to a soccer ball (the C60 molecule is often referred to as a bucky ball) [2]. A few years later, another form of carbon nanostructures, carbon nanotubes (CNTs), were discovered by Sumio Iijima of the NEC Corporation [1] from soot of an arc discharge apparatus. Carbon nanotubes (CNTs) are tubular structures with walls of hexagonal carbon (graphite structure), diameters as small as 0.7 nm, and lengths up to several microns, usually capped at one end by fullerene. The interest in CNTs in the last decade across the world for potential applications has been extraordinary due to their remarkable physical and mechanical properties [3–7]. Their mechanical properties are impressive by typically exhibiting stiffness and strength in the ranges of 230 to 725 GPa and 1.5 to 4.8 GPa, respectively [8]. Pristine carbon nanotubes are also extremely conductive. Due to their one-dimensional (1D) nature, charge carriers can travel through nanotubes without scattering, resulting in ballistic transport. The absence of scattering means that Joule heating is minimized so that CNTs can carry very large current densities of up to 10^9 mA/cm^2 [9]. In addition, carrier mobilities as high as 10^5 cm^2/Vs have been observed in semiconducting CNTs [10]. These properties have led to wide-ranging investigations of their potential for future electronics, field emitter devices, sensors, electrodes, high strength composites, storage of hydrogen, and so forth.

Currently, CNTs are usually synthesized by carbon-arc discharges [11], laser ablation of carbon targets [12], thermal-chemical vapor deposition (CVD) [13], and plasma-enhanced chemical vapor deposition (PECVD) [14]. Of these various growth methods, it has been shown [15–17] that CVD growth is the most suitable process to produce arrays of CNTs for electronic devices, sensors, field emitters, and other applications where controlled growth is needed. More recently, plasma-enhanced CVD (PECVD) has been studied for its ability to produce vertically aligned nanotubes at low temperatures.

Atmospheric-pressure plasmas (APPs) have been investigated as a possible replacement for low-pressure plasma processing in a number of areas, including semiconductor and flat panel display processing. In particular, many researchers have studied CNT growth using APP sources because of key advantages over vacuum (i.e., low-pressure) processing. APP processing minimizes the need for vacuum systems and enables continuous handling of large substrates. For this reason, many researchers have reported their own APP sources that are modified for CNT growth.

In this chapter, various APP sources that have been used for CNT growth are introduced. The APP technologies such as dielectric barrier discharges, plasma jet discharges, corona discharges, and so forth, that will be discussed in this chapter are not commercially available for CNT growth and are still under development in laboratories. The characteristics of the CNTs grown by the APP sources will also be discussed as functions of various growth conditions including growth temperature, plasma properties, and so on.

7.2 Atmospheric Pressure Plasma

7.2.1 Advantages of Atmospheric Pressure Plasma Processing

Plasmas are chemically active media. Depending on how they are activated and their working power, they can have low or very high "temperatures," known as cold or thermal plasmas, respectively. This wide temperature range enables various applications for plasma technologies: thin film processing for semiconductor or display manufacturing (surface etching, surface cleaning, and surface coatings), tissue sterilization, and carbon materials synthesis [18–21], to name a few. Among the different types of plasma sources, APP sources offer several advantages. APP processing eliminates the need for relatively expensive and complicated vacuum systems and components. Vacuum systems used for low-pressure plasma processing also require maintenance, making APP processing cost effective if the cost related to higher gas usage is not excessive. Also, for low-pressure processing, loadlocks and robotic assemblies must be used to shuttle materials in and out of vacuum. However, for APP processing, high processing speeds and even continuous processing of large area substrates by in-line processing is possible, thus facilitating industrial-level manufacturing. In addition, the size of the object that can be treated is not limited by the size of the vacuum chamber.

In recent years, there has been an increasing interest in APP processing such as etching, cleaning, and surface treatment [22–29] due to the possible advantages described above. To date, research on APP sources has made remarkable progress toward the development of stable plasmas for many different applications. Interestingly, APP processing is not only useful for industrial applications but is also reported to be useful for the synthesis of functional materials such as carbon nanotubes, diamond-like carbon, ultra-high-speed etching, silicon oxide nanoparticle deposition, and amorphous/crystalline silicon film deposition. All have been investigated extensively from the perspective of APPs because of both economical and technological reasons as mentioned above. In summary, major potential cost savings are associated with the in-line processing capability, which substantially reduces the substrate handling costs and leads to an increased throughput due to high processing rates. Capital cost savings for both equipment and line space (footprint), and the relative ease of integration are further benefits in comparison to low-pressure approaches.

7.2.2 Various Atmospheric Pressure Plasma Sources

For comparison purposes, it is convenient to classify the various types of APP sources by using a key parameter such as the frequency, f, of the voltage applied to generate the plasma: (1) direct current (DC) discharges, including steady-state or pulse-periodical regimes; (2) alternating current (AC) discharges; traditionally, f values up to 100 kHz; (3) radio frequency (RF) discharges, having f from 100 kHz to 100 MHz; and (4) microwave (MW) discharges, f >100 MHz. The differences in the plasma sources operated at different frequencies are shown in Tables 7.1, 7.2, and 7.3. As shown in Table 7.1, for DC or low-frequency operation, the plasma tends to have a low electron temperature and a low gas temperature while having a high breakdown voltage of a few kV. On the other hand, operation of a source at microwave frequencies of 2.54 GHz shows a high electron temperature and a high gas temperature as shown in Table 7.3.

Another way to classify APP sources is by the configuration that depends on the application. In the case of operation at a high frequency such as microwave frequency, due to high-energy particles in the plasma, torch-type plasma sources that produce a plasma flame in a gas flow are generally used while at DC or low frequencies, capacitively coupled plasma sources composed of two parallel electrodes are used. Regardless of the operation frequency, an inert gas such as Ar and He is generally used as the supply gas due to the lower breakdown voltage and ease of generation of glow discharge-like plasmas at atmospheric pressure. Among the various configurations used to generate atmospheric pressure plasmas, the most studied plasma source is a dielectric barrier discharge (DBD). These sources have a simple electrode configuration and are easily implemented to produce a plasma. In addition to DBDs and plasma torches, other types of plasma sources such as corona discharges and plasma jets have also been investigated. In the following sections, all of these widely used APP sources will be discussed in more detail.

7.2.2.1 Dielectric Barrier Discharges (DBDs)

One of the earliest reports of DBDs operated at atmospheric pressure was by Donohoe [30]. Donohoe used a large gap (cm) pulsed-barrier discharge in a mixture of helium and ethylene to polymerize ethylene [31]. Later, Kanazawa et al. [32] reported the development of a stable glow discharge at atmospheric pressure by using the DBD configuration shown in Figure 7.1. Normally, as shown in Figure 7.1, the source consists of two metal electrodes, with at least one electrode coated with a dielectric layer. To ensure stable plasma operation, the gap that separates the electrodes is limited to a few millimeters. Discharge gas flows in the gap and the discharge is ignited by means of a sinusoidal or pulsed [33] power source. Depending on the working gas composition, the voltage, and frequency excitation, the discharge can be either

TABLE 7.1

Characteristics of Atmospheric Pressure Direct Current (DC) and Low-Frequency Plasma Sources

Input Power	Source	Properties	Operating Conditions	Reference
DC 10 kHz	Arc discharge	T_g: 3500 K N_e: $10^{13} \sim 10^{14}$ cm^{-3} T_{rot}: 700 ~ 2500 K	Gas: air	99
50 Hz alternating current (AC)	Pin plane corona		First power 35 kV, 50 Hz Second power 4.5kV, 50 ~ 10^{50} Hz Gas: He, N$_2$, Ar, air	100
50 Hz ~ 10 kHz AC	Dielectric barrier discharge (DBD)	E_e: 1 ~ 10 eV N_e: $10^{14} \sim 10^{15}$ cm^{-3}	Vpp: 3 ~ 20 kV Gap distance: 0.2 ~ 5 mm Gas: air, He, N$_2$ Gas flow: 100 sccm	101
100 ~ 160 kHz AC	DBD jet	P: 500 W	Gas: air, N$_2$ Gas flow: 40 ~ 100 slm	102
20 kHz AC	Flexible tube		Gas: Ar Gas flow: 3 lpm Vrms: 8.5 kV Irms: 0.098 A	103
25 kHz pulse	DBD jet	T_e: 1 eV N_e: 4.3×10^{22} m^{-3}	Gas: Ar Vrms: 15 kV	104
20 kHz AC	Jet	T_g: 300 K T_e: 1.56 eV N_e: 1.8×10^{12}/cm^3	Hole diameter: 500 um Gas: air, N$_2$, He Gas flow: 5 lpm	105,106
50 Hz DC	Microplasma	T_{rot}: 700 ~ 1550 K T_{vib}: 4500 ~ 5000 K	Gas: air	107
DC	Micro DBD	N_e: $10^{14} \sim 10^{15}$ cm^{-3} E_e: 1 ~ 10 eV	Gas: He, N$_2$, Ar, Ne Ipp: 0.1 A	108

TABLE 7.2

Characteristics of Atmospheric Pressure Radio Frequency (RF) Plasma Sources

Input Power	Source	Properties	Operating Conditions	References
13.56 MHz RF	Dielectric barrier discharge (DBD)	T_{rot}: 490 ~ 630 K T_{vib}: 3000 ~ 3200 K	Gas: Ar P: 600 W	109–111
13.56 MHz	Atmospheric pressure plasma jet (APPJ)	T_g: <300°C N_e: 0.2 ~ 2 × 10^{11}/cm^3 E_e: 2 ~ 4 eV	Gas: He Gas flow: 50 lpm P: 3 ~ 30 Wcm^{-3}	112
13.56 MHz RF	Parallel plate	N_p: 4.4 × 10^{12} cm^{-3} T_e: 1.2 eV C_e: <10^6 cm^{-3} $N_{o\,atom}$: 7 × 10^{13} cm^{-3}	Gas: Ar, O$_2$	113
13.56 MHz RF	Capillary DBD	N_p: 5 × 10^{11} cm^{-3} (glow) N_p: 10^{15} cm^{-3} (filamentary) T_e: 1.3 eV (glow) T_e: 1.7 eV (filamentary)	Gas: Ar Gas flow: 40 ~ 100 slm	114
13.56 MHz RF	Fused hollow cathode (FHC)	T_{vib}: 4706 K	Gas: air, Ne Gas flow: 1000 sccm	115
144 MHz	Cold plasma inductively coupled plasma (ICP)	N_p: 10^{11} cm^{-3} N_e: 8 × 10^{14} cm^{-3} T_{exc}: 4000 ~ 4500 K	P: 50 W Gas: Ar Gas flow: 0.7 slm	116
13.56 MHz	Microplasma jet (MPJ)	T_g: <400 K N_e: 8 × 10^{20} m^{-3} (Ar) N_e: 7 × 10^{20} m^{-3} (He)	Gas: Ar + CH$_4$ Ar + C$_2$H$_2$ V_{rms}: 200 ~ 250 V I_{rms}: 0.4 ~ 0.6 A	117
13.56 MHz RF	APPJ	T_g: 25 ~ 200°C Charged particle density: 10^{11} ~ 10^{12} T_e: 1 ~ 2 eV	Gas: He, O$_2$	57
13.56 MHz RF	Needle	$T_{vib\&exc}$: 0.2 ~ 0.3 eV T_{rot}: few hundred K	Gas: He, Ar, N$_2$, O$_2$ V_{pp}: 200 ~ 500 V	118

TABLE 7.3

Characteristics of Atmospheric Pressure Microwave Plasma Sources

Input Power	Source	Properties	Operating Conditions	References
2.45 GHz	Microwave torch discharge (MTD)	T_e: 17,000 ~ 20,000K T_h:1,500 ~ 4,000K N_e:10^{20} ~ 10^{21} m^{-3}	Gas: N_2 Gas flow: 1 ~ 3 slm P: 100 ~ 400 W	119
2.45 GHz	Microwave plasma torch (MPT)	T_{exc}: 4,700 K N_e: 10^{21} m^{-3}	Gas: Ar P: 50 ~ 200 W	120
2.45 GHz	(H-Head)	$T_{substrate}$: 600°C	Gas: Ne, Ar P: 400 W	121
2.45 GHz	Microwave-induced pasma (MIP)	T_{exc}: 3010 ~ 4350 K T_{rot}: 2250 ~ 3550 N_e: 6.6 ~ 7.6 × 10^{14} cm^{-3}	Gas: Ar, O_2, air Gas flow: Ar:4 lpm O_2: 4 lpm P: 400 W	122, 123
2.45 GHz	Capillary tube	T_{exc}: 5200 K T_{rot}: 1700 K N_e: 4 × 10^{14} cm^{-3}	Gas: Ar Gas flow: 1 lpm P: 250 ~ 300 W	124
2.45 GHz	Microwave plasma torch (MPT)	N_e: 5 ~ 8 × 10^{14} cm^{-3}	Gas: N_2, Ar Gas flow N_2: 2.5 lpm, Ar: 12.5 lpm	125
2.45 GHz	Pulsed microwave plasma discharge	T_g: 3000 K	Gas: Ar, H_2 Gas flow Ar: 10 lpm, O_2: 0.15 lpm P: 600 W	126
2.45 GHz	Steam plasma torch (SPT)	T_g: 5000 K	Gas: Ar T_{steam}: 160°C	127
2.45 GHz	Axial injection torch (AIT)	T_e: 13,000 ~ 14,000 K T_h: 2,400 ~ 2,900 K N_e: 10^{21} m^{-3}	Gas: He Gas flow: 2 ~ 6 slm P: 100 ~ 2000 W	128

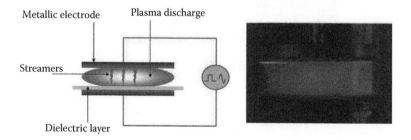

FIGURE 7.1

A dielectric barrier discharge. (Figure reprinted from C. Tendero et al. 2006. Atmospheric presure plasma: A review. *Spectrochimica Acta, Part B.* 61: 2–30. With permission.)

filamentary or glow-like [34,35]. After discharge ignition, charged particles are collected on the surface of the dielectric-covered electrode. This charge buildup creates a voltage drop that counteracts the applied voltage, leading to lower breakdown and a self-limiting discharge current. The plasma is generated through a succession of micro discharges, lasting for 10 to 100 ns, randomly distributed in space and time. These "streamers" are believed to be 100 μm in diameter and separated from each other by as much as 2 cm [36,37]. DBDs are sometimes confused with coronas because of these streamer discharges that may exhibit micro arcing. DBD sources typically contain a large number of these micro discharges. Under certain operating conditions, more homogeneous and diffuse-type discharges—that is, glow discharge-like plasmas—can be obtained [31,32,38,39], which are useful for industrial applications. Such glow discharge–like conditions are easily obtained in pure He or He-rich gas mixtures or, alternatively, with special electrode configurations and operating conditions in other gases.

7.2.2.2 Atmospheric Pressure Plasma Jets and Plasma Torches

Figure 7.2 shows a schematic diagram of an atmospheric pressure plasma jet (APPJ) operated with radio frequency (RF) power. The discharge has a coaxial configuration, with a gap between the inner and outer electrodes. The ionized gas from the plasma jet exits through a nozzle, where it is directed onto a substrate a few millimeters downstream. The inert gas flow typically runs through the gap of the coaxial structure, while the reactive gases are inserted through a capillary near the end of the inner electrode located near the nozzle. RF power at 13.56 MHz is generally used to operate the discharge. This APPJ source configuration has been used for many applications including polyimide, tungsten, tantalum, and silicon dioxide etching [41]. APPJ sources have also been frequently used for biomedical applications, including the induction of apoptosis in cancer cells [43–47]. APPJs have also been used to synthesize CNTs [48]. Overall, plasma jets are of great interest

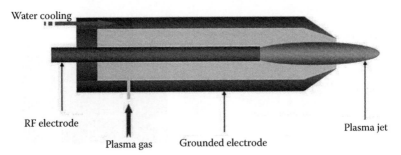

FIGURE 7.2
An atmospheric-pressure plasma jet for the deposition of silica films. (Figure reprinted from C. Tendero et al. 2006. Atmospheric pressure plasma: A review. *Spectrochimica Acta, Part B.* 61: 2–30. With permission.)

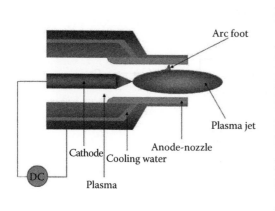
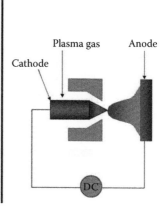

FIGURE 7.3

A cold plasma torch. (Figure reprinted from C. Tendero et al., 2006. Atmospheric pressure plasma: A review. *Spectrochimica Acta, Part B.* 61: 2–30. With permission.)

for technological applications ranging from thin film processing to biomedical applications.

The APP torch system shown in Figure 7.3 has similar characteristics to APPJs except that the plasma is generated between the tip of the center electrode and ground electrode near the exit of the torch. Many researchers have employed plasma torches for materials processing, including silicon etching [49], photoresist ashing [50], deposition of SiO_x or TiO_x films [51–54], treatment of vulcanized rubber [55], and the production of fullerenes [56].

Both APPJs and plasma torches generate reactive species in the discharge chamber at atmospheric pressure that are carried in a gas flow to a substrate. There is virtually no distinction in the operating concept between APPJs and plasma torches. However, APPJs are characterized by relatively low electron temperatures and gas temperatures, because gas molecules are dissociated between the electrodes in a micro discharge that is glow-like. In the case of plasma torches, a very high voltage of 10 ~ 50 kV is generally applied, and the reactive gas is dissociated in an arc-type discharge. Therefore, a typical atmospheric-pressure plasma torch tends to have a gas temperature of 3,000 ~ 20,000 K and a plasma density of $10^{16} ~ 10^{19}$ /cm^3, both of which are significantly higher than those found in APPJs [57].

7.2.2.3 Corona Discharges

A corona discharge is defined as a luminous glow localized in space around a sharp tip in a highly nonuniform electric field. Corona discharges are electrical discharges formed by ionization of a fluid surrounding a conductor, which occurs when the potential gradient at the sharp tip exceeds a certain value but is not sufficient to cause complete electrical breakdown or arcing

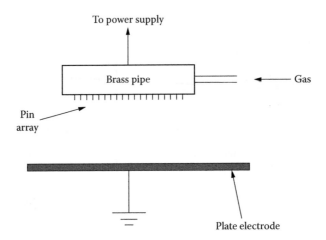

FIGURE 7.4
A corona discharge system. (Figure reprinted from Laroussi, M. 2002. Nonthermal decontamination of biological media by atmospheric-pressure plasmas: Review, analysis, and prospects. *IEEE Trans. Plasma Sci.* 30(4): 1409–1415. © 2002 IEEE. With permission.)

[58–60]. The Siemens research team [61] was the first to propose the use of a corona discharge to generate ozone for disinfecting water. This was the first report that applied plasmas for inactivation of microorganisms. Later, Menashi [62] used a pulsed RF-driven corona discharge to form a plasma at atmospheric pressures. This plasma could be described as a Townsend or negative glow discharge depending upon the field and potential distribution. By applying a high voltage to the sharp electrode, small localized discharges of very short duration can be observed in the gas gap of about 1 to 10 mm. A schematic diagram of a corona discharge system is shown in Figure 7.4. The system consists of a line of pins fastened to a power electrode. If a nonelectronegative gas such as He or Ar is used as the supply gas instead of air, the discharge is enhanced and can be operated at a relatively low voltage. The pin array is biased by a DC, AC, or pulsed power supply. The plasma usually extends about 0.5 mm from the metal tips. In the drift region outside this volume, charged species diffuse toward the planar electrode and are collected. Corona discharges in air are commonly used for ozone production [63] or for the activation of polymer surfaces before printing, pasting, or coating [64,65].

7.2.2.4 Characteristics of Atmospheric Pressure Discharges

Nonthermal atmospheric plasmas generated by these various discharge configurations are considered to possess certain common properties. The average electron temperature is fixed by ionization/recombination and attachment mechanisms. Figure 7.5 [66] shows the influence of the operating pressure on the transition from a glow discharge ($T_e > T_h$) to an arc discharge. In a low-pressure plasma, the gas temperature is lower than the electron

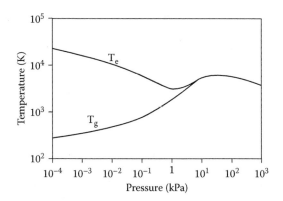

FIGURE 7.5
Evolution of the plasma temperature (electrons and heavy particles) with pressure in a mercury arc plasma. (Figure reprinted from Schütze, A., Jeong, James Y., and Hicks, Robert F. et al. 1998. The atmospheric-pressure plasma jet: A review and comparison to other plasma sources. *IEEE Trans. Plasma Sci.* 26(6): 1685–1694. © 1998 IEEE. With permission.)

temperature. The inelastic collisions between electrons and background gas atoms or molecules result in electronic excitation or ionization but do not lead to an increase in the gas temperature. When the pressure is higher, the collision frequency between electrons, ions, and neutral gas atoms/molecules increases. These collisions lead to both gas dissociation (i.e., inelastic) and heating (i.e., elastic). The difference between the electron temperature (T_e) and gas temperature (T_g) thus decreases. As the operating pressure approaches very high pressures (~1 atm), the kinetic energy of the electrons is transferred more effectively to the background gas atoms/molecules, and the two temperatures become similar (i.e., $T_e \sim T_g$). However, even for atmospheric-pressure plasmas, T_g can be close to room temperature for some of the discharge sources. A low gas temperature, T_g, is obtained by efficient heat conduction to the electrodes, or by limiting gas transit time in the discharge region. Figure 7.6 shows the range of gas temperatures and electron temperatures that are found in various atmospheric-pressure plasma sources. The gas temperatures of plasma jets and corona discharges are clearly low within the commonly used operating window. The electron temperature of plasma jets is estimated to be between 1 and 2 eV. However, there is a sufficient population of electrons at energies high enough to dissociate (and ionize) many molecules, including O_2, N_2, and He.

Table 7.4 shows typical breakdown voltages for the different plasma sources. The plasma jet has the lowest V_b value, even lower than that found for low-pressure glow discharges. The low electron temperature, T_e, shown in Figure 7.6 is related to the low V_b of 0.05 ~ 0.2 kV. Plasma jets possess a low T_e due to the source configuration; the center power electrode is spaced closely from the outer ground electrode in a coaxial fashion allowing the plasma to be ignited at low voltages. In the case of plasma torches, the plasma is ignited between

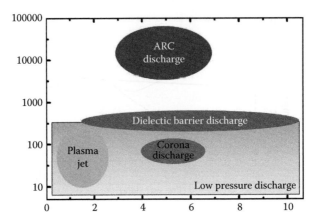

FIGURE 7.6
(See color insert.) Comparison of the gas and electron temperatures for different atmospheric-pressure plasmas versus low-pressure plasmas. (Figure reprinted from Schütze, A., Jeong, James Y., and Hicks, Robert F. et al. 1998. The atmospheric-pressure plasma jet: A review and comparison to other plasma sources. *IEEE Trans. Plasma Sci.* 26(6): 1685–1694. © 1998 IEEE. With permission.)

TABLE 7.4

Breakdown Voltages of Various Atmospheric-Pressure Plasmas (APPs) and a Low-Pressure Plasma

Source	V_b (kV)
Low-pressure discharge	0.2 ~ 0.8
Arc and plasma torch	10 ~ 50
Corona	10 ~ 50
Dielectric barrier discharge	5 ~ 25
Plasma jet	0.05 ~ 0.2

Source: Table reprinted from Andreas Schütze, James Y. Jeong, Robert F. Hicks, et al., The Atmospheric-Pressure Plasma Jet: A Review and Comparison to Other Plasma Sources. *IEEE Trans. Plasma Sci.* 26(6): 1685–1694. © 1998 IEEE.

the tip of the center electrode and a ground electrode near the exit of the torch, and the breakdown voltage is generally high. However, the breakdown voltage of atmospheric pressure plasma sources is always higher than several kV due to the reduced mean free path for electrons which limits their energy.

Table 7.5 shows typical densities of charged species in the different discharge sources. Except for the arc and plasma torch, all the plasma sources exhibit electron densities in the same range as low-pressure glow discharges. However, corona discharges show relatively lower plasma densities compared with other APP sources. The lower plasma density for the corona discharge is related to the formation of a localized discharge (filamentary discharge) near the pin electrode as compared to a large and uniform discharge over

TABLE 7.5

Density of Charge Species in Various Atmospheric-Pressure Plasma (APPs) and a Low-Pressure Plasma

Source	Plasma Density (cm⁻³)
Low-pressure discharge	$10^8 \sim 10^{13}$
Arc and plasma torch	$10^{16} \sim 10^{19}$
Corona	$10^9 \sim 10^{13}$
Dielectric barrier discharge	$10^{12} \sim 10^{15}$
Plasma jet	$10^{11} \sim 10^{12}$

Source: Table reprinted from Andreas Schütze, James Y. Jeong, Robert F. Hicks, et al., The Atmospheric-Pressure Plasma Jet: A Review and Comparison to Other Plasma Sources. *IEEE Trans. Plasma Sci.* 26(6): 1685–1694. © 1998 IEEE.

the electrode area. Other types of plasma sources that are larger in volume have plasma densities equal to or higher than low-pressure glow discharges. Therefore, the processing efficiency with APP sources can be at least as high if not higher than low-pressure plasmas under the appropriate operating conditions.

7.3 CNT Growth Process Using Atmospheric Pressure Plasmas

7.3.1 Comparison of Different Atmospheric Pressure Plasma Sources

As mentioned in the previous sections, remarkable progress has been made toward the formation of stable APP sources; this has led to tremendous interest in their applications. In addition to thin film processing, APPs are being explored for applications in energy, environment, and biomedicine. In recent years, APP systems have also been developed for growth and surface modification of materials. These types of APP systems are known to be beneficial for CNT synthesis because CNTs can be grown in an in-line process and without using a vacuum system. At present, the fabrication of CNTs at atmospheric pressure has been achieved using various atmospheric pressure discharge systems including DBDs, plasma jets, and coronas [67–72]. In the following sections, recent results for CNT fabrication using various APP sources and the characteristics of the CNTs will be discussed.

7.3.1.1 CNT Synthesis Using Nonequilibrium DBDs

DBDs are the most studied and implemented atmospheric-pressure discharge systems because of their simple electrode configuration, scalability, and overall discharge stability. In addition, due to an electrode configuration similar to low-pressure capacitively coupled plasmas (CCPs), a strong electric field is created near the substrate by the sheath voltage; therefore, vertically aligned and high-quality CNTs can be easily synthesized using a DBD. Many research teams including Nozaki et al. [70] have investigated CNT synthesis using the DBD plasma system at atmospheric pressure. Other researchers have used modified DBD systems to obtain higher plasma densities for more efficient growth of CNTs [73] or to obtain uniform gas distribution in addition to the high plasma density utilizing the gas distribution electrode similar to a conventional low-pressure CCP system [74]. The conventional DBD system and modified DBD system used for CNT synthesis will be discussed in the following sections in more detail.

7.3.1.1.1 CNT Synthesis Using Conventional DBD Systems

Nozaki et al. have investigated CNT synthesis using a conventional DBD operating at atmospheric pressure. Figure 7.7 shows the schematic diagram of the system [70]. A dielectric barrier layer of 500 μm thick alumina plate was installed on the top power electrode, and on the lower electrode a quartz plate covered with 20 nm thick Ni catalyst was used. This system is a typical DBD with at least one dielectric layer between the two metal electrodes facing each other. The top electrode was powered with AC frequency at 125 kHz, and the substrate electrode has a heater that can heat the substrate up to 600°C. The processing chamber was initially evacuated up to 10^{-3} Torr to remove impurities; the reactive gas mixture composed of $He/H_2/CH_4$ was then fed to bring the pressure up to atmospheric.

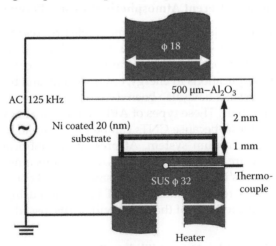

FIGURE 7.7
A conventional dielectric barrier discharge (DBD) system for carbon nanotube (CNT) synthesis used in investigation by Nozaki and colleagues. (Figure reprinted from Nozaki, T., Kimura, Y., and Okazaki, K. 2002. Carbon nanotubes deposition in glow barrier discharge enhanced catalytic CVD. *J. Phys. D: Appl. Phys.* 35: 2779–2784. With permission.)

Representative CNTs synthesized by the DBD system at 600°C are shown in Figure 7.8. As shown in the scanning electron microscopic (SEM) image in Figure 7.8 and transmission electron microscopic (TEM) image in Figure 7.9, the diameter of the synthesized CNTs was in the range of 40 ~ 50 nm and the CNT density was in the range of 10^9 ~ 10^{10} cm^{-2}. Also, as shown in Figure 7.9, the synthesized CNTs are not perfect hollow tubes and, instead, exhibit

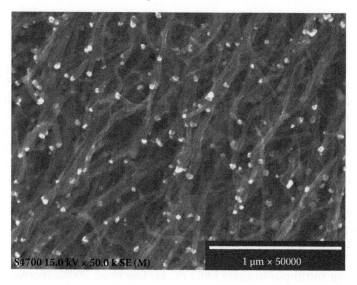

FIGURE 7.8
Scanning electron micrograph (SEM) image of carbon nanotubes (CNTs) deposited by a conventional dielectric barrier discharge (DBD) system. (Figure reprinted from Nozaki, T., Kimura, Y., and Okazaki, K. 2002. Carbon nanotubes deposition in glow barrier discharge enhanced catalytic CVD. *J. Phys. D: Appl. Phys.* 35: 2779–2784. With permission.)

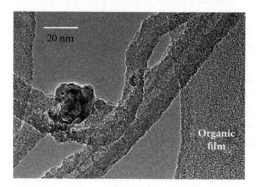

FIGURE 7.9
High-resolution transmission electron microscope (TEM) image showing detailed wall structure of multiwalled carbon nanotubes (CNTs). (Figure reprinted from Nozaki, T., Kimura, Y., and Okazaki, K. 2002. Carbon nanotubes deposition in glow barrier discharge enhanced catalytic CVD. *J. Phys. D: Appl. Phys.* 35: 2779–2784. With permission.)

bamboo-type structures. The imperfect CNTs obtained by the DBD system were related to defective growth caused by insufficient diffusion of carbon to the Ni catalyst metal surface. To obtain cleaner CNTs, a modified DBD system with a higher substrate temperature that can increase carbon diffusion efficiency on the metal catalyst surface or that with a higher plasma density which can increase the gas dissociation efficiency is required.

7.3.1.1.2 CNT Synthesis Using Multi-Pin-to-Plate-Type DBD Systems

The conventional DBD has a high breakdown voltage (30 kV/cm at air) and relatively low plasma density due to a high recombination rate at atmospheric pressure. This makes it difficult to use DBDs for processing applications other than surface treatment [75–79]. To obtain a higher plasma density, a modified DBD system composed of a multi-pin-powered electrode instead of a planar-powered electrode has been developed [80]. The pin shape of the multi-pin electrode was fabricated in the shape of a pyramid by machining the electrode surface. The multi-pin pyramid-shaped electrode resulted in a lower breakdown voltage because of the high localized electric field. In addition, the high density of pins on the electrode and diffusion of the discharge on the electrode surface allowed the microdischarges to merge and form a uniform glow discharge. Thus, the modified DBD not only has a lower breakdown voltage but also a higher plasma density compared to a conventional DBD system under similar discharge conditions.

Figure 7.10 shows the effect of number of pins on the discharge current density for a multi-pin-to-plate DBD (Y. H. Lee et al. 2005). Increasing the pin density on the powered electrode increased the discharge current at the same voltage, indicating a more efficient discharge (i.e., the plasma density increased with the pin density). Figure 7.11 shows the spectral intensity of lines corresponding to the optical emission of fluorine atoms, measured as a function of NF_3 flow rate in N_2, for a multi-pin-type powered electrode and a planar electrode DBD system, respectively. As shown, the modified DBD source with the multi-pin-to-plate electrode showed 2 ~ 3 times higher line intensity than the conventional DBD source indicating a higher gas dissociation rate.

The modified DBD source has been applied to CNT synthesis; Figure 7.12 shows a schematic diagram of the modified DBD system used for the growth of the CNTs (S. J. Kyung et al. 2005). As shown, the top electrode was composed of a multi-pin-type electrode covered by a dielectric barrier plate, and the bottom electrode was equipped with a heater for substrate heating. A gas mixture of $He/C_2H_2/NH_3$ was fed using a gas ring from the side, and the substrate temperature was varied from 400 ~ 500°C. The catalyst was prepared by sputter depositing a NiCr(10 nm)/Cr(100 nm) thin film on a glass substrate and exposing the film to a He(10 slm)/NH_3(150 sccm) atmospheric pressure plasma for 3 min at 500°C in the same processing chamber to increase the surface area and form nanosized NiCr alloy particles. CNTs synthesized for 3 minutes by the modified DBD after the pretreatment are

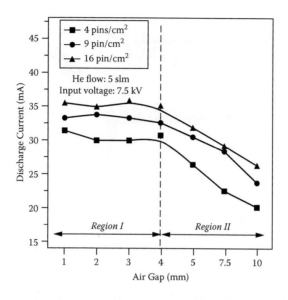

FIGURE 7.10

Discharge current measured as a function of the air gap between the electrodes for various pin densities at an applied voltage of 7.5 kV and a He flow rate of 5 slm. (Figure reprinted from Lee, Y. H., and Yeom, G. Y. 2005. Characteristics of a pin-to-plate dielectric barrier discharge in helium. *J. Kor. Phys. Soc.* 47(1): 74–78. With permission.)

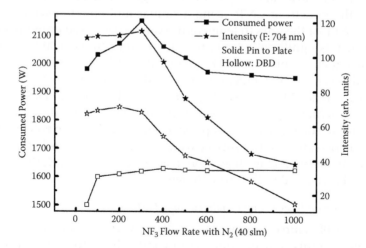

FIGURE 7.11

Effect of NF_3 flow rate in a NF_3(0 ~ 1 slm)/N_2 (40 slm) gas mixture on power consumption and relative optical emission intensity of F (704 nm) measured by optical emission spectroscopy (OES) for multi-pin-to-plate-type dielectric barrier discharge (DBD) and a conventional DBD. (Figure reprinted from Kyung, S. J., Maksym, V., and Yeom, G. Y. et al. 2007. Growth of carbon nanotubes by atmospheric pressure plasma enhanced chemical vapor deposition using NiCr catalyst. *Surf. Coat. Tech.* 201: 5378–5382. With permission.)

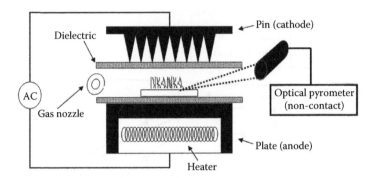

FIGURE 7.12
A pin-to-plate-type atmospheric pressure plasma-enhanced chemical vapor deposition (AP-PECVD) system for carbon nanotube (CNT) synthesis. (Figure reprinted from Kyung, S. J., Voronko, M., and Yeom, G. Y. 2005. Growth and field emission properties of multiwalled carbon nanotubes synthesized by pin-to-plate type atmospheric pressure plasma enhanced chemical vapor deposition. *J. Korean Phys. Soc.* 47(5): 824–827. With permission.)

shown in Figure 7.13 (SEM images) and Figure 7.14 (TEM images). As the substrate temperature is increased from 400°C to 500°C, the length of the CNTs increased from 1 ~ 3 μm while the CNT diameter decreased from 80 to 40 nm. In comparison to CNTs synthesized by a conventional DBD system at 600°C [70], the CNTs grown at 500°C by the modified DBD system showed a higher-quality structure despite the lower growth temperature. Another study on the growth of CNTs using a similar multi-pin-to-plate-modified DBD showed that the growth temperature can be decreased even further to 450°C, which is significantly lower than the temperature of 600°C required for a conventional DBD system (Kyung et al. 2005). The higher quality of CNTs and lower growth temperature obtained with multi-pin-to-plate-modified DBDs is related to the discharge efficiency. The modified DBD is characterized by a higher plasma density and more efficient gas dissociation at the same input voltage. The higher plasma density may lead to enhanced ion bombardment on the catalyst surface which can remove defects in the CNTs, while gas dissociation may allow more reactive carbon to reach the catalyst surface and increase the CNT growth rate.

7.3.1.1.3 CNT Synthesis Using Capillary-Type DBD Systems

A modified DBD source composed of capillary-type electrodes has also been investigated. In this source, a number of parallel holes were formed in the dielectric barrier layer to increase the plasma density [74] and to distribute the feed gas more uniformly [82]. Figure 7.15 shows a schematic diagram of the capillary-type modified DBD used to grow CNTs [74]. The perforated dielectric barrier layer on the top electrode charges during discharge operation and limits current flow to the electrode similar to the dielectric barrier layer in conventional DBD sources. By using perforated holes on the

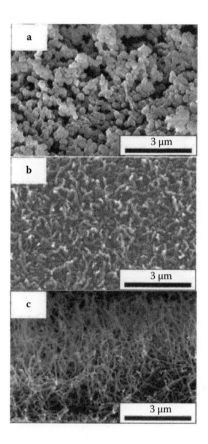

FIGURE 7.13
Scanning electron micrograph (SEM) image of carbon nanotubes (CNTs) grown with He(10 slm)/C₂H₂(210 sccm) /NH₃(270 sccm) on NiCr(10 nm)/Cr(100 nm)/glass substrate for 3 minutes at different substrate temperatures, after the He(10 slm)/NH₃(150 sccm) plasma pretreatment at 500°C for 3 min: (a) 400°C, (b) 450°C, and (c) 500°C. (Figure reprinted from Kyung, S. J., Voronko, M., and Yeom, G. Y. 2005. Growth and field emission properties of multiwalled carbon nanotubes synthesized by pin-to-plate type atmospheric pressure plasma enhanced chemical vapor deposition. *J. Korean Phys. Soc.* 47(5): 824–827. With permission.)

dielectric barrier surface, the feed gas can be more uniformly distributed. In particular, high aspect ratio holes in the dielectric barrier layer permit electrons to be accelerated in a large electric field, and a high-density plasma is obtained [82].

Figure 7.16 shows a SEM image of CNTs synthesized by the capillary-type modified DBD system at 400°C. Prior to the growth of CNTs, the glass substrate was sputter deposited with a Ni/Cr catalyst layer and exposed to the capillary-type modified DBD at 400°C using He (6 slm)/NH₃ gas (90 sccm). After the pretreatment, CNTs were synthesized at 400°C by flowing a gas mixture of C₂H₂/He/N₂ through the capillary holes. As shown in Figure 7.16, multiwalled CNTs could be grown even at a temperature of 400°C, possibly

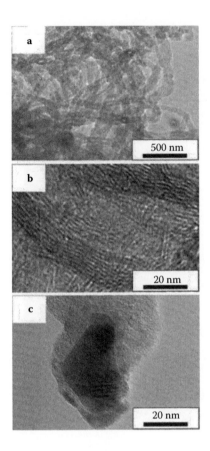

FIGURE 7.14
Transmission electron micrograph (TEM) images of carbon nanotubes (CNTs) synthesized and grown at 500°C by using a modified dielectric barrier discharge (DBD) system. (a) Low magnification TEM image of CNTs, (b) high-resolution TEM (HRTEM) image of the CNT body, and (c) the tip of CNT. (Figure reprinted from Kyung, S. J., Voronko, M., and Yeom, G. Y. 2005. Growth and field emission properties of multiwalled carbon nanotubes synthesized by pin-to-plate type atmospheric pressure plasma enhanced chemical vapor deposition. *J. Korean Phys. Soc.* 47(5): 824–827. With permission.)

due to the high plasma density obtained by the capillary-type DBD source. Therefore, by using modified DBD sources such as multi-pin-to-plate-type DBDs and capillary-type DBDs, the minimum growth temperature for CNTs can be decreased by about 100 ~ 150°C while also obtaining higher-quality CNTs in terms of their structural properties.

7.3.1.1.4 Vertically Aligned CNTs Using DBD Systems

As mentioned in the previous sections, DBD systems have an electrode configuration consisting of at least one dielectric barrier layer inserted between two electrodes which is similar to that of a low-pressure CCP system. Therefore, when AC or RF voltage is applied, a large sheath voltage is developed on

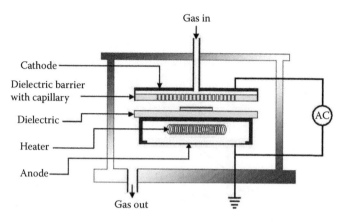

FIGURE 7.15

A capillary dielectric discharge-type atmospheric pressure plasma-enhanced chemical vapor deposition (AP-PECVD) system. (Figure reprinted from Kyung, S. J., Lee, Y. H., and Yeom, G. Y. et al. 2006. Deposition of carbon nanotubes by capillary-type atmospheric pressure PECVD. *Thin Solid Film* 506–507: 268–273. With permission.)

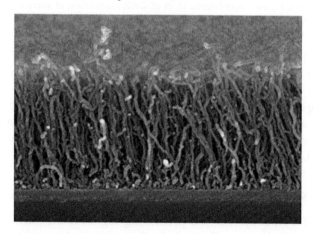

FIGURE 7.16

Scanning electron micrograph (SEM) image of carbon nanotubes (CNTs) grown on Ni (5 nm)/ Cr (100 nm)/Si substrates after He/NH$_3$ plasma pretreatment at 400°C for 5 min with a He/ NH$_3$/C$_2$H$_2$ plasma. (Figure reprinted from Kyung, S. J., Lee, Y. H., and Yeom, G. Y. et al. 2006. Deposition of carbon nanotubes by capillary-type atmospheric pressure PECVD. *Thin Solid Film* 506–507: 268–273. With permission.)

each electrode during each voltage cycle and, due to the short mean free path compared to a low-pressure system, the sheath distance is less than 10 ~ 100 μms [132–134] at atmospheric pressure. When a few kV of AC or RF voltage is applied between the electrodes, a very high voltage can be induced in the very short sheath distance, perpendicular to the substrate, allowing vertically aligned CNTs to be synthesized. Nozaki et al. [83] used a DBD system shown in Figure 7.17 to grow CNTs. As shown in Figure 7.18, vertically aligned CNTs

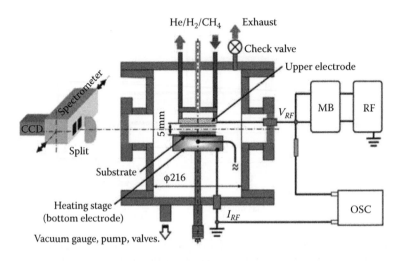

FIGURE 7.17
Atmospheric pressure radio frequency discharge reactor system. (Figure reprinted from Nozaki, T., Ohnishi, K., and Kortshagen, U. et al. 2007. Fabrication of vertically aligned single-walled carbon nanotubes in atmospheric pressure non-thermal plasma CVD. *Carbon* 45: 364–374. With permission.)

could be easily observed by SEM, and the directional growth of CNTs was related to the strong electric field formed by the sheath perpendicular to the substrate.

DC biasing can further improve the aspect ratio of vertically aligned CNTs. Kyung et al. applied a DC voltage on the substrate electrode in a capillary-type modified DBD [74]. By applying a DC bias voltage of –1.2 kV, as shown in Figure 7.19, highly aligned vertical CNTs could be obtained.

7.3.1.2 CNT Synthesis Using Atmospheric Pressure Plasma Jets and Torches

In the case of a DBD system, the processing region is the same as the discharge region. In the case of atmospheric pressure plasma jets (APPJs) or atmospheric plasma torches, the processing region and the discharge region are separated. Reactive species dissociated in the discharge region are transported to the substrate by a strong gas flow. In general, APPJs consist of a glow discharge or filamentary discharge and the gas temperature is generally low. In contrast, plasma torches are arc like and the gas temperature is generally high. CNT growth studies have been performed with both of these types of discharges [84–87].

CNT growth has been carried out by a plasma torch operated at microwave frequency. The high frequency and high electric field formed in the source region resulted in a very high plasma density and efficient gas dissociation. Figure 7.20 shows the schematic diagram of the plasma torch system used to synthesize the CNTs [85]. The microwave power was fixed at 1 kW and

FIGURE 7.18
Scanning electron micrograph (SEM) image of vertically aligned CNTs synthesized at 700°C and 30 min annealing (pretreatment) conditions. (Figure reprinted from Nozaki, T., Ohnishi, K., and Kortshagen, U. et al. 2007. Fabrication of vertically aligned single-walled carbon nanotubes in atmospheric pressure non-thermal plasma CVD. *Carbon* 45: 364–374. With permission.)

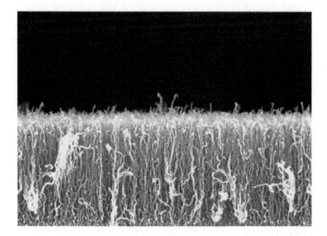

FIGURE 7.19
Scanning electron micrograph (SEM) image of vertically aligned carbon nanotubes (CNTs) grown on Ni (5 nm)/Cr (100 nm)/Si substrates after the He/NH$_3$ plasma pretreatment at 400°C for 5 min with a He/NH$_3$/C$_2$H$_2$ plasma by biasing the substrate with the direct current (DC) voltage of –1.2 kV. (Figure reprinted from Kyung, S. J., Lee, Y. H., and Yeom, G. Y. et al. 2006. Deposition of carbon nanotubes by capillary-type atmospheric pressure PECVD. *Thin Solid Film* 506–507: 268–273. With permission.)

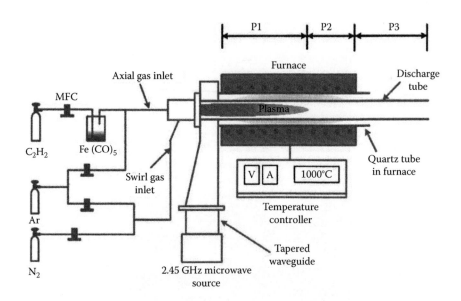

FIGURE 7.20

The microwave discharge atmospheric-pressure plasma (APP) torch system used for carbon nanotube (CNT) synthesis. (Figure reprinted from Hong, Y. C., and Uhm, H. S. 2005. Production of carbon nanotubes by microwave plasma torch at atmospheric pressure. *Phys. Plasmas* 12: 0535041-6. With permission.)

C_2H_2 gas was used as the carbon source gas. Bubbling the C_2H_2 gas through iron pentacarbonyl liquid ($Fe(CO)_5$) introduced the catalyst precursor in the plasma jet. In addition, Ar and N_2 were added to the gas flow to aid dissociation of the carbon source. The coating on the inside wall of the quartz tube was peeled off using a brush in order to collect the CNTs. SEM images of synthesized CNTs are shown in Figure 7.21 for growth temperatures ranging from 700°C to 1000°C. As the growth temperature is increased, the CNT density increases and the CNTs change from a bent shape to a more aligned and straight shape structure. The average diameter and growth rate of CNTs as a function of growth temperature are shown in Figure 7.22. The average diameter of CNTs increased from 47.78 to 111.9 nm and the growth rate increased from 0.74 to 2.5 μm/min as the growth temperature increased from 700°C to 1000°C, suggesting that carbon diffusion through the catalyst particles was enhanced with increasing growth temperature.

Jašek et al. [87] also used a microwave plasma torch system for CNT synthesis. In their experiments, Ar (1000 sccm)/H_2 (300 sccm)/CH_4 (50 sccm) gas mixtures were used to generate the plasma and the substrate was loaded near the tip of the plasma torch. Thin films of Fe were deposited on a SiO_2 on Si substrate as the catalyst material (Si/SiO_2/Fe (10 nm)). Prior to the growth of CNTs, the substrate was pretreated by annealing in a H_2 or NH_3 environment at 970 K. Figure 7.23 shows a SEM image of CNTs grown on Si/SiO_2/Fe (10 nm) using a Ar(1000 sccm)/H_2(300 sccm)/CH_4(50 sccm) plasma

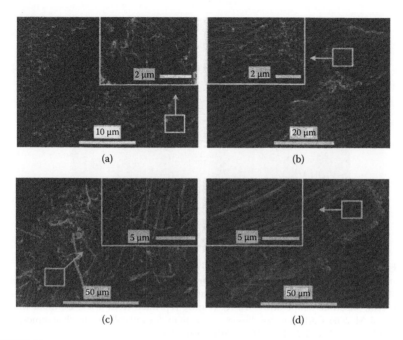

(a) (b)

(c) (d)

FIGURE 7.21

Scanning electron micrograph (SEM) images of carbon nanotubes (CNTs) obtained from an atmospheric microwave plasma torch. All the CNTs were synthesized at a mixture of 12.5 lpm argon and 2.5 lpm nitrogen as a swirl gas with 4.5×10^{-2} C_2H_2 gas. Images are obtained for CNTs grown at a furnace temperature of (a) 700°C, (b) 800°C, (c) 900°C, and (d) 1000°C. (Figure reprinted from Hong, Y. C., and Uhm, H. S. 2005. Production of carbon nanotubes by microwave plasma torch at atmospheric pressure. *Phys. Plasmas* 12: 0535041-6. With permission.)

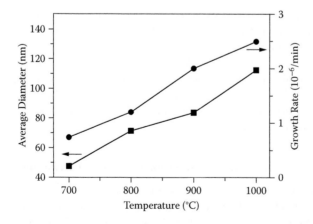

FIGURE 7.22

Average diameter and growth rate of the carbon nanotubes (CNTs), corresponding to Figure 7.21. (Figure reprinted from Hong, Y. C., and Uhm, H. S. 2005. Production of carbon nanotubes by microwave plasma torch at atmospheric pressure. *Phys. Plasmas* 12: 0535041-6. With permission.)

FIGURE 7.23
Scanning electron micrograph (SEM) of carbon nanotubes (CNTs) deposited on $Si/SiO_2/$ Fe substrate by a microwave plasma torch system (gas flow of Ar 1000 sccm, H_2 300 sccm, CH_4 50 sccm, substrate temperature 970 K). (Figure reprinted from Jašek, O., Eliáš, M., and Kadlečíková, M. 2006. Carbon nanotubes synthesis in microwave plasma torch at atmospheric pressure. *Mater. Sci. Eng. C* 26: 1189–1193. With permission.)

torch at 970°K. Highly aligned CNTs could be observed with diameters of 10 ~ 20 nm and lengths of approximately 50 μm. Thus, CNTs with very high aspect ratios could be obtained. In addition, the CNT density was also high in the temperature range of 870 ~ 1020 K. The growth of dense, vertical CNTs could be related not only to the high growth temperature but also to the high dissociation rate of the carbon source by the microwave plasma. The vertical alignment of the nanotubes is also caused by the highly dense growth. When CNTs are grown close to each other, the repulsive force between growing CNTs caused by van der Waals interactions results in the vertical growth of CNT bundles, known as the "crowding effect." Figure 7.24 shows TEM images of the CNTs grown by a microwave plasma torch. A small number of defects on the CNT sidewall structure can be observed. The TEM image also shows that the CNT diameter is similar to the metal catalyst particle size, indicating the importance of pretreatment of the metal catalyst to yield thin nanotubes with a high aspect ratio.

7.3.1.3 CNT Synthesis Using Microplasma or Corona Discharges

A few researchers have reported the synthesis of CNTs using a microplasma [88] or corona discharge [89]. A schematic diagram of the microplasma system studied by Chiang et al. [88] is shown in Figure 7.25. Using a plasma reactor composed of a metal capillary tube electrode and a counterelectrode mesh, and forming an atmospheric pressure discharge in Ar gas and metal-locene vapor (ferrocene ($Fe(C_5H_5)_2$ and/or nickelocene ($Ni(C_5H_5)_2$), bimetallic

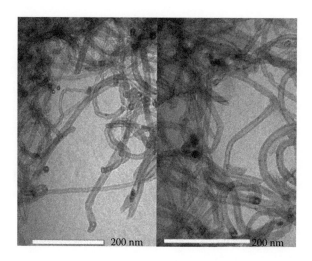

FIGURE 7.24
TEM images of CNTs grown by a microwave plasma torch.

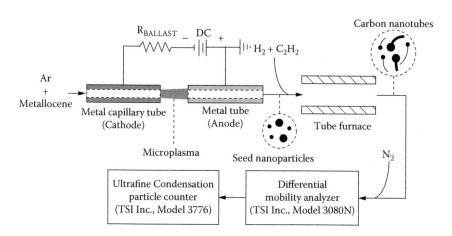

FIGURE 7.25
Microplasma reactor and tubular flow furnace used for catalytic, gas-phase growth of carbon nanotubes. (Figure reprinted from Lia, M. W., Hu, Z., and Tian, Y. L. et al. 2004. Low-temperature synthesis of carbon nanotubes using corona discharge plasma at atmospheric pressure. *Diamond Rel. Mater.* 13: 111–115. With permission.)

catalyst particles of less than 3 nm were generated. These catalyst particles were transferred to a tube furnace by the gas flow and CNTs were synthesized by the reaction of the catalyst particles with C_2H_2/H_2 gas. Figure 7.26 shows TEM images of the CNTs synthesized by the microplasma system and, as shown, 5 ~ 10 nm thick multiwall CNTs grown on a few nanometer-sized catalyst particles could be observed.

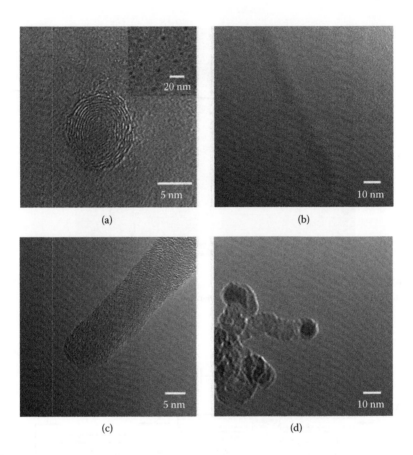

FIGURE 7.26
High-resolution transmission electron microscopy (HRTEM) images of carbon nanostructures grown in a tube furnace on (a) Fe particles at 500°C, (b) Fe particles at 600°C, (c) Ni particles at 500°C, and (d) Ni particles at 600°C. Experimental conditions: 0.5 sccm C_2H_2 + 50 sccm H_2. (Figure reprinted from Lia, M. W., Hu, Z., and Tian, Y. L. et al. 2004. Low-temperature synthesis of carbon nanotubes using corona discharge plasma at atmospheric pressure. *Diamond Rel. Mater.* 13: 111–115. With permission.)

Lia et al. [89] studied the synthesis of CNTs using a corona discharge. In general, a voltage higher than a few kV is required to ignite a corona discharge, resulting in an arc-type or a filamentary-type micro-discharge at the electrode tip. The arc-like properties lead to an electron temperature as high as 5 eV and facile dissociation of gas atoms/molecules that are known to be beneficial for CNT synthesis [90,91]. Figure 7.27 shows the schematic diagram for a corona discharge system used to synthesize CNTs. To synthesize well-ordered carbon nanotubes, an anodic aluminum oxide (AAO) template was used to mask the substrate. The channels in the AAO template were coated with a thin cobalt catalyst layer by electrodeposition. A CH_4/H_2 gas (1:10 ratio, total 22 sccm) mixture was used as the feed for CNT growth. Figures 7.28

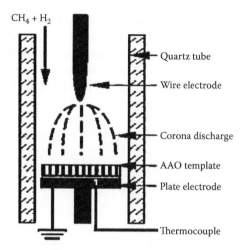

FIGURE 7.27
A corona discharge reactor used for carbon nanotube (CNT) growth with an anodized aluminum oxide (AAO) template. (Figure reprinted from Lia, M. W., Hu, Z., and Tian, Y. L. et al. 2004. Low-temperature synthesis of carbon nanotubes using corona discharge plasma at atmospheric pressure. *Diamond Rel. Mater.* 13: 111–115. With permission.)

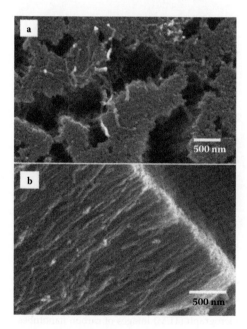

FIGURE 7.28
Scanning electron micrograph (SEM) images of carbon nanotubes (CNTs) grown by the corona discharge in Figure 7.27. (a) A top view: note that some carbon nanotubes appeared at the surface. (b) A cross-sectional view. (Figure reprinted from Lia, M. W., Hu, Z., and Tian, Y. L. et al. 2004. Low-temperature synthesis of carbon nanotubes using corona discharge plasma at atmospheric pressure. *Diamond Rel. Mater.* 13: 111–115. With permission.)

and 7.29 show SEM and TEM images, respectively, of CNTs grown by the corona discharge. CNTs were vertically aligned by the AAO template. The length and diameter of CNTs were 1 μm and 40 nm, respectively, confirming the presence of multiwalled CNTs. Defects on the sidewalls of the nanotubes were identified by Raman spectroscopy. Interestingly, CNTs could be grown

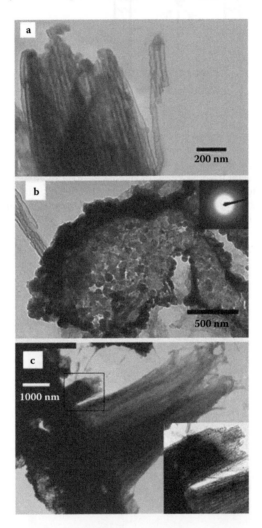

FIGURE 7.29
Transmission electron micrograph (TEM) images of carbon nanotubes (CNTs) grown by the corona discharge in Figure 7.27 after removing the anodic aluminum oxide (AAO) template. (a) A bundle of carbon nanotubes, (b) amorphous carbon, and (c) an ensemble of carbon nanotubes, the inset is an enlarged image from the rectangle labeled. (Figure reprinted from Lia, M. W., Hu, Z., and Tian, Y. L. et al. 2004. Low-temperature synthesis of carbon nanotubes using corona discharge plasma at atmospheric pressure. *Diamond Rel. Mater.* 13: 111–115. With permission.)

in a corona discharge at temperatures lower than 200°C. The high average electron temperature of 5 eV may allow methane gas to be dissociated (dissociation of hydrogen from methane requires 4.48 eV) in the gas phase and produce radicals that can react with and etch the cobalt catalyst to nucleate nanotubes at low temperatures. However, the precise reason for CNT growth at such low temperatures requires further investigation.

7.3.2 The Effect of Operating Conditions on CNT Synthesis

7.3.2.1 Raman Analysis of As-Grown CNTs

One of the analytical tools available for characterizing the structure of CNTs is Raman spectroscopy. Two distinctive peaks are observed in Raman spectroscopic data including the D-band peak at 1340 ~ 1345 cm^{-1} that is related to the defect-induced dispersive peak and the G-band peak at 1574 ~ 1580 cm^{-1} that is related to the crystalline hexagonal graphite structure. In addition, other peaks such as the 2D band (2691 cm^{-1}), E1g + D (2918 cm^{-1}), 2E2g (3224 cm^{-1}), and 2D + E2g (4263 cm^{-1}) are also often observed (the exact peak locations are slightly different depending on the growth condition, growth equipment, etc.) [92]. Figure 7.30 shows a typical Raman spectrum for multiwalled CNTs synthesized by a DBD system at

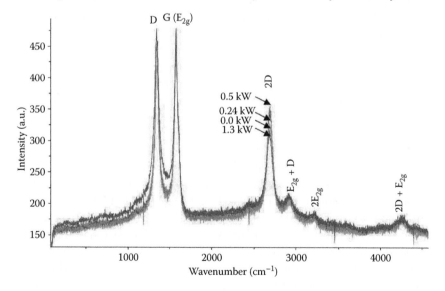

FIGURE 7.30

Raman spectra of multiwalled carbon nanotubes (MWCNTs) and MWCNTs treated in an air-dielectric barrier discharge (DBD) atmosphere. (Figure reprinted from Okpalugo, T. I. T., Papakonstantinou, P., and Brown, N. M. D. 2005. Oxidative functionalisation of carbon nanotubes in atmospheric pressure dielectric barrier discharge (APDBD). *Carbon* 43: 2951–2959. With permission.)

atmospheric pressure. In addition to the D-band and G-band peaks, other peaks related to graphite bonding are also observed. Among the various peaks, most researchers are interested in the ratio of the intensity of the D-band peak to the G-band peak (I_D/I_G) which is a measure of the degree of defects in the synthesized CNTs.

7.3.2.1.1 *Effect of Substrate Temperature on I_D/I_G Ratio*

In general, the energy supplied to the substrate through substrate heating or plasma power not only affects the diffusion rate of carbon on the catalyst surface but also affects the degree of defects in the as-grown CNTs. The effect of substrate temperature on the I_D/I_G ratio in CNTs grown by a capillary-type modified DBD system is shown in Figures 7.31a and 7.31b [74,93]. A gas mixture composed of $C_2H_2/NH_3/N_2/He$ was used as the carbon feedstock and NiCr was used as the catalyst. Increasing the substrate temperature resulted in a decrease in the ratio of I_D/I_G indicating that the defect density was decreased. The same effect of substrate temperature on defects in as-grown CNTs was also observed by Hong et al. [85] and Nozaki et al. [83]

7.3.2.1.2 *Effect of Substrate Biasing on I_D/I_G Ratio*

Figure 7.32 shows a Raman spectrum for CNTs grown using a capillary type modified DBD system with and without DC biasing of the substrate at –1.2 kV [74]. The DC biasing is expected to increase the plasma density near the substrate and enhance carbon diffusion on the metal catalyst or increase gas dissociation. In addition, the sheath voltage at the substrate surface is increased, enhancing ion bombardment. Although the specific role of ion bombardment on CNT growth is not well understood, ions could increase the substrate temperature or increase the mobility of carbon, both of which should reduce defects in the CNTs.

7.3.2.1.3 *Gas Mixture Effect on I_D/I_G Ratio*

The effect of gas composition on defects in as-grown CNTs using a DBD system at 600°C was investigated by Nozaki et al. [83]. Figure 7.33 shows that increasing the ratio of H_2 to CH_4 increased the peak intensities of both the D-band and the G-band peak, indicating an increase in the CNT growth rate. In addition, the G-band peak intensity was found to increase with respect to the D-band peak intensity as the ratio of $H_2:CH_4$ was increased. As previously reported by Lia et al. [89], the increase in the CNT growth rate is believed to be related to an increase in the density of activated hydrogen atoms in the discharge which can prevent recombination of carbon into hydrocarbon radicals. The defect density in the as-grown tubes may be reduced by the hydrogen atoms as well, which promotes carbon diffusion on the metal catalyst. Atomic hydrogen may also reduce the oxidized catalyst at a low temperature or react and etch the nanotubes during growth.

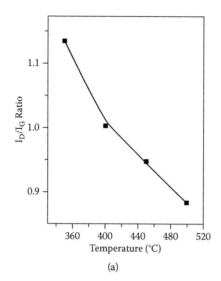

(a)

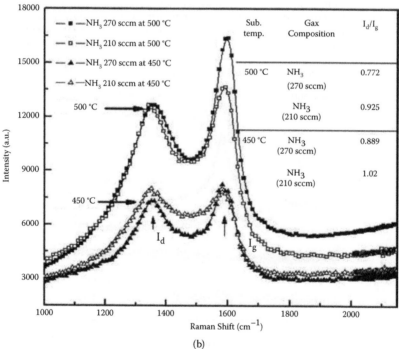

(b)

FIGURE 7.31

(a) I_D/I_G ratio from Fourier transform (FT)-Raman spectra of carbon nanotubes (CNTs). [1v] (S. J. Kyung et al. 2007) and (b) FT-Raman spectra of CNTs grown at 450 and 500°C and with the NH_3 gas rate of 210 and 270 sccm in He(10 slm)/C_2H_2(210 sccm)/NH_3 for 3 min. (Figure reprinted from Lee, Y. H., Kyung, S. J., and Yeom, G. Y. 2006. Characteristic of carbon nanotubes synthesized by pin-to-plate type atmospheric pressure plasma enhanced chemical vapor deposition at low temperature. *Carbon* 44: 807–823. With permission.)

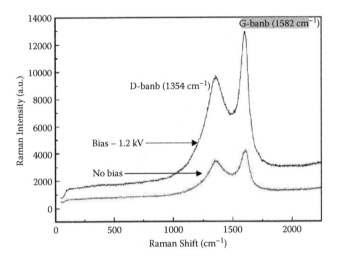

FIGURE 7.32
Raman spectra of carbon nanotubes (CNTs) grown with and without biasing the substrate during growth in a capillary-type modified dielectric barrier discharge (DBD) system. (Figure reprinted from Kyung, S. J., Lee, Y. H., and Yeom, G. Y. et al. 2006. Deposition of carbon nanotubes by capillary-type atmospheric pressure PECVD. *Thin Solid Film* 506–507: 268–273. With permission.)

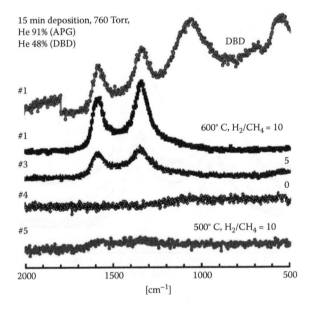

FIGURE 7.33
Raman spectroscopic data measured as a function of gas composition ratio of H_2/CH_4: (1) dielectric barrier discharge (DBD) at 600°C and $H_2/CH_4 = 10$, (2) ~ (4) atmospheric pressure glow discharge (APG) at different H_2/CH_4 ratios, and (5) APG at 500°C. (Figure reprinted from Nozaki, T., Ohnishi, K., and Kortshagen, U. et al. 2007. Fabrication of vertically aligned single-walled carbon nanotubes in atmospheric pressure non-thermal plasma CVD. *Carbon* 45: 364–374. With permission.)

7.3.2.1.4 Summary of the Effect of Various Process Conditions on I_D/I_G Ratio

The ratio of I_D/I_G obtained for different discharge systems, gas compositions, substrate temperatures, and different catalysts are summarized in Table 7.6. The data were compared with those obtained by thermal CVD. The I_D/I_G ratio is clearly affected by the various process parameters and by controlling these parameters carefully, high-quality CNTs similar to thermal CVD can be obtained using atmospheric pressure plasmas.

7.3.2.2 Effect of Catalyst Preparation on CNT Synthesis

The metal catalyst is an important aspect of CNT growth because the size of the catalyst controls the CNT diameter. Various methods have been investigated to control the size of the catalyst including pretreatment of a metal catalyst film, varying the catalyst film thickness, depositing nanoparticles directly on a substrate, or dissociating a catalyst precursor vapor to homogenously nucleate nanoparticles in the gas phase.

7.3.2.2.1 Effect of Pretreatment on the CNT Synthesis

To obtain nano-sized catalyst particles for CNT growth, a thin film of metal is often deposited on a substrate that needs to be cracked. In the case of thermal CVD of CNTs, relatively high process temperatures of 800 ~ 1000°C are used and the metal film is easily broken up into nanoparticles without any pretreatment steps. On the other hand, PECVD of CNTs at low pressures requires pretreatment of the metal films because low temperatures are used during growth. This is also true for CNT growth with APPs because the growth temperature is generally in the range of 400 ~ 600°C, similar to that of low-pressure PECVD.

The effect of pretreatment on metal films has been investigated (Figure 7.34) [74]. Thin films of Ni/Cr (5 nm/100 nm) were sputter deposited on soda lime glass and pretreated for 5 min using various conditions. The surface morphology was then characterized by SEM for the following conditions (a) as is, (b) 400°C heat, (c) He/NH$_3$ APP, and (d) He/NH$_3$ APP + 400°C heat [APP condition: NH$_3$ (90 sccm)/He (6000 sccm)]. Heating the film to 400°C resulted in an average surface roughness measured by atomic force microscope (AFM) of 2450 nm as compared to the original film roughness of 1498 nm. When the catalyst was pretreated with a He/NH$_3$ atmospheric pressure plasma in combination with heating at 400°C, the surface morphology showed smaller and more uniform particles with an average surface roughness measured by AFM of 71.4 nm. Therefore, more effective catalyst preparation was achieved by pretreatment using an APP and heating than heating alone. The decrease in the apparent metal catalyst particle size may be related to increased surface diffusion of Ni due to surface bombardment by ions or excited atoms and the removal of oxygen on the catalyst surface by hydrogen atoms from NH$_3$ dissociation. Thus, dense and high aspect

TABLE 7.6

Ratios of I_D/I_G Obtained for Different Discharge Systems, Gas Compositions, Substrate Temperatures, and Different Catalysts

Reference	Temperature (°C)	Gas Composition	Catalyst	I_D/I_G Ratio	Atmospheric-Pressure Plasma (APP) Source
70	600	CH_4/H_2	20 nm Ni film	1.0	Dielectric barrier discharge (DBD)
72	700	C_2H_4/H_2	10–40 nm Ni film	1.2	Thermal chemical vapor deposition (CVD)
72	800	C_2H_4/H_2	10–40 nm Ni film	0.6	Thermal CVD
85	700	$C_2H_2/Ar/N_2$	Iron pentacarbonyl	1.12	Microwave plasma torch
85	1000	$C_2H_2/Ar/N_2$	Iron pentacarbonyl	0.45	Microwave plasma torch
73	500	$C_2H_2/He/NH_3$	10 nm Ni/Cr film	0.772	Pin to plate type DBD
131	600	$CH_4/H_2/He$	Ni	1.5	Microwave discharge
20	350	$C_2H_2/He/NH_3$	10 nm Ni/Cr film	1.135	Capillary type DBD
20	500	$C_2H_2/He/NH_3$	10 nm Ni/Cr film	0.882	Capillary type DBD

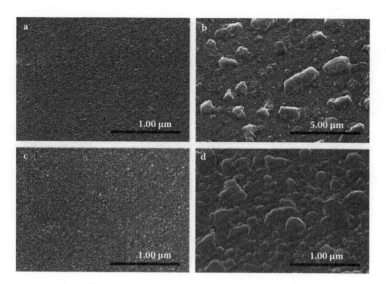

FIGURE 7.34
Effect of various pretreatments on the surface morphology of the Ni catalyst observed by a field emission scanning electron microscope (FE-SEM). (a) The sputtered Ni surface (b) after heating at 400°C, (c) after a He/NH$_3$ atmospheric pressure plasma treatment at room temperature, and (d) after the plasma treatment while heating at 400°C. (Figure reprinted from Kyung, S. J., Lee, Y. H., and Yeom, G. Y. et al. 2006. Deposition of carbon nanotubes by capillary-type atmospheric pressure PECVD. *Thin Solid Film* 506–507: 268–273. With permission.)

ratio CNTs can be grown with the appropriate pretreatment conditions (see Figures 7.17 and 7.34).

7.3.2.2.2 Effect of Ni Catalyst Layer Thickness

The effect of Ni catalyst film thickness (Ni on 100 nm thick Cr layer) on the CNT growth characteristics has been studied. Figure 7.35 shows SEM images of CNTs grown on a Ni catalyst film with thicknesses of (a) 0.5 nm, (b) 2.5 nm, (c) 5 nm, (d) 10 nm, (e) 20 nm, and (f) 40 nm using an APP generated with a NH$_3$/C$_2$H$_2$/He/N$_2$ gas mixture [20]. As shown in Figure 7.35g, initially, the CNT growth rate increased with the catalyst film thickness; however, further increase of the film thickness (>10 nm) resulted in a decrease in the CNT growth rate. In general, thinner metal films produce smaller size nanoparticles and the CNT growth rate is higher because a smaller number of carbon atoms are required to nucleate nanotubes. On the other hand, the decrease in the CNT growth rate when the catalyst layer is thinner than 10 nm may be due to surface oxidation because the catalysts are exposed to air before NH$_3$ pretreatment and CNT growth.

7.3.2.2.3 Nano-Size Catalyst Particle for SWCNTs

To synthesize single-walled CNTs (SWCNTs), nano-sized catalysts are required. Fe/Co catalysts with a diameter of 1 to 2 nm have been deposited

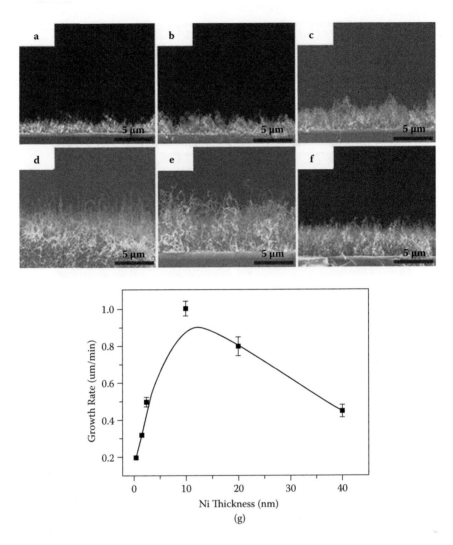

FIGURE 7.35
Scanning electron microscope (SEM) images of carbon nanotubes (CNTs) grown at different NiCr thicknesses: (a) 0.5 nm, (b) 2.5 nm, (c) 5 nm, (d) 10 nm, (e) 20 nm, and (f) 40 nm. All samples were grown at 500°C for 8 min, and the growth rates as a function of the NiCr thickness are shown in (g). (Figure reprinted from Kyung, S. J., Maksym, V., and Yeom, G. Y. et al. 2007. Growth of carbon nanotubes by atmospheric pressure plasma enhanced chemical vapor deposition using NiCr catalyst. *Surf. Coat. Tech.* 201: 5378–5382. With permission.)

on Al_2O_3/Si wafers by dip coating [94]. In addition to the small-sized catalysts, remote-like plasmas are also required to minimize or eliminate damage to the CNTs by ion bombardment. Atmospheric-pressure plasmas are well suited because the ion energy is reduced at high pressures, similar to remote-type plasmas. Using a DBD-type atmospheric pressure radio frequency discharge (APRFD) in a $He/H_2/CH_4$ gas mixture with nano-sized Fe/Co particles, Nozaki et al. [83] successfully synthesized SWCNTs.

Figure 7.36 shows TEM (a,b,c) and high-resolution TEM micrographs (d,e,f) of SWCNTs fabricated with Fe/Co nano-sized particles in a DBD. At low power (~40 W), SWCNT bundles were not observed (Figure 7.36a). At higher powers (~60 W), well-oriented SWCNT bundles, and a small number of isolated double-walled carbon nanotubes (DWCNTs) with diameters

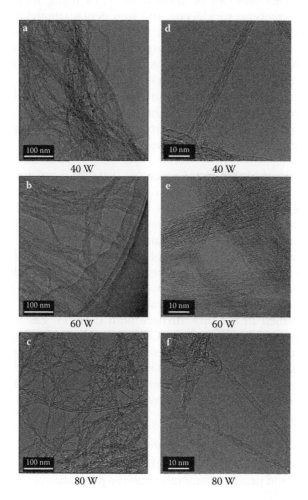

FIGURE 7.36

Transmission electron micrographs (TEMs) (a,b,c) and high-resolution TEM micrographs (d,e,f) of single-walled carbon nanotubes (SWCNTs) fabricated on Al_2O_3/Si wafer dip coated with Fe/Co nano-sized particles using a dielectric barrier discharge (DBD). (Figure reprinted from Nozaki, T., Ohnishi, K., and Kortshagen, U. et al. 2007. Fabrication of vertically aligned single-walled carbon nanotubes in atmospheric pressure non-thermal plasma CVD. *Carbon* 45: 364–374. With permission.)

of approximately 5 nm were observed (Figure 7.36b). Figure 7.36c shows SWCNT samples obtained at 80 W. The SWCNTs separate into much smaller bundles at this plasma power than those found in Figure 7.36a,b. In addition, DWCNTs and catalyst particles were easily observed. MWCNTs with diameters of 10 nm are also present in Figure 7.36c.

7.3.2.2.4 Homogeneous Catalyst Synthesis by Floating Catalyst Method

Floating metal catalysts can be formed in APPs using metal precursor vapors such as ferrocene, nicklocene, and iron pentacarbonyl as studied by Hong et al. [85] and Chiang et al. [88]. By using APPs such as microwave plasma torches or microplasmas, floating metal catalyst particles are formed by homogeneous nucleation. Microplasmas are miniaturized atmospheric pressure glow discharges that contain energetic electrons that permit nonthermal decomposition of the vapor precursors to nucleate nanoparticles. CNTs are catalyzed from the metal nanoparticles by flowing a carbon-containing gas mixture such as C_2H_2/H_2 together with the metal precursor or by transporting the metal nanoparticles by gas flow (Ar) into a heated tube furnace.

Size distributions of Ni and Fe floating catalyst particles formed by a microplasma and nanotubes grown in the tube furnace are shown in Figure 7.37 [88]. The size distributions were obtained in real time using a cylindrical differential mobility analyzer (DMA) (model 3080N, TSI, Inc.) and ultrafine condensation particle counter (CPC) (model 3776, TSI, Inc.). As shown, the geometric mean diameter of the Ni particles was 3.11 nm, while that of the Fe particles was 2.87 nm. The particle density (particles/cm^3) was higher for the Ni particles. Nanotubes were found to grow from Fe nanoparticles at 500°C, and Ni particles were catalyzed at even 400°C. TEM images of the CNTs formed in the tube furnace with C_2H_2/H_2 gas are shown in Figure 7.26. The activation energy for CNT growth on the catalyst particles was determined from aerosol measurements to be 117 kJ/mol for Fe and 73 kJ/mol for Ni, similar to results from Kim et al. [95]. The size of the particle did not significantly affect the activation energy for CNT growth. Therefore, at the same processing conditions, Ni catalysts were found to have a lower activation energy than Fe catalysts.

7.3.2.2.5 Summary of the Effect of Catalysts on CNT Synthesis

The effect of catalyst preparation methods, type of atmospheric pressure plasma sources, and growth temperature on the properties of CNTs are summarized in Table 7.7.

7.3.2.3 Posttreatment of CNTs by Using Atmospheric Pressure Plasmas

Atmospheric pressure plasmas can also be useful in surface treatment of CNTs to clean the nanotube surface, modify or functionalize the CNT surface, and so forth [96]. Han et al. used a DBD-type atmospheric pressure

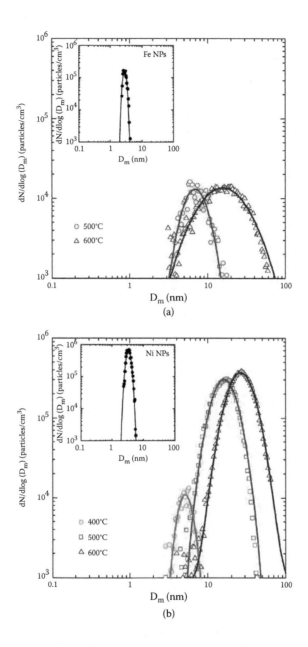

FIGURE 7.37
Temperature-dependent aerosol -size distributions of carbon nanostructures grown on (a) Fe and (b) Ni catalyst particles. Log-normal distribution fits are shown. The insets show the size distributions of the nanoparticles (NPs) synthesized in the microplasma reactor (experimental conditions: 10 sccm ferrocene + 90 sccm Ar for Fe NPs and 20 sccm nickelocene + 80 sccm Ar for Ni NPs). (Figure reprinted from Chiang, W. H., and Mohan Sankaran, R. 2007. Microplasma synthesis of metal nanoparticles for gas-phase studies of catalyzed carbon nanotube growth. *Appl. Phys. Lett.* 91: 1215031-3. With permission.)

TABLE 7.7

Effects of Catalyst Preparation Methods in Addition to the Atmospheric Pressure Plasma Sources and Temperature on the Carbon Nanotube (CNT) Characteristics

Reference	Synthesis Equipment	Catalyst (Preparation Method)	Treatment Process for Catalyst	Diameter of CNTs	Temperature (°C)
85	Microwave plasma torch	Iron pentacarbonyl (C_2H_2 gas bubbling)	—	47.78 nm	700
85	Microwave plasma torch	Iron pentacarbonyl (C_2H_2 gas bubbling)	—	111.9 nm	1000
83	Atmospheric pressure radio frequency discharge	Fe/Co 1 ~ 2 nm (dip-coating)	700°C in He/H_2 atmosphere	1 ~ 1.5 nm (single-walled carbon nanotubes)	700
14	Capillary type modified dielectric barrier discharge (DBD)	Ni 5 nm/Cr 100 nm (sputtering)	NH_3/He atmospheric discharge + 400°C heating	20 ~ 50 nm	400
131	Microwave-excited atmospheric-pressure plasma	Ni catalyst	—	20 nm	600
89	Atmospheric microplasma	Ferrocene (gas phase)	—	10 nm	600
89	Atmospheric microplasma	Nickelocene (gas phase)	—	10 nm	600
88	Corona discharge	Cobalt (electrodeposit)	—	40 nm	<200
70	DBD	Ni 20 nm (metal plating)	600°C in hydrogen atmosphere	40–50 nm	600
93	Pin-to-plate-type modified DBD	NiCr (10 nm)/Cr (100 nm)	NH_3/He atmospheric discharge + 450°C heating	40–45 nm	500
87	Microwave plasma torch	Fe 10 nm	—	10–20 nm	697

plasma to treat MWCNTs synthesized by a CVD system and increase the hydrogen storage efficiency. Due to the large surface area and the hollow structure of CNTs, MWCNTs are suitable for hydrogen gas storage. Han et al. synthesized MWCNTs on SiO_2 substrates via thermal CVD by introducing ferrocene ($Fe(C_5H_5)_2$) and xylene (C_8H_{10}) vapor mixed with Ar and H_2 at 800°C. The synthesized MWCNTs were high quality with a closed cap structure. However, for hydrogen storage, it would be better for CNTs to have an open cap and a defective structure almost like a nanoporous structure. For this reason, the MWCNTs were posttreated by an APP to remove the cap and induce defects. Nanotubes were exposed for 15 min to an atmospheric pressure plasma (2% O_2 in He) operated with AC voltage (3 kV peak-to-peak). Figures 7.38a through 7.38c show SEM and TEM images of the MWCNTs before the posttreatment. Figures 7.38d through 7.38f show SEM and TEM images of MWCNTs after the posttreatment. As shown in Figures 7.38a and 7.38b, the CNTs were multiwalled with lengths of 200 to 400 μm and diameters of 20 to 50 nm. From the high-resolution TEM image in Figure 7.38c, the MWCNTs were found to have a closed cap before treatment. In comparison,

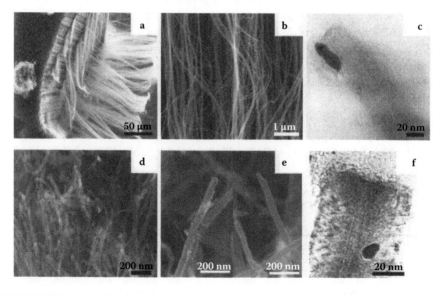

FIGURE 7.38

(a) and (b) Scanning electron micrograph (SEM) images of multiwalled carbon nanotubes (MWCNTs) prepared by a floating catalyst method using ferrocene/xylene. (c) Transmission electron micrograph (TEM) image of the MWCNT tip before an atmospheric-pressure plasma treatment; the nanohole is closed with catalysts and graphite layers. (d) and (e) SEM images of MWCNTs after plasma treatment, and (f) TEM image of the MWCNT tip after plasma treatment; the tip part and the nanotube wall are damaged by plasma. (Figure reprinted from Han, K. S., Kim, H. S., and Kang, J. K. et al. 2005. Atmospheric-pressure plasma treatment to modify hydrogen storage properties of multiwalled carbon nanotubes. *Appl. Phys. Lett.* 86: 2631051-3. With permission.)

after posttreatment by an atmospheric pressure plasma, the MWCNTs exhibited an open cap structure (Figures 7.38d through 7.38f). In addition, Raman spectroscopy showed an increase in the I_D/I_G ratio after posttreatment indicating damage to the nanotubes by the plasma treatment. Hydrogen storage characteristics of the nanotubes were compared before and after treatment by collecting hydrogen desorption spectra as a function of temperature. CNTs were first exposed to 99.999% hydrogen gas at 60 atm and 300 K for 12 hours. Then, the CNTs were heated from 80 K to 723 K at a rate of 3 K/min. As shown in Figure 7.39a, the non-plasma-treated MWCNTs showed a hydrogen discharge of about 4.9 wt% at 100 ~ 150 K, whereas as shown in Figure 7.39b, the plasma-treated MWCNTs showed a hydrogen discharge of about 5.1 wt% at 100 ~ 150 K in addition to 0.6 wt% at 300 K. Therefore, the defect sites introduced on the nanotube walls in addition to the opening of the cap structure by the postplasma treatment resulted in improved hydrogen storage capacity.

Posttreatment of screen-printed CNTs by atmospheric pressure plasmas has been investigated by Kyung et al. [97]. Screen-printed CNTs are used as the electron emission material in field emission displays, and to activate the tip of the screen printed CNTs, a tape activation method is used [98]. Surface treatment by an APP can improve the characteristics of the screen-printed CNTs after tape activation. A multi-pin-to-plate-type modified DBD was used to treat screen-printed CNTs in He/N$_2$ gas for 30 s. Figures 7.40a and 7.40b show cross-sectional SEM images of screen-printed CNTs before and after posttreatment by the plasma, respectively. As shown in Figure 7.40a,

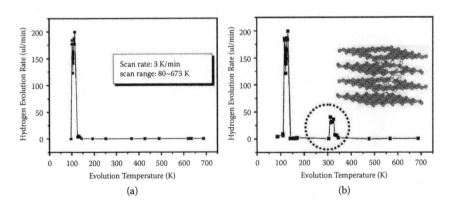

FIGURE 7.39
Hydrogen desorption spectra (a) before and (b) after atmospheric-pressure plasma treatment where molecules shown in the inset describe the desorption behavior near room temperature of hydrogen initially stored in the nanopores. Hydrogen and carbon atoms are in white and gray colors, respectively. (Figure reprinted from Han, K. S., Kim, H. S., and Kang, J. K. et al. 2005. Atmospheric-pressure plasma treatment to modify hydrogen storage properties of multiwalled carbon nanotubes. *Appl. Phys. Lett.* 86: 2631051-3. With permission.)

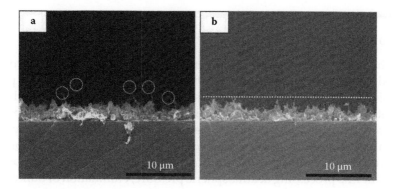

FIGURE 7.40
Scanning electron micrograph (SEM) images of screen-printed carbon nanotube (CNT) sample (a) after tape activation and (b) after the He/N$_2$ plasma treatment for 10 s after the tape activation. (Figure reprinted from Kyung, S. J., Park, J. B., and Yeom, G. Y. et al. 2007. The effect of atmospheric pressure plasma treatment on the field emission characteristics of screen printed carbon nanotubes. *Carbon* 45: 649–654. With permission.)

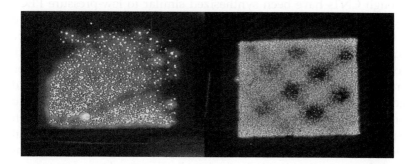

FIGURE 7.41
Field emission characteristics of the screen printed CNTs before (a) and after (b) treatment by an atmospheric pressure plasma. (Figure reprinted from Kyung, S. J., Park, J. B., and Yeom, G. Y. et al. 2007. The effect of atmospheric pressure plasma treatment on the field emission characteristics of screen printed carbon nanotubes. *Carbon* 45: 649–654. With permission.)

the screen-printed CNTs showed irregular heights after tape activation. This irregular surface profile produces hot spots that emit excessive electrons that can result in nonuniform emission and decrease the lifetime of the display. In comparison, as shown in Figure 7.40b, posttreatment by a He/N$_2$ plasma produces more uniform CNTs in terms of the height by removing the ends of the protruding nanotubes. Field emission characteristics of the screen-printed CNTs before and after treatment by the atmospheric pressure plasma are shown in Figures 7.41a and 7.41b, respectively, and confirm that the field emissions are more uniform and more stable after plasma treatment.

7.4 Conclusions

CNTs have been synthesized by researchers using many different growth methods including low-pressure arc discharges, thermal CVD, low pressure PECVD, atmospheric pressure plasma enhanced CVD, and so forth, for applications such as semiconductor channel layers, field emission tips for field emission displays, scanning tunneling probe tips, gas-detecting sensor materials, and so forth. Among the various growth methods, atmospheric pressure plasma-enhanced CVD offers several advantages such as the possibility of in-line processing at atmospheric pressure, elimination of bulky and expensive vacuum equipment, and so forth. Various types of atmospheric pressure plasmas have been developed to synthesize CNTs including DBDs, modified DBDs, plasma jets and plasma torches, microplasmas, and corona discharges. By optimizing the processing conditions, not only MWCNTs but also SWCNTs have been successfully synthesized at low temperature (compared to thermal CVD).

Although CNTs have been synthesized similar to low-pressure PECVD, damage to the nanotube surface is still observed in some cases, possibly due to ion bombardment. Atmospheric pressure plasma sources that can deposit high-quality CNTs uniformly over large areas are also under investigation. In addition, a basic understanding of the discharge chemistry, the interaction mechanism between the metal catalyst and atmospheric pressure discharge, the mechanism for high-quality CNT synthesis at low temperature, and source of damage to the CNT surface are needed. Therefore, before atmospheric pressure plasmas can be commercially used for CNT synthesis, additional understanding and optimization of the growth conditions is required. A more convenient industrial application may be surface treatment of CNTs where, because of simpler physics and chemistry, significant progress has already been made.

References

1. Iijima, S. 1991. Helical microtubules of graphitic carbon. *Nature* 354:56–58.
2. Kroto, H. W., Heath, J. R., and O'Brien, S. C. et al. 1985. C60: Buckminsterfullerene. *Nature* 318:162–163.
3. Saito, R., Dresselhaus, G., and Dresselhaus, M. S. 1998. *Physical Properties of Carbon Nanotubes*. London: Imperial College Press.
4. Meyyappan, M., and Srivastava, D. 2003. *Carbon Nanotubes: Handbook of Nanoscience, Engineering, and Technology*. Boca Raton, FL: CRC Press.
5. Dresselhaus, M. S., Dresselhaus, G., and Avouris Ph. 2001. *Carbon Nanotubes*. Berlin: Springer.

6. Grimes, C. A., Mungle, C., and Eklund, P. C. et al. 2000. The 500 MHz to 5.50 GHz complex permittivity spectra of single-wall carbon nanotube-loaded polymer composites, *Chem. Phys. Lett.* 319: 460–464.

7. Files, B. S., and Mayeaux, B. M. 1999. Carbon nanotubes. *Adv. Mater. Proc.* 156: 47–49.

8. Callister, W. D. 2003. *Materials Science and Engineering, an Introduction.* New York: Wiley.

9. Wei, B. Q., Vajtai, R., and Ajayan, P. M. 2001. Reliability and current carrying capacity of carbon nanotubes. *Appl. Phys. Lett.* 79(8): 1172–1174.

10. Dürkop, T., Kim, B. M., and Fuhrer, M. S. 2004. Properties and applications of high-mobility semiconducting nanotubes. *J. Phys.: Condens Matter* 16(18): R553–R580.

11. Iijima, S., Ajayan, P. M., and Ichihashi, T. 1992. Growth model for carbon nanotubes. *Phys. Rev. Lett.* 69: 3100–3103.

12. Thess, A., Lee, R., and Smalley, R. E. et al. 1996. Crystalline ropes of metallic carbon nanotubes. *Science.* 273: 483–487.

13. Yudasaka, M., Kikuchi, R., and Yoshimura, S. et al. 1995. Specific conditions for Ni catalyzed carbon nanotube growth by chemical vapor deposition. *Appl. Phys. Lett.* 67: 2477–2479.

14. Huang, Z. P., Xu, J. W., and Provencio, P. N. et al. 1998. Growth of highly oriented carbon nanotubes by plasma-enhanced hot filament chemical vapor deposition. *Appl. Phys. Lett.* 73: 3845–3847.

15. Kong, J., Soh, H. T., and Dai, H. et al. 1998. Synthesis of individual single-walled carbon nanotubes on patterned silicon wafers. *Nature* 395: 878–881.

16. Delzeit, L., Chen, B., and Meyyappan, M. et al. 2001. Multilayered metal catalysts for controlling the density of single-walled carbon nanotube growth. *Chem. Phys. Lett.* 348: 368–374.

17. Li, J., Papadopoulos, C., and Moskovits, M. et al. 1999. Highly-ordered carbon nanotube arrays for electronics applications. *Appl. Phys. Lett.* 75: 367–369.

18. Kogelschatz, U. 2004. Atmospheric-pressure plasma technology. *Plasma Phys. Control. Fusion* 46: B63–75.

19. Kunhardt, E. E. 2000. Generation of large-volume, atmospheric-pressure, non-equilibrium plasmas, *IEEE Trans. Plasma Sci.* 28: 189–200.

20. Kyung, S. J., Maksym, V., and Yeom, G. Y. et al. 2007. Growth of carbon nanotubes by atmospheric pressure plasma enhanced chemical vapor deposition using NiCr catalyst. *Surf. Coat. Tech.* 201: 5378–5382.

21. Kyung, S. J., Lee, Y. H., and Yeom, G. Y. 2006. Field emission properties of carbon nanotubes synthesized by capillary type atmospheric pressure plasma enhanced chemical vapor deposition at low temperature. *Carbon* 44: 1530–1534.

22. Deng, X. T., Shi, J. J., and Kong, M. G. et al. 2007. Protein destruction by atmospheric pressure glow discharges, *Appl. Phys. Lett.* 90: 0139031-3.

23. Deng, X. T., Shi, J. J., and Kong, M. G. 2007. Protein destruction by a helium atmospheric pressure glow discharge: Capability and mechanisms. *J. Appl. Phys.* 101: 0747011-3.

24. Yang, Z. S., Shirai, H., and Hasegawa, Y. et al. 2007. Synthesis of Si nanocones using rf microplasma at atmospheric pressure. *Thin Solid Film* 515: 4153–4158.

25. Chen, Q., Zhang, Y., and Ge, Y. et al. 2005. Atmospheric pressure DBD gun and its application in ink printability. *Plasma Sources Sci. Technol.* 14: 670–675.

26. Laroussi, M., and Lu, X. 2005. Room-temperature atmospheric pressure plasma plume for biomedical applications. *Appl. Phys. Lett.* 87: 1139021-3.
27. Kyung, S. J., Park, J. B., and Yeom, G. Y. 2006. Improvement of field emission from screen-printed carbon nanotubes by He/(N₂,Ar) atmospheric pressure plasma treatment. *J. Appl. Phys.* 100: 1243031 4.
28. Nojima, H., Park, R. E., and Takiyama, K. et al. 2007. Novel atmospheric pressure plasma device releasing atomic hydrogen: Reduction of microbial-contaminants and OH radicals in the air, *J. Phys. D: Appl. Phys.* 40: 501–509.
29. Kyung, S. J., Lee, Y. H., and Yeom, G. Y. 2006. Deposition of carbon nanotubes by capillary-type atmospheric pressure PECVD. *Thin Solid Films* 506: 268–273.
30. Donohoe, K. G. 1976. *The development and characterization of an atmospheric pressure non-equilibrium plasma chemical reactor.* Ph.D. dissertation. Pasadena: California Institute of Technology.
31. Donohoe, K. G., and Wydeven, T. 1979. Plasma polymerization of ethylene in an atmospheric pressure discharge. *J. Appl. Polymer Sci.* 23: 2591–2601.
32. Kanazawa, S., Kogoma, M., and Okazaki, S. et al. 1988. Stable glow plasma at atmospheric pressure. *J. Appl. Phys. D: Appl. Phys.* 21: 838–840.
33. Salge, J. 1996. Plasma-assisted deposition at atmospheric pressure. *Surf. Coat. Technol.* 80: 1–7.
34. Massines, F., and Gouda, G. 1998. A comparison of polypropylene-surface treatment by filamentary, homogeneous and glow discharges in helium at atmospheric pressure. *J. Phys. D: Appl. Phys.* 31: 3411–3420.
35. Moon, S. Y., Choe, W., and Kang, B. K. 2004. A uniform glow discharge plasma source at atmospheric pressure. *Appl. Phys. Lett.* 84(2): 188–190.
36. Eliasson, B., and Kogelschatz, U. 1991. Non-equilibrium volume plasma chemical processing. *IEEE Trans. Plasma Sci.* 19: 1063–1077.
37. Salge, J. 1996. Plasma-assisted deposition at atmospheric pressure. *Surf. Coat. Technol.* 80: 1–7.
38. Ammelt, E., Schweng, D., and.Purwins, H. -G. 1993. Spatio-temporal pattern formation in a lateral high-frequency glow discharge system. *Phys. Lett. A* 179: 348–354.
39. Breazeal, W., Flynn, K. M., and Gwinn, E. G. 1995. Static and dynamic two-dimensional patterns in self-extinguishing discharge avalanches. *Phys. Rev. E.* 52: 1503–1515.
40. Babayan, S. E., Jeong, J. Y., and Hicks, R. F. et al. 1998. Deposition of silicon dioxide films with an atmospheric pressure plasma jet. *Plasma Source Sci. Technol.* 7(3): 286–288.
41. Jeong J. Y., Babayan, S. E. and Selwyn. G. S. 1998. Etching materials with an atmospheric-pressure plasma jet. *Plasma Source Sci. Technol.* 7(3): 282–285.
42. Kim, K. G., Choi, J. D., and Yang, S. S. et al. 2011. Atmospheric-pressure plasma-jet from micronozzle array and its biological effects on living cells for cancer therapy. *Appl. Phys. Lett.* 98: 0737011-3.
43. Kim, G. C., Kim, G. J., and Lee, J. K. et al. 2009. Air plasma coupled with anti-body-conjugated nanoparticles: A new weapon against cancer. *J. Phys. D: Appl. Phys.* 42: 0320051-5.
44. Dobrynin, D., Fridman, G., and Fridman, A. 2009. Physical and biological mechanisms of direct plasma interaction with living tissue. *New J. Phys.* 11: 1150201-20.

45. Kim, G. J., Kim, W., and Lee, J. K. et al. 2010. DNA damage and mitochondria dysfunction in cell apoptosis induced by nonthermal air plasma. *Appl. Phys. Lett.* 96: 0215021-3.
46. Kim, J. Y., Kim, S. O., and Li, J. et al. 2010. A flexible cold micro-plasma jet using biocompatible dielectric tubes for cancer therapy. *Appl. Phys. Lett.* 96: 2037011-3.
47. Kim, S. J., Chung, T. H., and Leem, S. H. et al. 2010. Induction of apoptosis in human breast cancer cells by a pulsed atmospheric pressure plasma jet. *Appl. Phys. Lett.* 97: 0237021-3.
48. Hahn, J., Han, J. H., and Suh, J. S. 2004. New continuous gas-phase synthesis of high purity carbon nanotubes by a thermal plasma jet. *Carbon* 42: 877–883.
49. Koinuma, H., Ohkubo, H., and Hayashi, S. et al. 1992. Development and application of a microbeam plasma generator. *Appl. Phys. Lett.* 60(7): 816–817.
50. Inomata, K., Koinuma, H., and Shiraishi, T. et al. 1995. Open air photoresist ashing by a cold plasma torch: Catalytic effect of cathode material. *Appl. Phys. Lett.* 66(17): 2188–2190.
51. Inomata, K., Ha, H., and Koinuma, H. et al. 1994. Open air deposition of SiO_2 film from a cold plasma torch of tetramethoxysilane-H_2–Ar system. *Appl. Phys. Lett.* 64(1): 46–48.
52. Ha, H., Inomata, K., and Koinuma, H. 1995. Plasma chemical vapor deposition of SiO_2 on air-exposed surfaces by cold plasma torch. *J. Electrochem. Soc.* 142(8): 2726–2730.
53. Ha, H., Yoshimoto, M., and Ishiwara, H. 1996. Open air plasma chemical vapor deposition of highly dielectric amorphous TiO_2 films. *Appl. Phys. Lett.* 68(21): 2965–2967.
54. Ha, H., Moon, B. K., and Koinuma, H. et al. 1996. Structure and electric properties of TiO_2 films prepared by cold plasma torch under atmospheric pressure. *Mater. Sci. Eng.* B41(1): 143–147.
55. Lee, B. J., Kusano, Y., and Koinuma, H. et al. 1997. Oxygen plasma treatment of rubber surface by the atmospheric pressure cold plasma torch. *Jpn. J. Appl. Phys.* 36(5A): 2888–2891.
56. Inomata, K., Aoki, N., and Koinuma, H. 1994. Production of fullerenes by low temperature plasma chemical vapor deposition under atmospheric pressure. *Jpn. J. Appl. Phys.* 33(2A): L197–199.
57. Schütze, A., Jeong, James Y., and Hicks, Robert F. et al. 1998. The atmospheric-pressure plasma jet: A review and comparison to other plasma sources. *IEEE Trans. Plasma Sci.* 26(6): 1685–1694.
58. Goldman, M., and Sigmond, R. S. 1982. Corona and insulation. *IEEE Trans. Elect. Insulation* EI-17(2): 90–105.
59. Raizer, Y. P. 1991. *Gas Discharge Physics*. New York: Springer-Verlag.
60. Meek, J. M., and Craggs, J. D. 1953. *Electrical Breakdown of Gases*. London: Oxford University Press.
61. Siemens, W., and Poggendorfs, Ann. 1857. Ueber die elektrostatische Induction und die Verzögerung des Stroms in Flaschendrähten. *Phys. Chem.* 102: 66–122.
62. Menashi, W. P. 1968. *Treatment of surfaces*. U. S. Patent 3 383 163.
63. Kogelschatz, U., and Eliasson, B. 1995. Ozone generation and applications. In: *Handbook of Electrostatic Processes*. New York: Marcel Dekker, pp. 581–605.
64. Gerstenberg, K. W. 1991. A reactor for plasma polymerization on polymer films. *Mater. Sci. Eng.* A139: 110–119.

65. Greenwood, O. D., Boyd, R. D., and Badyal, J. P. S. et al. 1995. Atmospheric silent discharge versus low-pressure plasma treatment of polyethylene, polypropylene, polyisobutylene, and polystyrene. *J. Adhesion Sci. Technol.* 9(3): 311–326.

66. Boulos, M. I., Fauchais, P., and Pfender, E. 1994. *Thermal Plasmas: Fundamentals and Applications.* New York: Plenum Press.

67. Chen, Ch., Perry, W. L., and Phillips, J. et al. 2003. Plasma torch production of macroscopic carbon nanotube structures. *Carbon* 41: 2555–2560.

68. Li, M., Hu, Z., and Tian, Y. et al. 2004. Low-temperature synthesis of carbon nanotubes using corona discharge plasma at atmospheric pressure. *Diamond Relat. Mater.* 13: 111–115.

69. Smiljanic, O., Stansfield, B. L., and Deesilets, S. et al. 2002. Gas-phase synthesis of SWNT by an atmospheric pressure plasma jet. *Chem. Phys. Lett.* 356: 189–193.

70. Nozaki, T., Kimura, Y., and Okazaki, K. 2002. Carbon nanotubes deposition in glow barrier discharge enhanced catalytic CVD. *J. Phys. D: Appl. Phys.* 35: 2779–2784.

71. Yuan, L., Li, T., and Saito, K. 2003. Growth mechanism of carbon nanotubes in methane diffusion flames. *Carbon* 41: 1889–1896.

72. Choi, G. S., Cho, Y. S., and Kim, D. J. et al. 2002. Carbon nanotubes synthesized by Ni-assisted atmospheric pressure thermal chemical vapor deposition. *J. Appl. Phys.* 91: 3847–3854.

73. Kyung, S. J., Voronko, M., and Yeom, G. Y. 2005. Growth and field emission properties of multiwalled carbon nanotubes synthesized by pin-to-plate type atmospheric pressure plasma enhanced chemical vapor deposition. *J. Korean Phys. Soc.* 47(5): 824–827.

74. Kyung, S. J., Lee, Y. H., and Yeom, G. Y. et al. 2006. Deposition of carbon nanotubes by capillary-type atmospheric pressure PECVD. *Thin Solid Film* 506–507: 268–273.

75. Chang, J. S., Lawless, P. A., and Yamamoto, T. 2001. Corona discharge processes, *IEEE Trans. Plasma Sci.* 19: 1152–1166.

76. Roth, J. R. 2001. *Industrial Plasma Engineering.* Bristol: Institute of Physics.

77. Brown, S. C. 1994. *Basic Data of Plasma Physics.* Heidelberg: Springer, American Vacuum Society Classics.

78. Schmidt, M., and Schoenbach, K. H. 2001. *Low Temperature Plasma Physics.* Germany: Wiley-VCH.

79. Rakshit, A. B., Stock, H. M. P., and Twiddy, N. D. et al. 1978. Some ion-molecule reaction rate coefficient measurements at 300 and 100 K in a temperature variable flowing-afterglow apparatus. *J. Phys. B* 11: 4237–4247.

80. Lee, Y. H., Kyung, S. J., and Yeom, G. Y. 2005. The effect of N_2 flow rate in He/O_2/N_2 on the characteristics of large area pin-to-plate dielectric barrier discharge. *Jpn. J. Appl. Phys.* 44(2): L78–L81.

81. Lee, Y. H., and Yeom, G. Y. 2005. Characteristics of a pin-to-plate dielectric barrier discharge in helium. *J. Kor. Phys. Soc.* 47(1): 74–78.

82. Yi, C. H., Lee, Y. H., and Yeom, G. Y. 2003. Characteristic of a dielectric barrier discharges using capillary dielectric and its application to photoresist etching. *Surf. Coatings Technol.* 163–164: 723–727.

83. Nozaki, T., Ohnishi, K., and Kortshagen, U. et al. 2007. Fabrication of vertically aligned single-walled carbon nanotubes in atmospheric pressure non-thermal plasma CVD. *Carbon* 45: 364–374.

84. Hahn, J., Jung, H. Y., and Suh, J. S. 2004. Selective synthesis of high-purity carbon nanotubes by thermal plasma jet. *Carbon* 42: 3024–3027.

85. Hong, Y. C., and Uhm, H. S. 2005. Production of carbon nanotubes by microwave plasma torch at atmospheric pressure. *Phys. Plasmas* 12: 0535041-6.

86. Chen, C. K., Perry, W. Lee, and Phillips, J. 2003. Plasma torch production of macroscopic carbon nanotube structures. *Carbon* 41: 2555–2560.

87. Jašek, O., Eliáš, M., and Kadlečíková, M. 2006. Carbon nanotubes synthesis in microwave plasma torch at atmospheric pressure. *Mater. Sci. Eng. C* 26: 1189–1193.

88. Chiang, W. H., and Mohan Sankaran, R. 2007. Microplasma synthesis of metal nanoparticles for gas-phase studies of catalyzed carbon nanotube growth. *Appl. Phys. Lett.* 91: 1215031-3.

89. Lia, M. W., Hu, Z., and Tian, Y. L. et al. 2004. Low-temperature synthesis of carbon nanotubes using corona discharge plasma at atmospheric pressure. *Diamond Rel. Mater.* 13: 111–115.

90. Eliasson, B., and Kogelschatz, U. 1991. Nonequilibrium volume plasma chemical processing. *IEEE Trans. Plasma Sci.* 19: 1063–1077.

91. Chang, J. S., Lawless, P. A., and Yamamoto, T. 1991. Corona discharge processes. *IEEE Trans. Plasma Sci.* 19: 1152–1166.

92. Okpalugo, T. I. T., Papakonstantinou, P., and Brown, N. M. D. 2005. Oxidative functionalisation of carbon nanotubes in atmospheric pressure dielectric barrier discharge (APDBD). *Carbon* 43: 2951–2959.

93. Lee, Y. H., Kyung, S. J., and Yeom, G. Y. 2006. Characteristic of carbon nanotubes synthesized by pin-to-plate type atmospheric pressure plasma enhanced chemical vapor deposition at low temperature. *Carbon* 44: 807–823.

94. Murakami, Y., Miyauchi, Y., and Maruyama, S. et al. 2003. Direct synthesis of high-quality single-walled carbon nanotubes on silicon and quartz substrates. *Chem. Phys. Lett.* 377: 49–54.

95. Kim, S. H., and Zachariah, M. R. 2006. In-flight kinetic measurements of the aerosol growth of carbon nanotubes by electrical mobility classification. *J. Phys. Chem. B* 110: 4555–4562.

96. Han, K. S., Kim, H. S., and Kang, J. K. et al. 2005. Atmospheric-pressure plasma treatment to modify hydrogen storage properties of multiwalled carbon nanotubes. *Appl. Phys. Lett.* 86: 2631051-3.

97. Kyung, S. J., Park, J. B., and Yeom, G. Y. et al. 2007. The effect of atmospheric pressure plasma treatment on the field emission characteristics of screen printed carbon nanotubes. *Carbon* 45: 649–654.

98. Vink, T. J., Cillies, M., and Van De Laar, H. W. J. J. et al. 2003. Enhanced field emission from printed carbon nanotubes by mechanical surface modification. *Appl. Phys. Lett.* 83: 3552–3554.

99. Risacher, A., Larigaldie, S., and Picard, L. et al. 2007. Active stabilization of low-current arc discharges in atmospheric-pressure air. *Plasma Sources Sci. Technol.* 16: 200–209.

100. Akishev, Y. S., Demyanov, A. V., and Trushkin, N. I. et al. 1991. Modeling and applications of silent discharge plasmas. *IEEE Trans. on Plasma Sci.* 19: 309–323.

101. Wagnera, H. E., Brandenburga, R., and Behnk, J. F. et al. 2003. The barrier discharge: Basic properties and applications to surface treatment. *Vacuum* 71: 417–436.

102. Panousis, E., Clement, F., and Marlin, L. et al. 2006. An electrical comparative study of two atmospheric pressure dielectric barrier discharge reactors. *Plasma Sources Sci. Technol.* 15: 828–839.
103. Hong, Y. C., Cho, S. C., and Uhm, H. S. 2007. A long plasma column in a flexible tube at atmospheric pressure. *Phys. Plasmas* 14: 0745021-4.
104. Forster, S., Mohr, C., and Viol, W. 2005. Investigations of an atmospheric pressure plasma jet by optical emission spectroscopy. *Surf. Coat. Technol.* 200: 827–830.
105. Hong, Y. C., and Uhm, H. S. 2007. Air plasma jet with hollow electrodes at atmospheric pressure. *Phys. Plasmas* 14: 0535031-5.
106. Hong, Y. C., and Uhm, H. S. 2006. Microplasma jet at atmospheric pressure. *Appl. Phys. Lett.* 89: 2215041-3.
107. Massines, F., Segur, P., and Ricard, A. et al. 2003. Physics and chemistry in a glow dielectric barrier discharge at atmospheric pressure: Diagnostics and modeling. *Surf. Coat. Technol.* 174: 8–14.
108. Kogelschatz, U. 2007. Applications of microplasmas and microreactor technology. *Contrib. Plasmas Phys.* 47: 80–88.
109. Moon, S. Y., Choe, W., and Kang, B. K. 2004. A uniform glow discharge plasma source at atmospheric pressure. *Appl. Phys. Lett.* 84: 188–190.
110. Moon, S. Y., Rhee, J. K., and Choe, W. et al. 2006. α, γ, and normal, abnormal glow discharge modes in radio-frequency capacitively coupled discharges at atmospheric pressure. *Phys. Plasmas* 13: 0335021-6.
111. Moon, S. Y., Han, J. W., and Choe, W. 2006. Control of radio-frequency atmospheric pressure argon plasma characteristics by helium gas mixing. *Phys. Plasmas* 13: 0135041-4.
112. Park, J. Y., Henins, I., and Hicks, R. F. et al. 2000. An atmospheric pressure plasma source. *Appl. Phys. Lett.* 76: 288–290.
113. Moravej, M., Yang, X., and Babayan, S. E. et al. 2006. A radio-frequency nonequilibrium atmospheric pressure plasma operating with argon and oxygen. *J. Appl. Phys.* 99: 0933051-5.
114. Balcon, N., Aanesland, A., and Boswell, R. 2007. Pulsed RF discharges, glow and filamentary mode at atmospheric pressure in argon. *Plasma Sources Sci. Technol.* 16: 217–225.
115. Barankova, H., and Bardos, L. 2002. Fused hollow cathode cold atmospheric plasma source for gas treatment. *Catalysis Today* 72: 237–241.
116. Ichiki, T., Koidesawa, T., and Horiike, Y. 2003. An atmospheric-pressure microplasma jet source for the optical emission spectroscopic analysis of liquid sample. *Plasma Sources Sci. Technol.* 12: S16–S20.
117. Gil, A. Y., Focke, K., and Keudell, A. V. et al. 2007. Optical and electrical characterization of an atmospheric pressure microplasma jet for Ar/CH_4 and Ar/C_2H_2 mixtures. *J. Appl. Phys.* 101: 1033071-8.
118. Stoffels, E., Flikweert, A. J., and Kroesen, G. M. W. et al. 2002. Plasma needle: A non-destructive atmospheric plasma source for fine surface treatment of (bio) materials. *Plasma Sources Sci. Technol.* 11: 383–388.
119. Jasinski, M., Mizeraczyk, J., and Chang, J. S. et al. 2002. CFC-11 destruction by microwave torch generated atmospheric-pressure nitrogen discharge. *J. Phys. D: Appl. Phys.* 35: 2274–2280.
120. Stonies, R., Schermerand, S., Broekaert, J. A. C. et al. 2004. A new small microwave plasma torch *Plasmas Sources Sci. Technol.* 13: 604.

121. Bardos, L., and Barankova, H. 2005. Characterization of the cold atmospheric plasma hybrid source *J. Vac. Sci. Technol. A* 23(4): 933–937.
122. Moon, S. Y., and Choe, W. 2006. Parametric study of atmospheric pressure microwave-induced Ar/O_2 plasmas and the ambient air effect on the plasma. *Phys. Plasmas.* 13: 1035031-6.
123. Moon, S. Y., Choe, W., and Choi, J. J. et al. 2002. Characteristics of an atmospheric microwave-induced plasma generated in ambient air by an argon discharge excited in an open-ended dielectric discharge tube. *Phys. Plasmas* 9(9): 4045–4051.
124. Hong, Y. C., and Uhm, H. S. 2006. Atmospheric-pressure hybrid plasma with combination of ac and microwave. *Appl. Phys. Lett.* 89: 2515021-3.
125. Hong, Y. C., and Uhm, H. S. 2005. Production of carbon nanotubes by microwave plasma torch at atmospheric pressure. *Phys. Plasmas* 12: 0535041-6.
126. Cho, S. C., Hong, Y. C., and Uhm, H. S. 2007. Surface treatment of aluminum sheets by pulsed microwave plasma discharge at atmospheric pressure. *Jpn. J. Appl. Phys.* 46(6A): 3583.
127. Uhm, H. S., Kim, J. H., and Hong, Y. C. 2007. Disintegration of water molecules in a steam-plasma torch powered by microwaves. *Phys. Plasmas* 14: 0735021-6.
128. Rodero, A., Quintero, M. C., and Gameroet, A. et al. 1996. Preliminary spectroscopic experiments with helium microwave induced plasma produced in air by use of a new structure: The axial injection torch. *Spectrochimica Acta Part B.* 51: 467–479.
129. Temdero, C., Tixier, C., and Leprince, P. et al. 2006. Atmospheric pressure plasmas: A review. *Spectrochimica Acta, Part B.* 61: 2–30.
130. Laroussi, M. 2002. Nonthermal decontamination of biological media by atmospheric-pressure plasmas: Review, analysis, and prospects. *IEEE Trans. Plasma Sci.* 30(4): 1409–1415.
131. Matsushita, A., Nagai, M., and Zaima, S. 2004. Growth of carbon nanotubes by microwave-excited non-equilibrium atmospheric-pressure plasma. *Jpn. J. Appl. Phys.* 43(1): 424–425.
132. Park, J., Henins, I., Herrmann, H. W., Selwyn, G. S., and Hicks, R. F. 2001. Discharge phenomena of an atmospheric pressure radio-frequency capacitive plasma source. *J. Appl. Phys.* 89(1): 20–28.
133. Shi, J. J., Liu, D. W., and Kong, M. G. 2006. Plasma stability control using dielectric barriers in radio-frequency atmospheric pressure glow discharges. *Appl. Phys. Lett.* 89: 081502-3.
134. Shi, J. J., and Kong, M. G. 2005. Expansion of the plasma stability range in radio-frequency atmospheric-pressure glow discharges. *App. Phys. Lett.* 87: 201501-3.

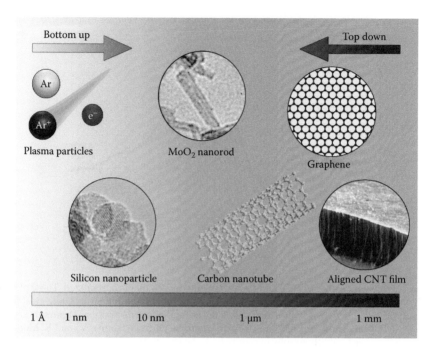

FIGURE P.1
Plasma processing of materials at different length scales. (Courtesy of Uwe Kortshagen, A. Chandra Bose, Davide Mariotti, and Liming Dai.)

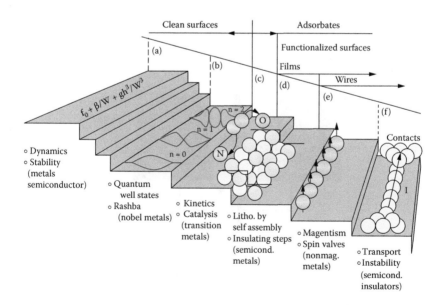

FIGURE 1.2
Representation of potential functionalizations of vicinal surfaces. (Reprinted from *Journal of Physics: Condensed Matter*, Vol. 21, Tegenkamp, "Vicinal surfaces for functional nanostructures," pp. 013002-2, Copyright (2009) with permission from IOP Publishing.)

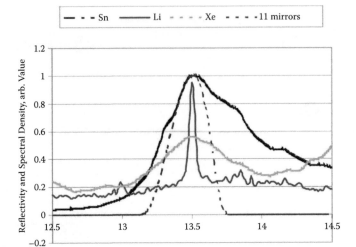

FIGURE 2.4
The spectral density plots of Li, Sn, and Xe are presented. Each plot is normalized. These plots reveal why Li is ideal as a mono-energetic light emitter, and why Xe and Sn are more ideal for their output. The fourth plot, 11 mirrors, shows the normalized reflectivity of 11 mirrors each made of Sn/Mo bilayer Bragg reflectors. (Figure reproduced from Banine, V., and R. Moors, Plasma sources for EUV lithography exposure tools. *Journal of Physics D: Applied Physics*, 2004. 37(23): p. 3207–3212. With permission.)

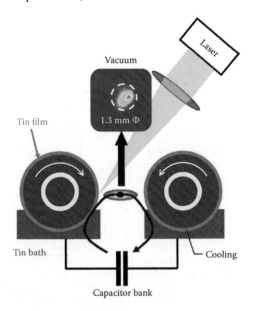

FIGURE 2.7
A rotating disk electrode source is one of a few different types of gas discharge produced plasmas. The two disks rotate, and are coated in a thin Sn layer, which replenishes the fuel. A laser is fired externally that causes a plume of Sn gas to gap the two electrodes. The potential on the electrodes is then discharged across the gap creating extreme ultraviolet (EUV) light. (Figure taken from Yoshioka, M. *Tin DPP Source Collector Module (SoCoMo): Status of beta products and HVM developments*, Extreme Ultraviolet (EUV) Lithography, 2010. 7636. San Diego, CA:SPIE. With permission.)

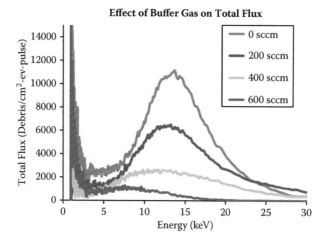

Effect of Buffer Gas on Total Flux

FIGURE 2.9

Measurements of total ion and neutral flux from an XTS 13-35 extreme ultraviolet (EUV) light source from 25° off axis are shown as a function of buffer gas flow rate. The buffer gas is injected between the EUV source and a foil trap as a method for mitigating the debris reaching the collector optics. The measured flux is predominately composed of Xe gas atoms. (Figure taken from Sporre, J., Detection of energetic neutral flux emanating from extreme ultraviolet light lithography sources, in *Department of Nuclear, Plasma, and Radiological Engineering*. 2010, University of Illinois at Urbana-Champaign. p. 126. With permission.)

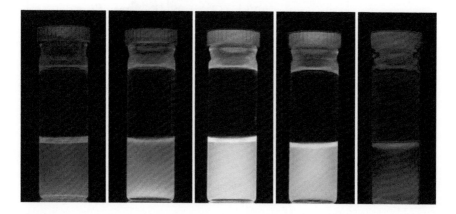

FIGURE 3.1

Photoluminescence of silicon nanocrystals that were synthesized in a two-stage plasma process: a first synthesis step and a subsequent etching step using a CF_4-etch chemistry. (From Pi, X. D., R. W. Liptak, J. D. Nowak et al. 2008b. Air-stable full-visible-spectrum emission from silicon nanocrystals synthesized by an all-gas-phase plasma approach. *Nanotechnology* 19: 245603. With permission.)

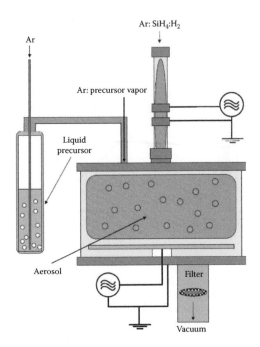

FIGURE 3.9
Two-stage plasma for the synthesis and plasma-aided surface grafting of organic surfactant molecules. (Reproduced from Mangolini, L., and U. Kortshagen. 2007. Plasma-assisted synthesis of silicon nanocrystal inks. *Adv. Mater.* 19: 2513–2519. With permission.)

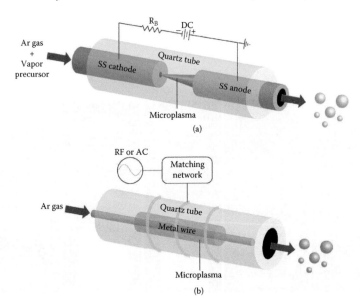

FIGURE 4.2
Possible configurations for microplasma synthesis of metal nanoparticles. (a) Vapor precursors such as organometallic compounds are dissociated in a direct current (DC) microplasma to form radicals that can homogeneously nucleate nanoparticles. (b) Solid metal wires are evaporated and sputtered by a radio frequency (RF) or alternating current (AC) microplasma to form vapors that condense and nucleate nanoparticles.

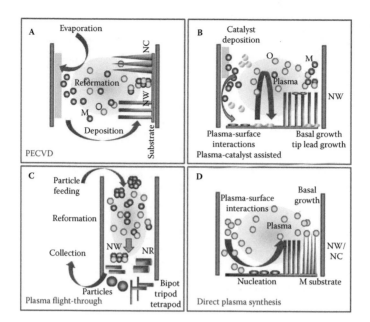

FIGURE 5.9
Different plasma-based approaches for NW synthesis: (a) plasma-enhanced chemical vapor deposition (PECVD); (b) plasma-catalyst-assisted; (c) plasma flight-through; and (d) direct plasma synthesis routes. (Reproduced with permission from Cvelber, U. Towards large-scale plasma-assisted synthesis of nanowires. *Journal of Physics D – Applied Physics*, 2011. 44:174014.)

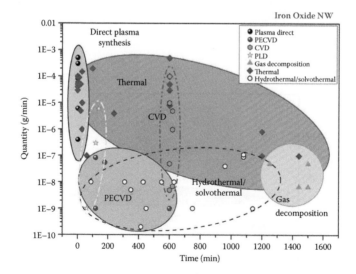

FIGURE 5.10
Time required for the synthesis of iron oxide nanowires (NWs) versus quantity produced for plasma-assisted syntheses in comparison with other methods (e.g., thermal, gas decomposition, pulse laser deposition [PLD], chemical vapor deposition [CVD], and hydrothermal/solvothermal. (Marked areas denote reported results for specific method, but can always be extended toward smaller quantities.) (Reproduced with permission from Cvelber, U. Towards large-scale plasma-assisted synthesis of nanowires. *Journal of Physics D – Applied Physics*, 2011. 44:174014.)

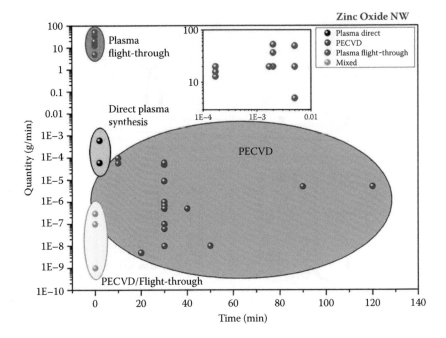

FIGURE 5.11

Time required for the synthesis of ZnO nanowires (NWs) versus quantity produced for plasma-assisted syntheses in comparison with other methods (e.g., thermal, gas decomposition, PLD, CVD and hydrothermal/solvothermal. (Marked areas denote reported results for specific method, but can always be extended toward smaller quantities.) (Reproduced with permission from Cvelber, U. Towards large-scale plasma-assisted synthesis of nanowires. *Journal of Physics D – Applied Physics*, 2011. 44:174014.)

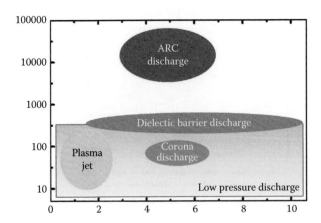

FIGURE 7.6

Comparison of the gas and electron temperatures for different atmospheric-pressure plasmas versus low-pressure plasmas. (Figure reprinted from Schütze, A., Jeong, James Y., and Hicks, Robert F. et al. 1998. The atmospheric-pressure plasma jet: A review and comparison to other plasma sources. *IEEE Trans. Plasma Sci.* 26(6): 1685–1694. © 1998 IEEE. With permission.)

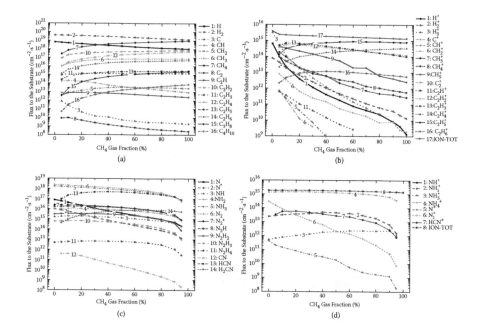

FIGURE 10.1

Calculated radially-averaged fluxes of the various species bombarding the substrate as a function of CH_4 fraction: (a) neutral species and (b) ions in a CH_4/H_2 gas mixture; (c) extra neutral species and (d) extra ions in a CH_4/NH_3 gas mixture. The operating conditions are 50 mTorr total gas pressure, 100 sccm total gas flow rate, 300W source power, 30W bias power at the substrate and 13.56 MHz operating frequency at the coil and at the substrate electrode. The substrate is heated to 550°C. (Reproduced from [42] with permission of Institute of Physics.)

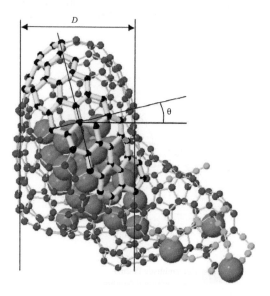

FIGURE 10.8

SWNT cap grown by means of the hybrid MD/MC model. Determination of the chiral angle $\theta = 14°$ and diameter $D = 11.45$ Å allows to assign a (12,4) chirality to the cap. (Reproduced from [167] with permission of the American Chemical Society.)

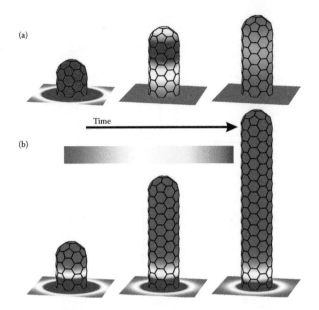

FIGURE 11.2
The progression of the growth of the single-walled carbon nanotubes (SWCNTs) in (a) neutral and (b) plasma (wide sheath) systems. The color represents the flux incident on the specific region from which a building unit (BU) will reach the base of the SWCNT; this quantity is proportional to the SWCNT growth rate. In neutral gas systems, BUs primarily deposit on the tips of the SWCNTs, and as the SWCNTs get longer, fewer BUs reach the base of the SWCNT, leading to slower, stifled growth. However, in a wide plasma sheath system, BU trajectories are directed toward the base of the SWCNTs, and a higher flux of BUs reaches the base of the SWCNT, leading to the rapid growth of long SWCNTs. (Figure and caption reproduced from Tam, E., and Ostrikov, K. 2008. Plasma-controlled adatom delivery and (re)distribution—Enabling uninterrupted, low-temperature growth of ultralong vertically aligned single walled carbon nanotubes. *Appl. Phys. Lett.*, 93(26): 261504. With permission.)

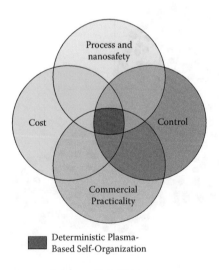

FIGURE 15.6
Venn diagram for the four main considerations in nanofabrication, deterministic plasma-based self-organization satisfies all criteria.

8

Structural Control of Single-Walled Carbon Nanotubes by Plasma Chemical Vapor Deposition

Rikizo Hatakeyama
Toshiaki Kato

CONTENTS

8.1 Freestanding Single-Walled Carbon Nanotubes Growth

The enormous potential of plasma chemical vapor deposition (CVD) for nanotube growth was first shown by Ren et al. in 1998 [1]. Since then, vertically and individually aligned multiwalled carbon nanotubes (MWNTs) have been grown by plasma CVD. Compared to carbon nanotubes (CNTs) grown by thermal CVD, which are spaghetti-like, entangled shapes, plasma CVD can produce well-aligned CNTs for direct integration in CNT-based nanoelectronics. However, CNTs produced by plasma CVD have been typically limited to MWNTs—that is, no one has succeeded in the growth of single-walled carbon nanotubes (SWNTs) which have superior electrical and optical characteristics. The growth of SWNTs by plasma CVD was first reported by our group in 2003 [2]. In plasmas, ions are accelerated by a potential drop created by space potentials between the plasma and substrate. The minimum value of this potential drop is determined by the electron temperature in the

219

plasma. Low electron temperature plasmas can significantly decrease the energy of ions coming to the substrate. Because the diffusion region in plasmas is known to have a very low electron temperature, we used a diffusion plasma to decrease the energy of ions attacking the catalyst to below a few eV. Figure 8.1a is a schematic illustration of a homemade diffusion plasma CVD system used in this study. A precise comparison between plasma CVD and thermal CVD under the same conditions is also possible with this system to clarify the effects of plasmas on the chirality distribution of SWNTs. Photoluminescence-excitation (PLE) mapping measurements are performed to characterize the chirality of SWNTs (Figure 8.1b). When we carry out the SWNT growth under the diffusion plasma region, SWNTs are grown on a flat substrate without using any catalyst support materials [3,4]. This shows that the critical element promoting catalyst aggregation is high energy ion bombardment. It is revealed that SWNTs grown by diffusion plasma CVD are well aligned and freestanding (i.e., all SWNTs are individually and vertically standing on the flat substrate). Figures 8.2a through 8.2c are a typical scanning electron microscope (SEM) image (Figure 8.2a), high magnification transmission electron microscope (TEM) image (Figure 8.2b), and Raman scattering spectra (Figure 8.2c) of freestanding SWNTs, respectively. Relatively high-quality SWNTs are found to be grown with the individually freestanding form. This freestanding alignment can be obtained by the

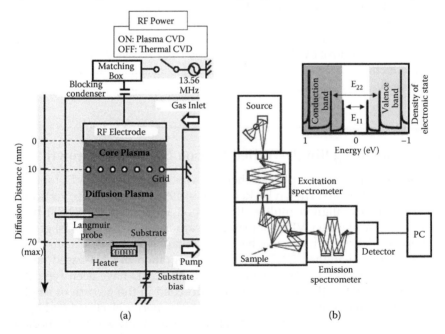

(a) (b)

FIGURE 8.1
(a) A diffusion plasma chemical vapor deposition (CVD) system. (b) Optical setup for the measurement of photoluminescence-excitation mapping.

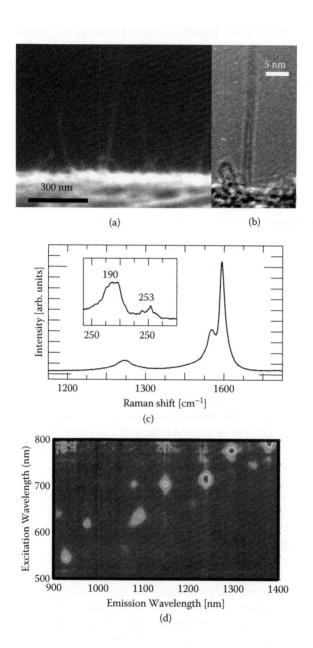

FIGURE 8.2

(a) Scanning electron microscope (SEM) and (b) transmission electron microscope (TEM) images of individually freestanding single-walled carbon nanotubes (SWNTs). (c) Raman scattering spectrum of individually freestanding SWNTs. Inset of (c) is emphasis of the radial breathing mode (RBM) region. (d) photoluminescence-excitation (PLE) map obtained from as-grown freestanding SWNTs without any dispersion process.

plasma sheath electric field. Based on our estimation, the rotation energy of the dipole moment in SWNTs is much higher than the thermal energy, which disturbs the tube alignment. This indicates that the individual SWNTs can be aligned along the electric field. Owing to their unique as-grown state, it is possible to directly detect photoluminescence (PL) spectra from the as-grown freestanding SWNTs on the substrate, as exemplified in Figure 8.2d of PLE map [5]. This is a remarkable advantage for optoelectronic applications and fundamental studies toward the diameter [6] and chirality control, which will be discussed later.

8.2 Chirality Distribution Control

Because the chirality of SWNTs directly determines their electronic and optical properties, the selective synthesis of SWNTs with desired chiralities is one of the major challenges in nanotube science and applications. Some progress has been made by silica supported CoMo [7] and zeolite supported FeCo [8] catalysts. FeRu [9] and FeNi [10] catalysts have also been developed to obtain narrow chirality distributions. We note that all of these results for narrow chirality distribution have been limited to magnetic catalysts. Because the existence of residual ferromagnetic catalysts particles in SWNTs is an important obstacle to research on intrinsic magnetic properties of SWNTs, the growth of SWNTs with nonmagnetic catalysts is indispensable. Despite reports of SWNT growth with nonmagnetic catalysts [11,12], the diameter and chirality (n,m) distributions have not been sufficiently controlled for fundamental studies and applications.

In this subsection, we discuss our recent achievements for narrow-chirality distributed growth of SWNTs from an Au catalyst by plasma CVD [13,14]. PLE mapping [15] is used to assign (n,m) of SWNTs grown from the Au catalyst at different H_2 concentrations (Figures 8.3a through 8.3c). The total pressure is kept at 50 Pa by adjusting the pumping efficiency of a rotary pump throughout the experiment. Lower H_2 concentrations (0 and 3 sccm) lead to larger diameter tubes and more widely (n,m)-distributed tubes with (6,5), (7,5), (7,6), (8,4), (8,6), and (8,7) (Figures 8.3a and 8.3b). On the other hand, at 7-sccm H_2 concentration, a narrow (n,m) distribution with a dominant peak corresponding to the (6,5) tube (Figure 8.3c) is achieved. The ultraviolet visible near-infrared (UV-vis-NIR) optical absorbance spectra of Au-catalyzed plasma CVD SWNTs grown at 7-sccm H_2 flow rate show one dominant peak in the first van Hove E_{11} range (900 to 1400 nm) corresponding to SWNTs with (6,5) chirality (Figure 8.3d). Because metallic SWNT peaks could not be observed in the UV-vis-NIR spectra (Figure 8.3d), the concentration of metallic SWNTs may be lower than conventional methods. The radial breathing mode (RBM) in the Raman spectra measured with 632.8 nm and

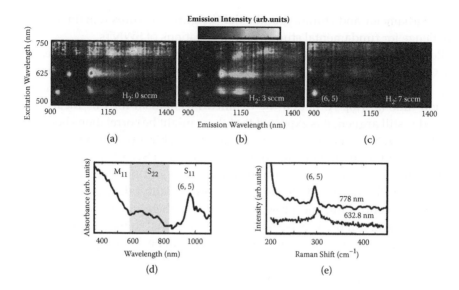

FIGURE 8.3

(a,b,c) Photoluminescence-excitation (PLE) maps of single-walled carbon nanotubes (SWNTs) grown from Au catalyst by plasma chemical vapor deposition (CVD) at (a) 0-sccm, (b) 3-sccm, and (c) 7-sccm H_2 flow rates, respectively. (d) Ultraviolet visible near infrared (UV-vis-NIR) spectrum of SWNTs grown from Au catalyst at 7-sccm H_2 flow rate and (e) Raman spectra of as-synthesized SWNTs grown from Au catalyst at 7-sccm H_2 flow rate using the excitation laser wavelengths of 632.8 nm and 778 nm.

778 nm lasers on the same SWNT sample also confirms the growth of the (6,5) SWNTs (Figure 8.3e). This is the first result reporting narrow chirality distribution of SWNTs grown from a nonmagnetic catalyst [14].

Based on the systematic investigations, it is revealed that low temperature and short time growth with Au-plasma CVD can avoid aggregation of catalyst particles during SWNT growth, which suppresses the growth of large diameter SWNTs, resulting in a narrow chirality distribution.

8.3 Length Control

The length of SWNTs is another very important factor that determines their properties. Recently, length-controlled SWNTs, especially very short-length (>100 nm) SWNTs, have attracted intense attention because of their unique features. However, compared to the study of tube diameter and chirality, very few efforts have been devoted to controlling the axial structure (i.e., the length of SWNTs). Furthermore, simultaneous control of both the radial (diameter and chirality) and axial (length) structures remain a major challenge for fundamental studies and applications of SWNTs.

Recent progress in *in situ* TEM observation during SWNT growth revealed that metal-catalyzed SWNT growth is initiated by the formation of a carbon cap structure on the surface of a catalytic nanoparticle with a certain incubation time (t_i) [16]. Although the detailed mechanism for this incubation period is still argued, it is expected that there might be correlations between the t_i and SWNT structures such as diameter and chirality. When we assume that the t_i of the small diameter (or specific chirality) SWNTs is shorter than that of the larger (or other chiralities) one, it should be possible to selectively grow the narrow-diameter (or narrow-chirality) distributed SWNTs by strictly controlling the growth time (t_g) at their initial growth stage. Because the length of SWNTs should be proportional to the growth time at the initial growth stage, the very short SWNTs can be obtained by adjusting the growth time. In the case of thermal CVD, the SWNT growth gradually starts and stops after the initiation of feeding and pumping the hydrocarbon gas, respectively. The growth time in thermal CVD includes some uncertainty, which makes it difficult to precisely control growth time. In the case of plasma CVD, on the other hand, reactive ions and radicals are the main species for nanotube growth, and SWNT growth is carried out only when a plasma is generated. This suggests that the growth time can be controlled by tuning an electric power supply used for plasma generation, and precise t_g control on the order of microsecond is possible. Based on this strategy, we attempted to grow narrow-diameter and narrow-chirality distributed short SWNTs by precisely adjusting the t_g with time programmed plasma CVD. Although plasma CVD is well-known for SWNT growth, there have not been any prior reports focusing on controlling the growth time.

The length distribution of SWNTs was carefully investigated by TEM and atomic force microscopy (AFM). In the case of relatively long time (~60 s), the length distribution of SWNTs is broad (Figure 8.4a). On the contrary, when the growth time is carefully controlled on the order of a few seconds, almost all the SWNTs are very short (<100 nm), and the distribution is also narrow, which was confirmed by direct TEM observation and AFM characterization (Figure 8.4b). This indicates that it is possible to directly grow short SWNTs by precisely adjusting the growth time with time programmed plasma CVD. Interestingly, when we check the PLE map of SWNTs grown for long (Figure 8.4c) and short (Figure 8.4d) times, in the case of short time growth, the main chirality species are limited to (7,6) and (8,4) (Figure 8.4d). This indicates that the chirality distribution at the initial growth stage is very narrow [17]. Consequently, it is possible to produce length-controlled SWNTs with narrow-chirality distribution by precisely adjusting the growth time with plasma CVD.

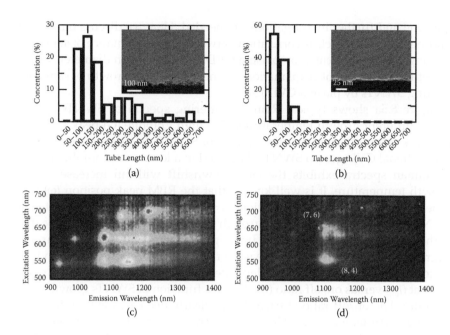

FIGURE 8.4

(a,b) Length distributions of SWNTs produced by long (a) and short (b) growth time. Inset shows the typical transmission electron micrograph (TEM) images of freestanding single-walled carbon nanotubes (SWNTs). (c,d) Photoluminescence-excitation (PLE) maps of SWNTs grown for long (c) and short (d) times.

8.4 Metallicity Control

Thin film transistors (TFTs) are one of the most promising applications of SWNTs. Although the high mobility and flexibility of SWNTs films can provide lots of opportunities to be utilized in various kinds of industrial applications, the low on/off current ratio in SWNT TFTs caused by the mixture of metallic and semiconducting SWNTs restricts the practical use of SWNTs in TFT applications. Recent progress in chemical separation enables us to fabricate good devices with on/off ratio: ~10^4 and effective gate mobility: ~52 cm²/Vs [18]. However, impurities and defects are sometimes introduced in chemically treated nanotubes during the separation process, which significantly decreases the device performance. Because as-grown SWNTs maintain the original high quality with low impurity concentration, the selective growth of semiconducting SWNTs is desirable. Dai et al. reported the preferential growth of semiconducting SWNTs by plasma CVD [19]. Although several similar reports with plasma CVD [20–22] and thermal CVD [23] have also been reported, the elucidation of this selective growth is still an open question and further investigations are needed. Here we discuss our recent findings that show a clear correlation between the performance of semiconducting devices fabricated by plasma CVD and their mean diameter, which might lead to a possible explanation for the preferential growth of semiconducting SWNTs by plasma CVD [24].

Figure 8.5a shows typical Raman scattering spectra of SWNTs grown under different growth temperatures. The high graphite (G)-peak to defect (D)-peak ratio indicates that the quality of SWNTs is comparable to other conventional CVD grown SWNTs. The RBM in a lower wave number region in Raman spectra exhibits the clear downshift with an increase in the growth temperature. It is well known that the RBM peak position (ω) and tube diameter (d) have a correlation: $\omega = 248/d$ (i.e., the mean diameter of SWNTs is found to increase with growth temperature). This seems to be due to the catalyst particle size effect. Higher growth temperatures cause particle aggregation and result in the increase of the particle size, which can produce larger-diameter SWNTs. A clear dependence is obtained from the plot of on current (I_{on}) versus on/off ratio (I_{on}/I_{off}) as a function of SWNTs growth temperature. The on/off ratio of each device clearly decreases with a decrease in the growth temperature (Figure 8.5b). The concentration of the working devices, which have on/off ratios greater than 5, is counted and plotted as a function of the growth temperature (Figure 8.5c). Noticeably, the working device concentration is only 2.5% in the case of 600°C (smaller-diameter SWNTs), whereas more than 90% of the devices work in the case of 800°C (larger-diameter SWNTs). The density of SWNTs grown under the different growth temperatures is almost the same.

In order to explain the dependence of the working device concentration on the SWNT diameter, devices were irradiated by an Ar plasma, and a defect formation rate is estimated from the current change before and after the plasma treatment. In the case of small-diameter SWNT devices, the on/off ratio does not change, and on and off currents significantly decrease after the Ar plasma irradiation, whereas the on/off ratio increases with an increase in the Ar plasma irradiation time and the off current depression is significant compared to that of the on current in the case of large-diameter SWNT devices. Based on these results, the following model can be developed to explain the dependence of the working device concentration on the diameter. Due to the curvature effect, small-diameter SWNTs are more unstable than large-diameter ones. Hence, both metallic and semiconducting SWNTs are easily deformed by the Ar plasma irradiation without any difference in the tube metallicity. On the other hand, in the weak curvature range, the dependence of the defect formation rate on a unique metallicity appears, which might correlate with the reactivity, binding energy between carbon and carbon, and healing process. This model is consistent with the selective etching of metallic SWNTs by gas phase reaction, which was previously reported [25]. Further detailed studies relating to the selective damage

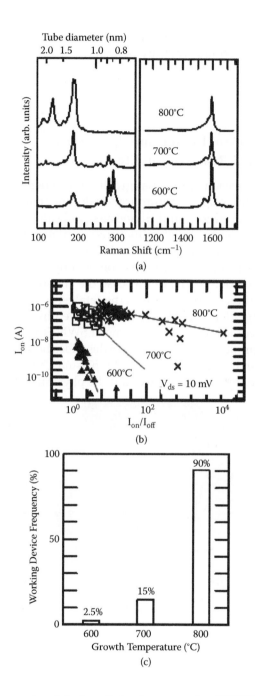

FIGURE 8.5
(a) Raman scattering spectra of single-walled carbon nanotubes (SWNTs) grown at different growth temperatures. (b) I_{on}-I_{on}/I_{off} plot of thin film transistors (TFTs) with SWNTs grown at different growth temperatures. (c) Histogram of working device concentration of TFTs with SWNTs grown at different growth temperatures.

of metallic SWNTs might provide the possible answer for the preferential growth of semiconducting SWNTs by plasma CVD.

8.5 Chirality Control by Bottom-Up Electric-Field-Assisted Reactive Ion Etching

The method of reactive ion etching has greatly contributed to modern micro-electronics manufacturing for ultra large-scale integrated (ULSI) circuits, where plasma-produced ions and dissociated radicals act as physical and chemical means, respectively, of miniaturizing materials and devices by cutting or chipping from the top down. Here let us propose a method of bottom-up electric-field-assisted reactive ion etching (BU-ERIE) for the precise chirality control of SWNTs. In this case the chemical etching proceeds such that hydrocarbons are etched from the catalyst surface, resulting in the suppression of large-diameter SWNT growth. The physical etching, on the other hand, proceeds such that small-diameter SWNTs are selectively etched due to the large-curvature effect. Figure 8.6 describes how specific chiralities (*n,m*) of SWNTs can be selectively grown by plasma CVD, where the chemical etching rate (i.e., hydrogen density) and the physical etching rate (i.e., ion

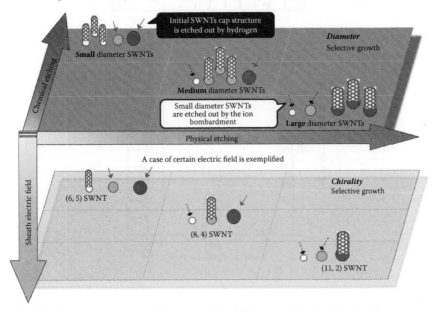

FIGURE 8.6
Approach to control single-walled carbon nanotube (SWNT) chirality by bottom-up electric-field-assisted reactive ion etching under plasma chemical vapor deposition (CVD).

irradiation energy) increase in the directions of the arrows, respectively. For instance, in the case of high chemical and low physical etching rates with weak electric fields, it is possible for small-diameter SWNTs to grow because the initial cap structure of intermediate-diameter SWNTs is etched by hydrogen while the cap structure of large-diameter SWNTs is not formed due to the low binding energy between the Au catalyst and carbon atoms. At equal chemical and physical etching rates, the intermediate-diameter SWNTs can grow because small-diameter SWNTs are etched by ion bombardment and the initial cap structure of relatively large-diameter SWNTs is etched by hydrogen. In addition, formation of large-diameter SWNT caps is suppressed. In the case of low chemical and high physical etching rates, it is possible for relatively large-diameter SWNTs to grow because the small- and intermediate-diameter SWNTs are etched by ion bombardment, and large-diameter SWNT cap structures are not formed due to a low binding energy. Because an electric field can modify the electronic state of the SWNT cap structure, the growth of specific chirality types of SWNTs might be enhanced by adding large electric fields (i.e., plasma sheath electric fields) during the nucleation period. For example, under strong electric fields, the growth of (6,5), (8,4), and (11,2) SWNTs might be selectively enhanced in the range of small-, intermediate-, and large-diameter SWNTs, respectively [26].

References

1. Ren, Z. F., Huang, Z. P., Xu, J. W., Wang, J. H., Bush, P., Siegal, M. P., and Provencio, P. N. 1998. *Science* 282: 1105.
2. Kato, T., Jeong, G. -H., Hirata, T., Hatakeyama, R., Tohji, K., and Motomiya, K. 2003. *Chem. Phys. Lett.* 381: 422.
3. Kato, T., Hatakeyama, R., and Tohji, K. 2006. *Nanotechnology* 17: 2223.
4. Kato, T., and Hatakeyama, R. 2008. *Appl. Phys. Lett.* 92: 031502.
5. Kato, T., and Hatakeyama, R. 2008. *J. Am. Chem. Soc.* 130: 8101.
6. Kato, T., Kuroda, S., and Hatakeyama, R. 2011. *J. Nanomater.* 2011: 490529.
7. Kitiyanan, B., Alvarez, W. E., Harwell, J. H., and Resasco, D. E. 2000. *Chem. Phys. Lett.* 317: 497.
8. Miyauchi, Y., Chiashi, S., Murakami, Y., Hayashida, Y., and Maruyama, S. 2004. *Chem. Phys. Lett.* 387: 198.
9. Li, X., Tu, X., Zaric, S., Welsher, K., Seo, W. S., Zhao, W., and Dai, H. 2007. *J. Am. Chem. Soc.* 129: 15770.
10. Chiang, W. H., and Sankaran, M. R. 2009. *Nat. Mater.* 8: 882.
11. Zhou, W., Han, Z., Wang, J., Zhang, Y., Jin, Z., Sun, X., Zhang, Y., Yan, C., and Li, Y. 2006. *Nano Lett.* 6: 2987.
12. Takagi, D., Homma, Y., Hibino, H., Suzuki, S., and Kobayashi, Y. 2006. *Nano Lett.* 6: 2642.

13. Ghorannevis, Z., Kato, T., Kaneko, T., and Hatakeyama, R. 2010. *Jpn. J. Appl. Phys.* 49: 02BA01.
14. Ghorannevis, Z., Kato, T., Kaneko, T., and Hatakeyama, R. 2010 *J. Am. Chem. Soc.* 132: 9570.
15. Bachilo, S. M., Strano, M. S., Kittrel, C., Hauge, R. H., Smalley, R. E., and Weisman, R. B. 2002. *Science* 298: 2361.
16. Hofmann, S., Sharma, R., Ducati, C., Du, G., Mattevi, C., Cepek, C., Cantoro, M., Pisana, S., Parvez, A., Cervantes-Sodi, F., Ferrari, A. C., Dunin-Borkowski, R., Lizzit, S., Petaccia, L., Goldoni, A., and Robertson, J. 2007. *Nano Lett.* 7: 602.
17. Kato, T., and Hatakeyama, R. 2010. *ACS Nano* 4: 7395.
18. Wang, C., Zhang, J., Ryu, K., Badmaev, A., Arco, L. G. D., and Zhou, C. 2009. *Nano Lett.* 9: 4285.
19. Li, Y., Mann, D., Rolandi, M., Kim, W., Ural, A., Hung, S., Javey, A., Cao, J., Wang, D., Yenilmez, E., Wang, Q., Gibbons, J. F., Nishi, Y., and Dai, H. 2004. *Nano Lett.* 4: 317.
20. Qu, L., Du, F., and Dai, L. 2008. *Nano Lett.* 8: 2682.
21. Kim, U. J., Lee, E. H., Kim, J. M., Min, Y. -S., Kim, E., and Park, W. 2009. *Nanotechnology* 20: 295201.
22. Mizutani, T., Ohnaka, H., Okigawa, Y., Kishimoto, S., and Ohno, Y. 2009. *J. Appl. Phys.* 106: 073705.
23. Ding, L., Tselev, A., Wang, J., Yuan, D., Chu, H., McNicholas, T. P., Li, Y., and Liu, J. 2009. *Nano Lett.* 9: 800.
24. Kato, T., and Hatakeyama, R. 2010. *J. Nanotechnol.* 2010: 256906.
25. Zhang, G., Qi, P., Wang, X., Lu, Y., Li, X., Tu, R., Bangsaruntip, S., Mann, D., Zhang, L., and Dai, H. 2006. *Science* 314: 974.
26. Hatakeyama, R., Kaneko, T., Kato, T., and Li, Y. F. 2011. *J. Phys. D: Appl. Phys.* 44: 174004.

9

Graphene Growth by Plasma-Enhanced Chemical Vapor Deposition (PECVD)

M. Meyyappan
Jeong-Soo Lee

CONTENTS

9.1 Introduction

Low-temperature plasma processing has become a popular technique of choice among the nanotechnology community in the last decade to grow carbon nanotubes (CNTs) and inorganic nanowires, for attaching functional molecular groups to the surface of nanostructures, and chemisorption-based hydrogen storage in CNTs [1]. In the past, diamond, diamond-like carbon, carbon nanotubes, and carbon nanofibers [2] have been grown by plasma-enhanced chemical vapor deposition (PECVD), and recently it was demonstrated that graphene can also be prepared using a low-temperature plasma. Research activities on growth and characterization of graphene and application development have increased exponentially since the discovery of its interesting properties in 2004 [3]. This chapter provides a background on graphene, its properties and applications, a brief description of conventional graphene preparation techniques, a discussion on the use of PECVD for the growth of graphene, and future challenges.

9.2 Properties and Characterization

Graphene is sp^2-bonded carbon atoms packed into a flat, one-layer thick, honeycomb crystal lattice. Graphene is the building block to create carbon nanotubes [4]; for example, a monolayer of graphene rolled up into a cylinder constitutes a single-walled carbon nanotube with a chiral vector (*n,m*), and multiple layers of graphene sheets rolled up into a hollow cylinder resemble a multiwalled carbon nanotube. Graphene exhibits interesting electrical, optical, mechanical, and thermal properties. It is a semimetal or zero band gap semi-conductor. Graphene layers with a finite width or the graphene nanoribbons (GNRs) exhibit varying electrical properties depending on configuration: for example, zigzag GNRs are metallic, whereas armchair GNRs can be metal-lic or semiconducting [5]. In the case of the latter, the band gap is inversely proportional to the GNR width. Ballistic electron transport characterized by the absence of scattering yields mobilities of the order of 15,000 cm^2/V.s at room temperature. A monolayer of graphene is opaque with only 2% of the white light absorbed. Exceptional mechanical strength has been predicted by simulations with a Young's modulus close to 1 TPa. Other attractive charac-teristics include high thermal conductivity, chemical inertness, and a high level of hydrophobicity. Recent reviews [5,6] provide a detailed discussion of the properties of graphene and current status of the literature.

Optical microscopy can be used to visualize and contrast the number of layers in graphene. Typically, graphene is deposited on or transferred to an insulating layer, and the contrast results are dependent on the wavelength of the light source. On SiO$_2$/Si substrates, the insulator thickness is chosen as 285 nm or 465 nm for observation [5–7]. Figure 9.1 shows an optical image at the edge of an as-grown graphene film on 300 nm SiO$_2$/Si. The region cov-ered with graphene is easily distinguishable from the bare SiO$_2$ spots based on different color. Transmission electron microscopy (TEM) and atomic force microscopy (AFM) are also commonly used to characterize graphene, but the most unambiguous approach to date has been Raman spectroscopy. A G peak located at about 1580 cm^{-1} and a two-dimensional (2D) peak at about 2700 cm^{-1} (also known as G′ peak) are the specific signatures for graphene. Any D peak at around 1350 cm^{-1} is an indication of defects in the sample. The shape, width, and position of the 2D peak and the intensity ratio of the 2D/G peaks are used to characterize the number of layers [8]. The 2D/G intensity ratio appears to be 2 or above for monolayer graphene, ~1 for bilayer gra-phene, and progressively smaller for three to five layers. Also, the 2D peak appears to shift toward higher wave numbers and broaden with an increase in the number of layers. Figure 9.2 shows a scanning electron microscopy

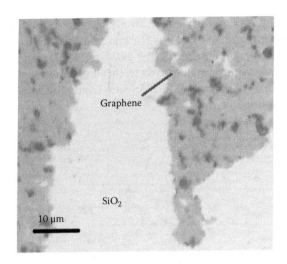

FIGURE 9.1
Optical image of a graphene film on 300 nm SiO_2/Si grown by chemical vapor deposition (CVD). (Image courtesy of Yunfan Zhao and Michael Oye.)

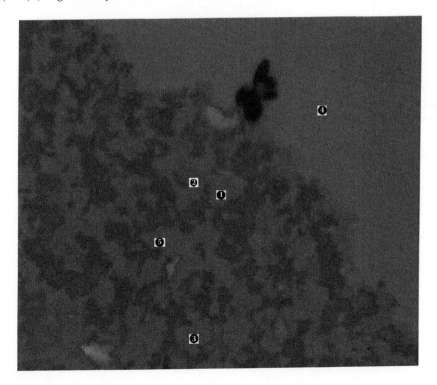

FIGURE 9.2
Scanning electron micrograph (SEM) of graphene film grown by chemical vapor deposition (CVD) on nickel catalyst. (Courtesy of Saebyuk Jeong.)

(SEM) image of a graphene film grown by chemical vapor deposition (CVD) using nickel catalyst, and Raman spectra taken at five different locations in Figure 9.2 are displayed in Figure 9.3. The 2D/G peak ratio is about 1 in all the locations indicating approximately two to three graphene layers. Figure 9.4 shows an AFM scan of a graphene film on SiO_2 from another CVD run showing a thickness of 1 to 2 nm.

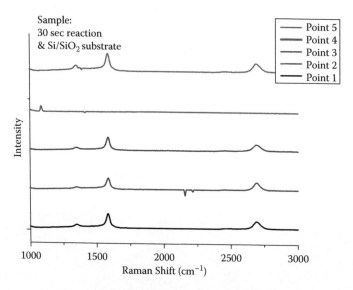

FIGURE 9.3
Raman spectra taken on five spots marked in Figure 9.2. (Courtesy of Saebyuk Jeong.)

FIGURE 9.4
Atomic force microscopy (AFM) scan of a graphene film on SiO_2 indicating 1 to 2 nm thickness. (Courtesy of Saebyuk Jeong.)

9.3 Applications

The interesting properties mentioned above have led to the exploration of numerous applications such as energy storage devices, transistors, chemical and biosensors, nanoelectromechanical systems, and others [9–26], just as it has been with carbon nanotubes previously [4]. One of the promising applications for graphene is in the construction of a supercapacitor [9–12]. Also known as an ultracapacitor that combines the high-power density of a capacitor with the high-energy density of a battery. The energy storage mechanism involves the capacitance arising from the charge accumulated at the electrolyte–electrode interface. This electrochemical double-layer (EDL) capacitance is different from the pseudocapacitance based on faradaic, redox reactions. EDL capacitors commonly use carbon-based electrodes including activated carbon, graphite, and CNTs. Stoller et al. [9] used chemically modified graphene (CMG) derived from the reduction of graphite oxide (GO) to construct supercapacitors. They mixed CMG with a binder material and shaped it into disk-like electrodes. The supercapacitor consists of a symmetric arrangement of two such electrodes backed by their collector metals and separated by a porous membrane. The electrodes are soaked in either aqueous or organic electrolytes. The supercapacitors by Stoller et al. showed 135 and 99 F/g for aqueous and organic electrolytes, respectively. The surface area of their CMG electrodes was 705 m^2/g, resulting in a specific surface capacitance of ~14 $\mu F/cm^2$. Du et al. [10] were able to obtain higher specific capacitances of ~28 $\mu F/cm^2$ using graphene nanosheets without performance degradation over 500 cycles. Besides pristine graphene sheets, graphene–polyaniline composites have also been considered and shown to have a specific capacitance as high as 480 F/g [11]. Commonly used collector metals such as aluminum may oxidize over time, thus causing performance deterioration. Ku et al. [12] showed that addition of a thin film of a noble metal (Pt or Au) on top of the collector metal improves the performance of graphene supercapacitors.

Qu et al. demonstrated the use of graphene as electrocatalyst for oxygen reduction in fuel cells [13]. This is usually done by Pt-based electrodes in the fuel cell literature, but the high cost and limited supply of Pt thus far has prevented commercialization of this technology. Qu et al. prepared nitrogen-doped graphene (N-graphene) by chemical vapor deposition from a mixture of ammonia and methane in argon. The N-graphene was shown to have superior electrocatalytic performance over undoped graphene as well as standard Pt electrodes. The CVD-grown graphene films are also ideal as transparent conducting electrodes in the fabrication of solar cells because they exhibit a low sheet resistance and high transmittance compared to indium tin oxide (ITO). Moreover, the flexibility of the graphene films is more suited for organic photovoltaics because bending often results in cracks in the case of ITO [14].

Graphene-based electrodes are useful for electrochemical biosensors as well [21–25]. Alwarappan et al. [21] showed the ability to detect dopamine, which is a neurotransmitter in humans, and monitoring its concentration is important to treat disorders such as Parkinson's disease. They showed that graphene electrodes are able to detect small concentrations of dopamine even in the presence of interferants such as serotonin and ascorbic acid. Graphene electrodes have also been demonstrated for glucose detection [25].

The interesting electronic properties of graphene have led to extensive investigations on the potential for the use of graphene as the conducting channel in complementary metal oxide semiconductor (CMOS) transistors and novel beyond-CMOS-type architectures [15–20]. An excellent review of this topic is given by Banerjee et al. [19]. Because graphene is a zero-gap material, opening up a band gap is the first big challenge to realize useful devices. Various approaches proposed in the literature to create a band gap include introduction of uniaxial strain, exploiting graphene–substrate interaction, creating nanoribbons of width 10 nm or less, and breaking the inversion symmetry in bilayer graphene through application of vertical electric fields. Other practical issues demanding serious solutions include contact engineering, suitable choice of substrate, and high-κ gate dielectic.

Recently, Xia et al. [15] showed a room-temperature on-off ratio of 100 for a bilayer graphene field effect transistor (GFET). Although most of the works involve transfer of graphene layers grown elsewhere in GFET fabrication, Kondo et al. demonstrated an *in situ,* patterned growth of graphene by CVD at 650°C using a 200 nm thick Fe catalyst [16]. This catalyst layer was later removed by wet etching after the formation of the source and drain contacts. Their graphene layers were able to support a current density of 10^8 A/cm^2, and the device transconductance was 22 mS/mm for a channel length of 3 μm. Kim et al. [26] fabricated a dual-gate GFET with Al_2O_3 as gate dielectric; however, the graphene layer grown on copper had to be transferred to the device platform as in many other efforts in the literature.

9.4 Graphene Preparation

Exfoliation was one of the earliest methods used [3] to produce single-layer graphene (SLG) and few-layer graphene (FLG) (2 to 10 layers). Graphite consists of graphene sheets held together by weak van der Waals forces that can be broken mechanically or chemically to release individual graphene sheets. Novoselov et al. [3] created small patterns of highly oriented pyrolytic graphite covered with a photoresist using oxygen plasma etching and then repeatedly used Scotch® tape to peel layers of graphene attached to the photoresist. Graphene has also been produced in bulk quantities by reducing graphite oxide using hydrazine hydrate.

CVD has become a popular technique to grow SLG and FLG films [27–31]. A controlled study by Yu et al. provides insight into the growth mechanism on thick nickel substrates [27]. These authors used thick nickel foils, first exposed to pure H_2 flow for an hour. This was followed by a 20 minute exposure of the substrate to a mixture of methane, hydrogen, and argon (0.15:1.0:2.0) at a total flow rate of 315 sccm and 1000°C. During this period, the hydrocarbon decomposes and diffuses into the nickel bulk. Then, a controlled cooling phase was used by removing the substrate downstream from the hot zone, which appears to be the critical step. Note here that there is no graphene during the heating phase, and as such, this is not a "deposition" process; instead, graphene forms as carbon atoms segregate out of nickel during cooling. If the cooling rate is slow, no graphene is found because carbon atoms continue to diffuse deeper into the bulk nickel. If the cooling rate is rapid, carbon atom segregation occurs but results in a defective graphite structure. A moderate cooling rate of the order 10°C/s results in a steady C segregation leading to graphene layer formation.

Several works have taken advantage of the above mechanism and prepared SLG and FLG films on Ni (100 to 300 nm) deposited on SiO_2 [28–31]. De Arco et al. [28] and Kim et al. [29] were able to prepare bilayer and few-layer graphene over a 4-inch wafer using this approach. When copper is used as substrate, the cooling rate-controlled segregation is not operative since the solubility of carbon atoms in copper is very low. Li et al showed growth of single, bi and three layer graphene on 25 micron thick copper foils using methane as feedstock at 1000° C and concluded that a surface-catalyzed process on copper surface is responsible for the growth [30].

9.5 Plasma-Enhanced Chemical Vapor Deposition (PECVD) of Graphene

The earliest report appears to be that by Wu et al. that describes the preparation of carbon nanowalls (CNWs) using a microwave PECVD [32]. Their chamber also consisted of a parallel plate setup with a DC bias on the lower electrode. They used Si, GaAs, stainless steel, and copper, all coated with one of NiFe, CoFe, FeMn, or CoCrPt catalysts at a thickness of 20 to 100 nm. The feedstock consisted of 40 sccm H_2 and 10 sccm methane. Interestingly, all the conducting substrates yielded only carbon nanotubes, while CNW growth was seen only on sapphire substrates. These were well-aligned carbon sheets of several nanometers thickness. The dependence of CNT versus CNF growth on the conductivity of the substrate is not clearly understood. Since the above report, several other studies have appeared on using various types of discharges to prepare graphene [33–44].

Hiramatsu et al. reported growth of CNWs using a radio frequency (RF) capacitively coupled methane discharge [33,42,43]. They injected atomic hydrogen produced from an inductive H_2 plasma into the hydrocarbon plasma. CNF growth occurred on Si, SiO_2, and sapphire substrates without any catalyst, which is different from the experience of Wu et al. described above [32]. The key here is the hydrogen injection, and no CNWs were seen without this added hydrogen. Figure 9.5 shows SEM images of these CNWs grown with atomic hydrogen injection, and the thickness is in the 1 to 10 nm range and sometimes with two layers near the edge. Figure 9.6 is a TEM image showing overlapping multilayered graphene domains with random

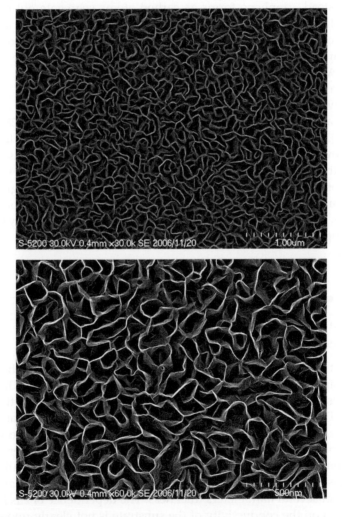

FIGURE 9.5
Scanning electron microscope (SEM) images of carbon nanowall growth with H injection at different magnifications. (Courtesy of Mineo Hiramatsu and Masaru Hori.)

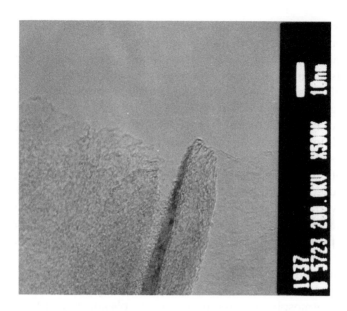

FIGURE 9.6
Transmission electron microscope (TEM) image of a carbon nanowall. (Courtesy of Mineo Hiramatsu and Masaru Hori.)

orientations. Wang et al. used an inductive plasma with methane concentrations of 5% to 100% in H_2 and observed CNW growth on substrates of Si, W, Mo, Zr, Ti, Hf, Nb, Ta, Cr, 304 stainless steel, SiO_2, and Al_2O_3, all without any catalyst [34,35]. Their characterization revealed one to two graphene layers with less than 1 nm in total thickness. No growth was observed at temperatures below 600°C, and CNW morphology was readily apparent between 630 and 830°C. The morphology did not depend on the nature of the substrate, an interesting feature considering that they used 12 different substrates. These authors proposed [36] that at the early stage, graphene layers grow parallel to the substrate, and the onset of vertical growth occurs after a 1 to 15 nm flat layer. At this point, sufficient force develops at the grain boundaries to curl up the leading edge of the top layers. Now, under the influence of the vertical electric field inside the plasma sheath, the high surface mobility of the carbon atoms and the electric-field-induced polarization of the graphitic layers help to keep the nanosheets growing in the vertical direction instead of piling thicker on the substrate. The authors claimed that the copious production of atomic H in the inductive plasma is key to obtaining CNWs [36]. The atomic hydrogen etches away amorphous carbon efficiently and prevents formation of secondary nuclei that might interfere with CNW growth. Also, cross-linking at the free edge of the growing sheets is eliminated, preventing the edges from getting thicker. This reasoning seems plausible because capacitive discharges may not produce as much atomic hydrogen as in an inductive discharge due to smaller electron densities. Hiramatsu et al. [33]

could get CNW growth only with atomic hydrogen injection into a capacitive hydrocarbon discharge, as mentioned earlier.

Chuang et al. [37] used an acetylene-ammonia mixture in contrast to all other efforts featuring methane and hydrogen. This feedstock also enabled CNWs on silicon substrates without any catalyst. Obraztsov et al. employed a DC discharge of methane and hydrogen and observed well-ordered graphite films of about 1.5 nm in thickness on 0.5 mm thick nickel substrates [38]. They kept their H_2 mole fraction of the inlet mixture at 92%. A remote plasma has also been used to grow single-layer and multilayer graphene films wherein the elimination of the electric field effect provides planar films [44]. Finally, Yuan et al. used a microwave plasma at 1.2 kW power with 10% methane in H_2 and grew graphene sheets at 800°C without any catalyst [41]. This was the only plasma-grown graphene in the literature that featured a very sharp, distinct 2D peak in the Raman spectra at 2653 cm^{-1}, providing evidence for a single-layer graphene.

Even though all of the above studies have focused on graphene growth on substrates, PECVD has also been used for bulk synthesis [45]. Dato et al. employed an atmospheric pressure microwave plasma reactor to synthesize graphene sheets using liquid ethanol as a precursor in an argon plasma. This approach creates an argon plasma in a 2.1 cm diameter quartz tube reactor at atmospheric pressure, into which an aerosol mixture of argon and ethanol is injected. The dissociated contents emerging from the plasma undergo rapid cooling, and the solid products collected in a filter consist of single-layer and bilayer graphene sheets [45].

The examples given above demonstrate that PECVD can be used to grow single-layer and multilayer graphene sheets, parallel or vertical to the growth substrate, with or without a catalyst and on a wide range of substrates. This versatility is truly appealing from an application development point of view. However, in order to make progress, several questions need to be answered, requiring further careful investigations. The list below is similar to the one raised in the case of PECVD of CNTs in 2003 at a similar stage in the evolution of carbon nanotubes [46]:

- What is the growth mechanism for PECVD of graphene?
- Thermal CVD results indicate carbon solubility during heating followed by segregation during cooling for nickel substrates, but catalytic activity for copper substrates. Is that also true for PECVD?
- A wide variety of substrates have been used in PECVD with almost substrate-independent characteristics. How can this be explained?
- Unlike a thermal process using methane at temperatures between 650 and 850°C, methane plasma will have numerous other stable hydrocarbons, atoms, and radicals with differing solubilities. What is the role of all these species?

- The hydrocarbon plasma would also feature ions with high kinetic energy and fast atoms. What are their roles, and how do they affect the growth?
- What is the mechanism behind the ability to obtain vertical graphene layers or the carbon nanowalls?
- What is the effect of dilution with hydrogen? Atomic hydrogen is thought to be an etchant to remove unwanted amorphous carbon impurity. Is there more on the utility of atomic hydrogen? Because DC and RF capacitive discharges differ from the high-density inductive and microwave discharges in plasma density (and hence their ability to dissociate H_2 and the hydrocarbon to produce H), would the recipes for dilution ratios (ratio of hydrocarbon to hydrogen) be different for these two classes of plasmas?
- Would the conventional wisdom of using plasmas for low-temperature processing apply to graphene growth? Can PECVD enable graphene preparation at temperatures below 500°C? How about room temperature growth?
- Issues related to large area growth and uniformity?
- How do PECVD samples differ from graphene prepared by other techniques including thermal CVD in terms of defects and other metrics?

Acknowledgments

The Division of IT-Convergence Engineering, World Class University (WCU) Program at POSTECH (South Korea) is acknowledged for support through the National Research Foundation of Korea funded by the Ministry of Science and Technology (R31-2008-000-10100-0). The authors acknowledge Mike Oye and Yunfan Zhao of NASA Ames, Professors Mineo Hiramatsu (Meijo University, Japan) and Masara Hori (Nagoya University, Japan) and Saebyuk Jeong (POSTECH) for providing unpublished figures.

References

1. M. Meyyappan, *J. Phys. D. Appl. Phys.* 44, xxx (2011).
2. M. Meyyappan, *J. Phys. D. Appl. Phys.* 42, 213001 (2009).
3. K. V. Novoselov et al., *Science.* 306, 666 (2004).
4. M. Meyyappan (Ed.), *Carbon Nanotubes: Science and Applications.* CRC Press, Boca Raton, FL (2004).

5. W. B. Choi , I. Lahiri, R. Seelaboyina, and Y. S. Kang, *Critical Rev. Solid State Mater. Sci.* 35, 52 (2010).
6. Y. Zhu, S. Murali, W. Cai, X. Li, J. W. Suk, J. R. Potts, and R. S. Ruoff, *Adv. Mater.* 22, 3906 (2010).
7. Z. H. Ni et al., *Nano Lett.* 7, 2758 (2007).
8. A. C. Ferrari et al. *Phys. Rev. Lett.* 97, 187401 (2006).
9. M. D. Stoller, S. J. Park, Y. Zhu, J. An, and R. S. Ruoff, *Nano Lett.* 8, 3498 (2008).
10. X. Du, P. Guo, H. Song, and X. Chen, *Electrochem. Acta.* 55, 4812 (2010).
11. K. Zhang, L. L. Zhange, X. S. Zhao, and J. Wu, *Chem. Mater.* 22, 1392 (2010).
12. K. Ku, B. Kim, H. Chung, and W. Kim, *Syn. Metals.* 160, 2613 (2010).
13. L. Qu, Y. Liu, J. B. Baek, and L. Dai, *ACS Nano.* 4, 1321 (2010).
14. L. G. De Arco, Y. Zhang, C. W. Schlenker, K. Ryu, M. E. Thompson, and C. Zhou, *ACS Nano.* 4, 2865 (2010).
15. F. Xia, D. B. Farmer, Y. M. Lin, and P. Avouris, *Nano. Lett.* 10, 715 (2010).
16. L. Liao, J. Bai, Y. Qu, Y. Huang, and X. Duan, *Nanotechnology.* 21, 015705 (2010).
17. D. Kondo, S. Sato, K. Yagi, N. Harada, M. Sato, M. Nihei, and N. Yokoyama, *Appl. Phys. Express.* 3, 025102 (2010).
18. S. Rumyantsev, G. Liu, W. Stillman, M. Shur, and A. A. Balandin, *J. Phys. Condens. Matter.* 22, 395302 (2010).
19. S. K. Banerjee, L. F. Register, E. Tutuc, D. Basu, S. Kim, D. Reddy, and A. H. Macdonald, *Proc. IEEE.* 98, 2032 (2010).
20. T. Palacios, A. Hsu, and H. Wang, *IEEE Commun. Magazine.* P122, 2032 (2010).
21. S. Alwarappan, A. Erdem, C. Liu, and C. Z. Li, *J. Phys. Chem. C.* 113, 8853 (2009).
22. S. He et al., *Adv. Funct. Mater.* 20, 453 (2010).
23. Y. Shao, J. Wang, H. Wu, J. Liu, I. A. Aksay, and Y. Lin, *Electroanalysis.* 22, 1027 (2010).
24. Y. Shao et al., *J. Mater. Chem.* 20, 7491 (2010).
25. P. Wu et al., *Electrochemica. Acta.* 55, 8607 (2010).
26. S. Kim et al., *Appl. Phys. Lett.* 94, 062107 (2009).
27. Q. Yu, J. Lian, S. Siriponglert, H. Li, Y. P. Chen, and S. S. Pei, *Appl. Phys. Lett.* 93, 113103 (2008).
28. L. G. De Arco, Y. Zhang, A. Kumar, and C. Zhoue, *IEEE Trans. Nanotechnol.* 8, 135 (2009).
29. K.S. Kim et al, *Nature.* 957, 706 (2009).
30. X. Li et al, *Science.* 324, 1312 (2009).
31. B. J. Lee, H. Y. Yu, and G. H. Jeong, *Nanoscale Res. Lett.* 5, 1768 (2010).
32. Y. Wu, P. Qiao, T. Chong, and Z. Shen, *Adv. Mater.* 14, 64 (2002).
33. M. Hiramatsu, K. Shiji, H. Amano, and M. Hori, *Appl. Phys. Lett.* 84, 4708 (2004).
34. J. J. Wang, M. Y. Zhu, R. A. Outlaw, X. Zhao, D. M. Manos, B. C. Holoway, and V. P. Mammana, *Appl. Phys. Lett.* 85, 1265 (2004).
35. J. J. Wang, M. Y. Zhu, R. A. Outlaw, X. Zhao, D. M. Manos, and B. C. Holoway, *Carbon.* 42, 2867 (2004).
36. M. Zhu, J. J. Wang, B. C. Holoway, R. A. Outlaw, K. Hou, V. Shutthanandan, and D. M. Manos, *Carbon* 45, 2229 (2007).
37. A. T. H. Chuang, B. O. Boskovic, and J. Robertson, *Diamond Relat. Mater.* 15, 1103 (2006).

38. A. N. Obraztsov, E. A. Obraztsova, A. V. Tyurnina, and A. A. Zolotukhin, *Carbon.* 45, 2017 (2007).
39. A. Malesevic, R. Kemps, L. Zhang, R. Erni, G. Van Tendeloo, A. Vanhulsel, and C. Van Haesendonck, *J. Optoelect. Adv. Mater.* 10, 2052 (2008).
40. N. G. Shang, P. Papakonstantinou, M. McMullan, M. Chu, A. Stamboulis, A. Potenza, S. S. Dhesi, and H. Marchetto, *Adv. Func. Mater.* 18, 3506 (2008).
41. G. D. Yuan, W. J. Zhang, Y. Yang, Y. B. Tang, Y. Q. Li, J. X. Wang, X. M. Meng, Z. B. He, C. M. L Wu, I. Bello, C. S. Lee, and S. T. Lee, *Chem. Phys. Lett.* 467, 361 (2009).
42. M. Hiramatsu and M. Hori, *Carbon Nanowalls*, Springer, New York (2010).
43. W. Takeuchi, K. Takeda, M. Hiramatsu, Y. Tokuda, H. Kano, S. Kimura, O. Sakata, H. Tajiri, and M. Hori, *Phys. Solidi A*, 207, 139 (2010).
44. G. Nandamuri, S. Roumimov, and R. Solanki, *Appl. Phys. Lett.* 96, 154101 (2010).
45. A. Dato, V. Radmilovic, Z. Lee, J. Phillips, and M. Frenklach, *Nano Lett.* 8, 2012 (2008).
46. M. Meyyappan, L. Delzeit, A. Cassel, and D. Hash, *Plasma Sources Sci. Technol.* 12, 205 (2003).

10

Modeling Aspects of Plasma-Enhanced Chemical Vapor Deposition of Carbon-Based Materials

Erik Neyts
Ming Mao
Maxie Eckert
Annemie Bogaerts

CONTENTS

10.1 Introduction

For many decades, gas discharge plasmas have played an important role in the deposition of thin films and nanostructured materials, in plasma-enhanced chemical vapor deposition (PECVD).[1-3] In PECVD, a gas discharge is applied to create reactive species (e.g., radicals, by electron impact dissociation of a molecular gas); these reactive species act as growth precursors for the deposition process. To improve the performance of the plasma growth processes, a clear understanding of the plasma and of the interaction with the growing films or nanostructures is needed. Computer simulations can be a very useful tool for this, because information can be obtained which is often difficult to access experimentally.

In this chapter, an overview is given of different modeling approaches that can be applied for PECVD of carbon-based materials. This includes two different aspects: modeling of the plasma and modeling of the deposition process. Plasma modeling includes both physical (e.g., plasma breakdown, electrical characteristics, electron heating, and fluid dynamics) and chemical aspects (i.e., creation and destruction of the various plasma species by chemical reactions). The latter can give indications on which species are important precursors for the deposition and how the plasma operating conditions can be tuned to optimize the deposition process. Therefore, in this chapter, we focus mainly on the plasma chemistry modeling.

The second aspect of PECVD modeling involves the deposition process (i.e., the interaction of the plasma with the growing film). This can be simulated by phenomenological, mechanistic models or by detailed atomistic descriptions of the growth process. For the film deposition modeling, input from the plasma simulations, such as the fluxes of species arriving at the substrate, is desirable. Vice versa, the plasma-surface interactions provide boundary conditions for the plasma simulations, such as sticking coefficients and surface reaction probabilities at the walls.

In our research group, we are active in both plasma chemistry modeling and modeling of thin film deposition, for different types of gas discharges and gas mixtures, and different materials. Therefore, the overview of the different modeling approaches will be illustrated by some examples from our own research. For a consistent picture, the examples shown here are all related to hydrocarbon plasmas and carbon-based films and nanostructured materials, but the same principles apply also to other gas mixtures and materials.

10.2 Plasma Chemistry Modeling for Plasma-Enhanced Chemical Vapor Deposition (PECVD)

There exist several plasma modeling approaches in the literature, including analytical models, zero-dimensional (0D) chemical kinetics models, one-dimensional (1D) or two-dimensional (2D) fluid approaches, Boltzmann models, Monte Carlo (MC) and particle-in-cell–Monte Carlo collision (PIC-MCC) simulations and hybrid approaches. These models will be explained below in a bit more detail. Table 10.1 gives a literature overview of the different modeling approaches applied to PECVD. Although the overview in this book chapter is mainly limited to hydrocarbon plasmas and carbon-related films, the table also summarizes modeling efforts for silicon-related films, to demonstrate that the work on carbon films is representative of PECVD in general.

It is clear from Table 10.1 that different modeling approaches have been used for PECVD, for different types of films, plasma reactors, gas mixtures, and pressures. This is not an exhaustive but rather an illustrative list. All these models focus on plasma chemistry. Surface chemistry is typically accounted for by a simple sticking model. Only a number of plasma models include detailed surface chemistry, as is clear from this table. The same literature overview is presented in a more concise way in Table 10.2, illustrating that most efforts for PECVD applications have been made by fluid models. Below, we will give a short overview of the different modeling approaches, presenting some examples from literature for hydrocarbon plasmas and explaining the strengths and weaknesses of these models for PECVD applications.

10.2.1 Analytical Modeling

In *analytical models* the plasma behavior is described with a number of analytical formulae, derived from plasma physics theory. This yields a fast description of the plasma, but it is, of course, an approximation, only valid under specified operating conditions. Gordillo-Vázquez and Albella developed a quasi-analytical space-time-averaged kinetic model for a radio-frequency (RF) inductively coupled plasma (ICP) in $C_2H_2/H_2/Ar$ used for nanocrystalline diamond (NCD) film deposition, with special focus on the underlying mechanisms driving the nonequilibrium plasma chemistry of C_2.[4–6] The authors suggest that the growth of NCD films under these conditions is very sensitive to the contribution of C_2 and C_2H species.[5] Another simplified analytical model has been developed by Dandy and Coltrin to simulate the diamond growth in a direct current (DC) arc jet reactor.[7] Both gas-phase reactions and surface processes for diamond film growth in a CH_4/H_2 plasma mixture were considered. The results from this simplified model agree with stagnation flow models in a wide range of operation conditions.[7]

TABLE 10.1

Overview of Different Models Applied for PE-CVD Applications of Carbon-Based Films and Nanostructures as well as Silicon-Based Films

Film Type	Plasma	Model	Working Gas (Pressure)	Surface Chemistry	Reference
Carbon-Based Films or Nanostructures					
Nanocrystalline diamond (NCD) film	Inductively coupled plasma (ICP)	Analytical model	$C_2H_2/H_2/Ar$	No	4–6
Diamond film	Direct current (DC) arc jet		CH_4/H_2 (1 atm)	Detailed surface chemistry	7
NCD film	Microwave (MW) plasma	Zero-dimensional (0D) global model	$Ar/H_2/CH_4$ (200 mbar)	No	8,9
Carbon nanotubes (CNTs)	ICP		CH_4/H_2, $Ar/CH_4/H_2$, CH_4/N_2 (few Torr)	Sticking model	10–12
a-C:H film	Electron cyclotron resonance (ECR) plasma		CH_4 (0.15 Pa)	Surface kinetics	13
Polypropylene treatment	DBD		N_2/O_2 (1 atm)	Yes	14,15
CNTs	DC plasma	One-dimensional (1D) fluid model	C_2H_2/NH_3 (8 Torr)	No	16–19
CNTs	Capacitively coupled plasma (CCP)		CH_4/H_2 (3 Torr)	Sticking model	20,21
Diamond-like carbon (DLC) film	CCP		CH_4 (0.14 Torr)	Sticking model	22
a-C:H film	CCP		C_2H_2 (0.3 Torr)	Sticking model	23–25
NCD film	MW plasma		CH_4/H_2 (10s Torr)	Sticking model	26
a-C:H film	CCP	Two-dimensional (2D) fluid model	CH_4 (0.14 Torr)	Sticking model	27
Diamond-like carbon (DLC) film	ICP		CH_4 (10 mTorr)	Detailed model	28
DLC film	DC plasma		CH_4/H_2 (1 atm)	Detailed model	29
CNTs	ICP		Ar/CH_4 (50 mTorr)	Sticking model	30,31

Film	Reactor	Model	Plasma chemistry	Surface chemistry	Ref.
DLC film	CCP	Particle-in-cell–Monte Carlo collision (PIC-MCC) simulations	CH_4/H_2 (0.14 Torr)	Sticking model	32
DLC film	CCP		CH_4 (few mTorr)	No	33
a-C:H film	CCP		CH_4 (50–300 mTorr)	No	34
Ultrananocrystalline diamond/nanocrystalline diamond [(U)NCD] film	MW or hot filament (HF) plasma	2D Hybrid model	$Ar/CH_4/H_2$ (10s Torr)	Sticking model	35–39
NCD film	DC arc jet		$Ar/CH_4/H_2$ (10s Torr)	Sticking model	40
CNTs	ICP		$CH_4(C_2H_2)/H_2(NH_3)$ (<50 mTorr)	Sticking model	41,42
Silicon-Based Films					
μc-Si:H film	CCP	Analytical model	SiH_4 (3 Torr)	Detailed	43
a-Si:H	CCP	0D Global model	Ar/SiH_4	Sticking model	44
a-Si:H/μc-Si:H	CCP		SiH_4	Simple model	45
a-Si:H	CCP	1D Fluid model	SiH_4/He (sub Torr)	Sticking model	46
a-Si:H	CCP		SiH_4/H_2 (0.14 Torr)	Sticking model	47
SiC	ICP		$C_3H_8/SiH_4/H_2$	Sticking model	48
SiNx film	CCP	2D Fluid model	$N_2/SiH_4/NH_3$	No	49
a-SiNx:H	CCP		SiH_4/NH_3	Sticking model	50
μc-Si:H	CCP		SiH_4/H_2	No	51
a-Si:H	CCP	PIC-MCC simulations	SiH_4/H_2 (<300 mTorr)	Sticking model	52
a-Si:H	Pulsed plasma		SiH_4	Sticking model	53
a-Si:H	CCP	2D Hybrid model	SiH_4/H_2	Sticking model	54

Notes: All of these models describe the plasma chemistry; some models also include a description of the surface chemistry, by either simple sticking models or more detailed surface kinetics, as indicated in column 5. The abbreviations are explained in the text.

TABLE 10.2

Summary of the Different Modeling Approaches Used for Describing the Plasma Chemistry for Plasma-Enhanced Chemical Vapor Deposition (PECVD) Applications of Different Carbon or Silicon-Based Materials

Film (Structure) Type	Analytical Method	0D Global Model	Fluid Model	PIC-MCC Simulations	Hybrid Model
a-C:H		13	23–25,27	34	
DLC	7		22,28,29	32,33	
(U)NCD	4–6	8,9	26		35–39
CNTs		10–12	16–21,30–31		41,42
Polypropylene		14,15			
a-Si:H		44,45	46,47	52,53	54
μc-Si:H	43		51		
SiC			48		
SiNx			59,50		

Note: The numbers in the table correspond to the references of these papers.

10.2.2 0D Chemical Kinetics or Global Modeling

In describing the plasma chemistry, a *0D chemical kinetics approach* is often applied. It is also called *global modeling* as it does not include spatial variations in the plasma. The method is based on solving rate equations for different plasma species (i.e., electrons, various types of molecules, ions, radicals, excited species). This approach can take into account a large number of different species and reactions without too much computational effort and is therefore of particular interest for describing the plasma chemistry.

Hassouni, Lombardi, and colleagues studied an $Ar/H_2/CH_4$ microwave (MW) discharge used for NCD deposition by means of experiments and 0D plasma chemistry modeling.[8,9] The authors suggest that "sp" species, especially C_2H_2, play a key role in the surface chemistry that governs the diamond growth.[8] As soot formation is often observed experimentally in this kind of plasma reactor, the mechanisms giving rise to soot particles, including both ionic and neutral pathways, were investigated.[9]

Several authors also presented 0D chemical-kinetics models of the plasma chemistry for carbon nanostructure (e.g., carbon nanotube, CNT) growth in ICPs, in various gas mixtures, such as CH_4/H_2,[10] $Ar/CH_4/H_2$,[11] and CH_4/N_2.[12]

A 0D plasma kinetics model incorporating detailed surface chemistry was developed by Möller[13] for the deposition of hydrogenated amorphous carbon (a-C:H) films from a low-pressure electron cyclotron resonance (ECR) plasma in CH_4. He predicted that direct ion incorporation, deposition from the adsorbed layer, and re-etching all contributed more or less equally to determine the carbon growth rate.

Finally, a very powerful and efficient 0D chemical kinetics model, called Global_kin, was developed by Dorai and Kushner and applied to investigate the gas phase and surface kinetics during humid-air dielectric barrier discharge (DBD) treatment of polypropylene.[14,15]

10.2.3 Fluid Modeling

It should be realized that a 0D model assumes a uniform plasma composition and does not explicitly consider the transport of the plasma species throughout the reactor. The transport of plasma species is accounted for in the *fluid approach*, in which the different plasma species are considered as separate fluids, and their behavior is described with continuity equations for mass, momentum, and energy (i.e., the first three moments of the Boltzmann transport equation). In practice, an energy balance equation is typically only applied for the electrons, because the other plasma species can be considered more or less in thermal equilibrium with the background gas. Moreover, the momentum conservation equations are commonly reduced to transport equations, based on the drift-diffusion approximation (i.e., transport by diffusion), and for the charged species, also by migration (drift) in the electric field. When the mass continuity and flux equations for the different species and the electron energy balance equation are solved simultaneously with the Poisson equation, a self-consistent electric field distribution can be calculated. The charged species densities, as obtained from the mass continuity equations, are inserted into the Poisson equation, which yields an electric field distribution that is in turn used in the charged species transport equations (migration-term). In this way, a self-consistent picture of the plasma behavior can be obtained.

Fluid approaches, both in 1D[16–21] and 2D,[30,31] have been applied to describe the plasma chemistry in various types of plasma reactors used for CNT and related carbon nanostructure growth, including DC plasmas,[16–19] capacitively coupled plasmas (CCPs),[20,21] and ICPs.[30,31] Various types of gas mixtures were considered in these models, based on either CH_4 or C_2H_2 as hydrocarbon growth precursors, mixed with H_2 or NH_3 as etchant gases, sometimes diluted with Ar. Typical calculation results include the density profiles and fluxes of the species bombarding the growing nanostructure (i.e., the growth precursors). It is reported that C_2H_2, C_2H_4, H, and CH_3 are major neutral species. Atomic H is important because it yields preferential etching of amorphous carbon phases, resulting in more "clean" CNT formation. Furthermore, the models yield information about the conversion (also called "decomposition rate") of the feedstock gases in the plasma.

Herrebout et al. carried out a comparison between a 1D[22] and 2D[27] fluid model for a CH_4 RF CCP used for the deposition of diamond-like carbon (DLC) or amorphous hydrogenated carbon (a-C:H) layers. The species densities and fluxes toward the substrate calculated with the 1D and 2D fluid model were in agreement.[27] Of course, a 2D fluid simulation can give additional information on the radial variation of the fluxes toward the substrate, and hence can make predictions on the uniformity of film deposition. The

C_2H_2 plasma chemistry was also investigated in our group by using a 1D fluid model, focusing on gas phase polymerization reactions.[23–25]

A 1D fluid model for NCD deposition from a MW plasma operating in CH_4/H_2 was developed by Hassouni and coworkers.[26] The authors found that the addition of CH_4 at low concentration (<5%) in the discharge did not lead to a significant change in the energy transfer process but significantly affected the electron temperature. The added CH_4 was rapidly converted to C_2H_2 near the gas inlet, and the latter was transported to the discharge region.

Bera and coworkers[28] developed a self-consistent 2D fluid model and applied it to an ICP operating in CH_4. The surface deposition/etching processes, involving adsorption–desorption, adsorption layer reaction, ion stitching, direct ion incorporation, and carbon sputtering, were also considered in this model. Some parametric studies provided an insight into the discharge and process characteristics for carbon thin film deposition and etching.

Finally, Farouk et al. applied the 2D fluid software CFD-ACE+ to investigate an atmospheric pressure CH_4/H_2 DC microplasma for thin film deposition.[29] The surface chemistry for the deposition model was taken into account. The authors found that H_3^+, CH_5^+, and $C_2H_5^+$ were the dominant ions; CH_2 and CH_3 were the main radicals; and C_2H_6, C_3H_8, C_2H_4, and C_2H_2 were also present at high density. These species were determined as the prominent growth species for DLC deposition at atmospheric pressure.

A fluid approach has a reasonable calculation time and is therefore also suitable for describing the plasma chemistry for PECVD processes. However, the plasma species are assumed to be more or less in equilibrium with the electric field (i.e., the energy gained by the electric field is more or less balanced by the energy lost through collisions). This is not completely true, especially for the electrons, which typically gain more energy from the strong electric field in the plasma than they lose by collisions. The nonequilibrium behavior of the plasma species is fully accounted for in kinetic models, which are described in the next section.

10.2.4 Kinetic Models

As mentioned above, kinetic models take into account the nonequilibrium behavior of the plasma species. This can be done by solving the full Boltzmann transport equation for every plasma species, instead of only solving the first (three) moments. This *Boltzmann model* is more accurate, but it becomes mathematically very complicated, especially if one tries to model the plasma in more than one dimension. Simplifications to the Boltzmann equation are possible but are only valid under certain conditions. In general, a Boltzmann model on its own is not used for describing complex plasma chemistries. Most often, a Boltzmann model is limited to describe the electron kinetics, and the results of it are subsequently used as input in a fluid model. By solving the electron energy distribution function (EEDF) with the Boltzmann equation for fixed electric field values and given gas compositions, and calculating the mean electron energy and electron impact collision rates from the EEDF, look-up tables can be created for the collision rates as a function of mean electron energy. In a fluid model, the mean electron energy is calculated from the electron energy balance equation, as mentioned above, and the corresponding collision rates are then interpolated from the look-up tables.

Another modeling approach, which is very accurate and moreover, mathematically straightforward, is *Monte Carlo (MC) simulations*, which treat the plasma species as individual particles. A number of superparticles are followed, which represent a large number of real particles (e.g., electrons), as defined by their "weight." During successive time-steps, the movement of these superparticles, under the influence of the electric field, is simulated with Newton's laws, and their collisions during every time-step (i.e., occurrence and kind of collision, new energy and direction after the collision) are treated with random numbers. By following a large number of superparticles in this way, their behavior can be statistically simulated. However, in order to reach statistically valid results, a large number of superparticles need to be followed, and hence long calculation times are required, especially for slow-moving species. Moreover, this modeling approach is not fully self-consistent, because the electric field, needed to simulate the species' trajectories, has to be given as input in the model. Therefore, MC models are often applied in combination with fluid models, in hybrid approaches, as will be described below.

To overcome the non-self-consistent character of the MC approach, it can also be incorporated into *particle-in-cell–Monte Carlo collision (PIC-MCC) simulations*. In this approach, the above description of the superparticles is complemented in every time-step with the solution of the Poisson equation (i.e., the charged species densities, necessary to solve the Poisson equation, are obtained from the positions of all [charged] superparticles at every time-step, as calculated with Newton's laws). In this way, a fully self-consistent description of the plasma behavior can be obtained. However, due to the coupling of the statistical description of the superparticles with the solution of the Poisson equation, the PIC-MCC simulations are even more time consuming and are therefore less suitable for describing a large number of different plasma species in complex plasma chemistries.

Nevertheless, a few groups have reported the use of a PIC-MCC approach for describing the plasma chemistry in hydrocarbon plasmas used for DLC deposition. Ivanov et al.[32] developed a 1D PIC-MCC model for a RF CCP in a mixture of CH_4 and H_2 used for DLC film deposition. The PIC-MCC model was compared with a 1D fluid model.[22] The nonstationary and nonlocal features of the EEDF were demonstrated in the PIC-MCC calculations. Another 1D PIC-MCC model for a CCP in CH_4 was developed by Proshina and coworkers.[33] The frequency and pressure dependence of the ion energy distribution function (IEDF) was studied, and a correlation between the IEDF and the experimental deposition rate of DLC films was analyzed. Finally, a two-dimensional PIC-MCC simulation of a CCP in CH_4 was carried out by Alexandrov and Schweigert.[34] Two regimes of discharge operation were observed in their simulations. Their calculation results showed agreement with experimental data from the literature.[55]

10.2.5 Hybrid Modeling

It is clear that every modeling approach has its own advantages and disadvantages. Therefore, different models can also be combined into a *hybrid modeling network*. For instance, for the energetic plasma species, such as the electrons that are not in equilibrium with the electric field, MC simulations (or the Boltzmann equation) can be applied, whereas for other plasma species, which can be considered more or less in thermal equilibrium, such as the heavy particles (ions, molecules, radicals), the much faster fluid approach is valid and, moreover, ensures a self-consistent description of the plasma behavior. In this way, the drawbacks of the different modeling approaches are avoided, and the benefits of the models are fully realized. Such a hybrid modeling approach is also suitable for describing the plasma chemistry.

A 2D hybrid model was developed by May, Ashfold, Mankelevich, and colleagues, to simulate the PECVD processes for NCD and ultrananocrystalline diamond (UNCD) growth.[35-39] The model consists of three parts that describe (1) the activation of the gas mixture, (2) the gas dynamics and chemical kinetics, and (3) the gas–surface interactions. The model is applied to both hot filament (HF) and microwave (MW) plasma CVD systems. The shape and size of the plasma ball are taken as input parameters, obtained from experiments. Also, a uniform absorbed power density and electron temperature were applied; the latter was obtained from a 0D plasma kinetics model, solving balance equations for given reduced electric fields. Thirty-five chemical species and around 300 reactions were considered in the model. Based on these model calculations, the authors could explain why in their experiments UNCD films can be grown much more easily in the MW plasma than in the HF CVD reactor. Furthermore, the model predicts that the densities of CH_3, C, and C_2H are greater than the C_2 density, suggesting that these species are more important precursors for UNCD growth.[35,37] Another 2D model was developed by the same group for a DC arc jet reactor, also used for microcrystalline diamond (MCD) and NCD deposition.[40] This model includes gas activation, expansion in the low-pressure reactor chamber, and reaction chemistry in the plasma and at the surface. C and CH are predicted to be the main radical species bombarding the growing (nano)diamond surface.

Kushner and coworkers also developed a very powerful 2D hybrid model, called the hybrid plasma equipment model (HPEM)[56] that describes the plasma by a combination of an electromagnetics module for the electric and magnetic fields in the plasma, a kinetic model (either Monte Carlo or Boltzmann) for the fast electrons, and a fluid model for the slow electrons and heavy particles. This model is mainly used to simulate the plasma conditions for etching applications (e.g., Zhang and Kushner[57]), although it has also been applied for physical vapor deposition (PVD) (e.g., Arunachalam,[58] Grapperhaus[59]). Another very powerful 2D hybrid model, called nonPDPSIM, was also developed by Kushner and coworkers[60] and has been used, among others, to study the continuous processing of polymers in repetitively pulsed atmospheric pressure discharges.[61]

We recently applied the HPEM to describe the detailed plasma chemistry in an ICP reactor used for CNT growth in four different gas mixtures—CH_4/H_2, C_2H_2/H_2, CH_4/NH_3, and C_2H_2/NH_3.[41] In the CH_4/H_2 plasma, 33 species (electrons, ions, radicals, and background neutrals) along with 58 electron impact reactions, 115 ion-neutral and 45 neutral-neutral reactions are taken into account. For the C_2H_2/H_2 plasma, 48 different species are considered, which take part in 105 reactions (i.e., 31 electron-impact reactions, 29 neutral-neutral, and 45 ion-neutral reactions). When NH_3 is used as dilution gas instead of H_2, an extra number of 22 species, 43 electron-impact reactions, 48 ion-neutral, and 67 neutral-neutral reactions were added to the model. All details about this plasma chemistry can be found in Mao and Bogaerts.[41] In a subsequent paper[42] a numerical parameter study was carried out with this model, varying the gas ratios, gas pressure, coil power, bias power, and substrate temperature. Typical calculation results included the power density deposited in the plasma, the gas temperature, the electron temperature and density, the densities of all different plasma species, as well as their fluxes to the substrate. The latter results are of special interest for CNT growth.

Figure 10.1 illustrates the calculated radially averaged fluxes of the various neutral species (a) and ions (b) bombarding the substrate as a function of CH_4 gas fraction in a CH_4/H_2 gas mixture, as well as the fluxes of the extra neutrals (c) and ions (d) included in the model for a CH_4/NH_3 gas mixture. It is observed from Figure 10.1a that C_2H_2, C_2H_4, and C_2H_6 are the dominant molecules regardless of the gas mixture ratios, in addition to the feedstock gases CH_4 and H_2. Because it was reported[62] that the unsaturated hydrocarbons, such as C_2H_2 and C_2H_4, have a higher reactivity than CH_4 on transition metal clusters, and decompose more easily at the catalyst surface than CH_4 at lower temperature, it is possible that C_2H_2 and C_2H_4 are the major precursors for CNT growth in a CH_4/H_2 plasma.

The primary radicals are H and CH_3 at all mixture ratios. Atomic hydrogen plays a key role in the dehydrogenation of the adsorbed hydrocarbons, enhancing the surface diffusion of carbon, and etching of amorphous carbon.[62] On the other hand, the CH_3 radicals are probably responsible for the amorphous carbon fraction in the CNTs.[62] As is clear from Figure 10.1a, the H flux decreases gradually, whereas the CH_3 flux increases at increasing CH_4 gas fraction. This suggests that low CH_4 gas fractions (i.e., until about 20%) will lead to "clean" conditions for CNT growth, as the amorphous phase, which might be present in the growing tube, will be etched away by the H-flux. In Matthews et al.,[63] it was also concluded, albeit for a MW-PECVD system, that an optimal ratio of H_2/CH_4 was obtained at 50:10, hence corresponding to less than 20% CH_4 in the gas mixture. Finally, it is clear that the

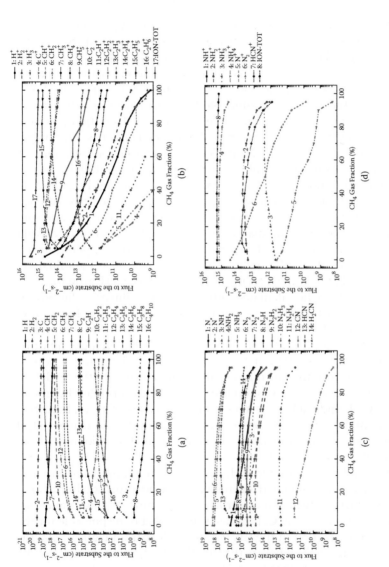

FIGURE 10.1

(See color insert.) Calculated radially-averaged fluxes of the various species bombarding the substrate as a function of CH$_4$ fraction: (a) neutral species and (b) ions in a CH$_4$/H$_2$ gas mixture; (c) extra neutral species and (d) extra ions in a CH$_4$/NH$_3$ gas mixture. The operating conditions are 50 mTorr total gas pressure, 100 sccm total gas flow rate, 300W source power, 30W bias power at the substrate and 13.56 MHz operating frequency at the coil and at the substrate electrode. The substrate is heated to 550°C. (Reproduced from [42] with permission of Institute of Physics.)

fluxes of C and C_2 are the lowest of all radicals. These observations confirm that the carbon source for CNT growth mainly arises from the decomposition of hydrocarbon molecules on the surface of the catalyst.[62]

As is apparent from Figure 10.1b, most ion fluxes show a rapid decrease as the CH_4 fraction increases, except for $C_2H_x^+$ ($x = 2, 3, 4, 5, 6$). The $C_2H_2^+$ and $C_2H_3^+$ fluxes exhibit a local maximum at 10% CH_4 gas fraction, while $C_2H_x^+$ ($x = 4, 5, 6$) rise with increasing CH_4 concentration. It is suggested in Mao and Bogaerts[42] that this may result in enhancement of physical sputtering and hence cause damage of the CNTs.

When NH_3 is applied as the dilution gas instead of H_2, some new species appear in the plasma, as is clear from Figures 10.1c and 10.1d, which show only the fluxes of the "extra species." The fluxes of the other species exhibit the same trends as in the CH_4/H_2 plasma; therefore, they are not shown here again. H_2 was found to be again the most dominant species, in spite of the fact that it is now not one of the feedstock gases. It appears to be mainly generated by decomposition of NH_3 at the conditions under study. Apart from the hydrocarbons C_2H_2, C_2H_4, and C_2H_6, N_2 and HCN also become predominant, in addition to the feedstock gases CH_4 and NH_3, as is clear from Figure 10.1c. Atomic hydrogen is again the most important radical, followed by CH_3, N, N_2H, and NH_2.

Finally, it is observed from Figure 10.1d that the total ion flux remains almost constant as a function of CH_4 fraction, as was also more or less the case in the CH_4/H_2 gas mixture (see Figure 10.1b). However, NH_4^+ now becomes the predominant ion when NH_3 is used as the additive. It is apparent that NH_4^+, NH_3^+, and NH_2^+ do not drop significantly and even rise (in the case of NH_3^+) upon increase of CH_4, or decrease of the NH_3 fraction. This indicates that NH_3 is more efficiently ionized at lower NH_3 fraction.

Similar results were also obtained in Mao and Bogaerts[42] for C_2H_2/H_2 and C_2H_2/NH_3 gas mixtures at varying gas ratios, although in this case some long-chain hydrocarbons were produced as well. In general it was concluded that a lower fraction of carbon-supplying gases (CH_4 and C_2H_2) compared to etching gases (H_2 and NH_3) will lead to more "clean" CNT growth conditions, which is in agreement with literature observations.[10,19,63–65] The same applies to a higher ICP power, a moderate gas pressure of about 100 mTorr (at least for single-walled [SW] CNTs), a high bias power (for aligned CNTs), and an intermediate substrate temperature.[42]

When modeling PECVD, it is interesting to realize that the HPEM does not only apply to the behavior of the plasma, but also to film deposition. Besides the modules mentioned above (i.e., electromagnetics module, electron kinetics module, and fluid part), some extra modules can be added in order to simulate deposition or etch processes at a substrate. These extra modules include a MC module for calculating the ion and neutral energy and angular distribution functions when bombarding the substrate, an analytical surface kinetics model, and a MC module for feature profile evolution.[66] Although these extra modules are mainly applied for simulating etch processes,[57] they are equally suitable for describing deposition processes. The same applies to the other hybrid model, nonPDPSIM, which is more suitable for atmospheric pressure processing.

10.3 Modeling the Deposition Process for PECVD

The fluxes of various plasma species impinging on a substrate, as calculated by the plasma modeling, can be used as input for the simulations of the plasma–surface interactions and hence for describing the growth process of thin films or nanostructured materials, in PECVD. Roughly, two different simulation approaches can be distinguished for describing thin film deposition or nanostructure growth mechanisms: an atomistic approach, providing detailed information on the interaction mechanisms of plasma species with the growing film or nanostructure, and a more macroscopic, phenomenological approach, which can give an overall illustrative view of the growth mechanisms. In the following, we will first explain a few of these phenomenological, mechanistic approaches presented in literature, before focusing on the detailed atomistic simulations of thin film and nanostructure growth processes, where some examples of our own research group will be given. Again, we focus only on carbon-based materials.

10.3.1 Mechanistic Modeling

There exist a variety of mechanistic models in the literature for describing thin film deposition[35–40,67–72] or carbon nanostructure growth.[73–88] Often, these models do not necessarily apply to PECVD, but apply to physical vapor deposition (PVD) (e.g., magnetron sputter deposition)[71,72] or chemical vapor deposition (CVD).[73–84] Therefore, we will not go into detail about these models. As an example, we give here only a brief literature overview of mechanistic models for the growth of CNTs (and related nanostructures).[73–88] Typically, these models are more or less based on the same processes, including adsorption and desorption of carbon species at the metal catalyst particle, surface and bulk diffusion, surface reaction on the catalyst and substrate, and nanostructure nucleation and growth. In general, these mechanistic models can be divided into three groups from simulation point of view: kinetic models,[73–80] multiphysics, multiphase integrated models,[81–84] and kinetic Monte Carlo (MC) models.[85–88] Kinetic simulations are particularly attractive because of their simplicity and reduced computational effort. In a kinetic simulation, all possible processes are described either with diffusion coefficients (for diffusion processes) or with rate constants (for reactions). Based on these parameters, a set of continuity equations is solved as a function of time, allowing the time-dependent growth rate and length of the CNTs to be predicted, along with the influence of processing parameters on these quantities. The multiphysics, multiphase models are based on the same principles, but they are integrated in a reactor model (for CVD) or a plasma model (for PECVD), to obtain self-consistent calculations. However, in this approach it is typically not so straightforward to express bulk diffusion processes.

Most of the models listed above are applied to CNT growth by CVD, but Ostrikov and coworkers specifically developed mechanistic models for the PECVD of carbon nanostructure growth.[78,79,85–88] The model presented by Denysenko and Ostrikov[78,79] for the PECVD of carbon nanofibers (CNFs) is a kinetic model that accounts for adsorption and desorption of C_2H_2 and H on the catalyst surface, surface and bulk diffusion, incorporation into a graphene sheet, as well as ion- and radical-assisted processes on the catalyst surface that are unique to a plasma environment. It was shown that plasma ions play a key role in the carbon precursor dissociation and surface diffusion, enabling low-temperature growth of carbon nanostructures. In Denysenko and Ostrikov,[79] the plasma heating effects were considered. The authors found that the calculated growth rates were in better agreement with the available experimental data than the results without heating effects. Finally, Levchenko, Ostrikov, and colleagues[85–88] presented an interesting multiscale MC/surface diffusion model for the plasma-based growth of carbon nanocone arrays on metal catalyst particles. The model includes the three main physical phenomena that play a key role in the nanostructure formation: (1) diffusion of adsorbed carbon atoms on the substrate surface toward the metal catalyst nanoparticles; (2) dissolution into the nanoparticle and eventually saturation of metal catalyst with carbon, resulting in nanocone nucleation and growth on top of the catalyst particle; and finally, (3) sputtering of the carbon nanocone with impinging carbon ions from the plasma. The model predictions suggested that the plasma parameters can effectively tailor the nanocone array properties and ultimately increase the array quality.[87]

As mentioned in Section 10.2.5, mechanistic models describing the surface kinetics (for deposition or etching) can also be integrated in hybrid plasma models. This is the case for HPEM and nonPDPSIM, developed by Kushner and coworkers, as was explained above (see more details in the literature[57,61]). Also the hybrid model developed by May, Mankelevich et al.[35–38] (see above) does not consider only the plasma processes, but also the plasma–surface interactions, more specifically for (U)NCD thin film deposition. It accounts for nine gas–surface reactions, involving the H-abstraction to form surface sites, and the subsequent reactions of H and hydrocarbon radicals with these surface sites. These reactions affect the gas composition near the surface.[35,38] For the case of UNCD deposition, the renucleation mechanisms (i.e., the creation of surface defects which change the growth direction and act as a renucleation site) were discussed in detail.[35] It was stated that the measured film morphology can be rationalized based on competition between H atoms,

CH_3 radicals, and other CH_x species reacting with dangling bonds on the surface.[38] Further, a general mechanism for the deposition of MCD and NCD from CH_4/H_2 gas mixtures and for UNCD films from $Ar/CH_4/H_2$ gas mixtures was proposed, which is consistent with published experimental data.[38]

It should be noted, however, that the gas surface reactions included in this model involve only those for which thermodynamic and kinetic data were available (i.e., only for H, H_2, C_2H_2, and the CH_x species, and not for C_2, C_2H, and higher hydrocarbon species).[38] While the simulations reach long-time scales (i.e., (U)NCD growth is simulated in real time,[89]), this method is limited by the requirement that a complete catalog of all relevant transitions and their rate constants has to be known in advance. Therefore, the completeness of this catalog depends on the intuition of the scientist applying the method and the availability of the kinetic data. Furthermore, not all transition mechanisms are known. In May et al.,[89] the questions that arise during the construction of the catalog are presented.

In general it can be concluded that the mechanistic modeling can be very instructive to obtain more insights into the growth processes of thin films or nanostructured materials, but it provides a more qualitative picture, and it depends strongly on the availability and correctness of reaction rate coefficients. The need for input data on surface reaction rate coefficients is a general weak point of the mechanistic models. They need to be obtained from experiments, or from detailed atomistic descriptions, such as ab initio (i.e., first principles) methods or classical molecular dynamics (MD) simulations. These methods are computationally more expensive, but they allow investigation of the evolution of the system without any assumptions regarding (reaction) mechanisms, as will be explained below.

10.3.2 Detailed Atomistic Simulations: General Overview

In solid-state physics, one of the most popular first principles methods is density functional theory (DFT). Within DFT, the energy of the investigated system is completely determined by the electronic charge density. DFT allows the calculation of the energy of many-body systems, and, therefore, enables the search for their energetically most probable configuration.

DFT calculations have been applied by various scientists to find probable reaction mechanisms of (U)NCD thin film deposition.[90–94] For instance, the role of the CH_3 radical as the main species contributing to diamond growth within the "standard growth mechanism" of diamond has been elucidated by DFT calculations.[91,92]

Although DFT calculations provide very valuable information about probable reaction mechanisms, the DFT calculation time becomes excessively long for systems containing more than about 100 atoms. In order to extend the length scale of first principles calculations, hybrid quantum-mechanical–molecular mechanics (QM/MM) methods have been developed.[95,96] In QM/MM, the essential part of the system (i.e., the atoms that are involved in the investigated reaction) is treated by means of, for example, DFT, and the surrounding atoms by means of classical dynamics. QM/MM calculations have confirmed the prominent role of the CH_3 radical in the standard growth mechanism of diamond.[95,96]

Other simplified approaches, such as DFT-based tight binding (DFTB), might alleviate the restriction of the limited system size; however, both the number of atoms and the attainable time scale remain one or two orders of magnitude smaller than what can be achieved with classical MD. Therefore, whereas quantum mechanical calculations provide valuable kinetic data and they have a decreasing computational cost (due to cheaper and faster processors), MD simulations are believed to continue to have an important role for the exploration of unknown reaction mechanisms.[97]

In classical MD simulations, the movement of all the atoms in the system is calculated as a function of time by Newton's laws, where the force is obtained as the negative of the gradient of the interatomic interaction potential. Hence, it is a deterministic method, which does not depend on *a priori* assumptions, in contrast to the mechanistic models described above. However, the reliability of the classical MD calculations strongly depends on the quality of the interatomic interaction potential.

Numerous interatomic potentials have been developed over the years. In the following paragraphs, we introduce some of the more well-known types relevant for plasma–surface interactions. The simplest types of potentials are pair potentials. To this category belong, for example, the Lennard-Jones potential, which is often applied for liquids, and the Morse potential. Its functional form can be written as follows:

$$V(r_{ij}) = 4\varepsilon \left[\left(\frac{\sigma}{r_{ij}} \right)^{12} - \left(\frac{\sigma}{r_{ij}} \right)^{6} \right].$$

Although many-body effects are not accounted for, pair potenials have also been developed for describing SiO_2 (e.g., the BKS potential).[98] This type of potentials, which are of the Born-Meyer type, also take into account long-range ionic interactions. Their functional form is

$$V(r_{ij}) = \frac{q_i q_j e^2}{4\pi\varepsilon_0 r_{ij}} + A_{ij} \exp(-b_{ij} r_{ij}) - \frac{C_{ij}}{r_{ij}^6}.$$

The advantages of this approximation include its simplicity and the fact that it can describe certain bulk properties such as the radial distribution function (RDF) or density reasonably well. The disadvantage is that it does not account for directionality of the covalent Si-O bonds and cannot describe nonbulk properties, such as the surface structure.

The earliest attempt to describe chemistry in covalent materials is the Stillinger-Weber (SW) potential, which includes a three-body term.[99] The SW potential is written as

$$V = \frac{1}{2} \sum_{i,j} \phi(r_{ij}) + \sum_{i,j,k} f(r_{ij}) f(r_{ik}) \left(\cos\theta_{jik} + \frac{1}{3} \right)^2$$

where $\phi(r_{ij})$ is a pair potential and $f(r)$ are cut off functions limiting the interaction range to a present distance. Although this potential allows for chemical bonds to be formed and broken, the SW potential fails when applied to nontetrahedral polytypes (e.g., liquid silicon or surface structures). Various researchers extended the SW potential (e.g., for Si-O-Cl and Si-O-F). Further extensions have allowed inclusion of ionic effects in order to describe, for example, SiO_2.[100]

A major improvement for realistically describing covalent materials was the development of the Tersoff-type potentials.[101] The Tersoff potential is written as

$$V = \frac{1}{2} \sum_{i,j} \phi_{rep}(r_{ij}) + \frac{1}{2} \sum_{i,j} B_{ij} \phi_{att}(r_{ij})$$

In this form, $\phi_{rep}(r_{ij})$ and $\phi_{att}(r_{ij})$ denote repulsive and attractive pair potentials, respectively, and B_{ij} is the bond-order function. It is a function of the coordination G_{ij} of the ij bond, which is given by

$$G_{ij} = \sum_{k} f(r_{ik}) g(\theta_{jik}) h(r_{ij} - r_{ik}).$$

In these potentials, the total energy of the system is written as a sum over pair contributions (taken as Morse potentials), in which the attractive component is modified by a many-body term. This many-body term is a function of the local binding topology for each atom. Corresponding to valence shell electron pair repulsion (VSEPR) theory, an angular function is included to allow open structures. Parametrizations have been developed for C, Si, BN, among others. Brenner extended the original Tersoff potential to describe hydrocarbons, accounting for the different chemistry of C and H, overbinding of radicals, and conjugation effects.[102]

Metals are very often described by embedded atom method (EAM) potentials.[103] In EAM, the potential energy of any atom is composed of a pairwise term and an embedding function:

$$V = \frac{1}{2} \sum_{i,j} \phi(r_{ij}) + F(\rho_{ij}).$$

The embedding function is a function of the local electron density associated with each atomic site:

$$\rho_i = \sum_j f_i(r_{ij}).$$

Therefore, EAM originates from DFT, which states that the energy of a solid is a unique function of the electron density. Although relatively simple, EAM potentials allow a relatively accurate description of metal bulk properties, as well as defects, surfaces, and impurities.

A universal type of potential is the reactive force field ("ReaxFF").[104] Here, the total system energy is a sum of several partial energy terms; these include energies related to lone pairs, undercoordination, overcoordination, valence and torsion angles, conjugation, hydrogen bonding, as well as van der Waals and Coulomb interactions:

$$E_{system} = E_{bond} + E_{lp} + E_{over} + E_{under} + E_{val} + E_{pen} + E_{C2} + E_{tors} + E_{H-bond} + E_{vdWaals} + E_{Coulomb}$$

Because Coulomb and van der Waals interactions are calculated between every pair of atoms, the ReaxFF potential describes not only covalent bonds but also ionic bonds and the whole range of intermediate interactions. Furthermore, as the ReaxFF potential takes polarization into account, it should be able to describe CNT growth in a charged environment such as plasmas, where polarization effects are caused by nonuniform microscopic distributions of charges and fields in the vicinity of nanostructures.[105]

In our research group, we make use of different potentials, depending on the requirements. We use the Brenner potential for hydrocarbons for describing amorphous carbon thin film deposition or (U)NCD growth. For CNT growth, the interaction with the metal catalyst particle needs to be accounted for, which is not included in the Brenner potential. Therefore, we use the ReaxFF potential. For the growth of metal oxide thin films, such as TiO_2 or complex oxides, for example, $Mg_xM_yO_z$ (with M = Al, Y, Cr), we use a more simplified potential—that is, a classical pairwise ionic potential, based on Coulomb interaction, short-range repulsion, and Van der Waals attraction.[106] The potential is based on only three parameters. In spite of the simplicity of this potential, the results obtained were found to be in agreement with experimental data for magnetron sputter-deposition.[107]

In the following, we will show some examples of MD simulations for the growth of different types of materials, obtained from our own research group. We will limit ourselves to carbon-based materials, as they were deposited by PECVD, in contrast to the metal oxide thin films, which were grown by either PVD (magnetron sputter-deposition)[107] or CVD.[108] Also the coupling with MC calculations, needed to handle the longer time scale behavior of surface relaxation in the case of, for example, diamond deposition or CNT growth, will be briefly explained.

10.3.3 Molecular Dynamics (MD) Simulations for Amorphous Hydrogenated Carbon (a-C:H) Thin Film Deposition

Amorphous hydrogenated carbon (a-C:H) can be characterized as a mixture of sp^2 and sp^3 carbon, containing a considerable fraction of hydrogen. Various plasma sources can be used to deposit a-C:H, including DC glow discharges, rf CCPs, ICPs,[109] and so forth, as well as remote sources, such as remote MW-PECVD or the expanding thermal plasma (ETP). The ETP source is of particular interest. Although the growth is purely chemical (i.e., without ion bombardment), films of medium density (1.8 g.cm^{-3}) and hardness (14 GPa) could be obtained at very high rates (up to 70 nm.s^{-1}).[110] Below, we describe some of our simulation efforts to unravel the plasma–surface interactions and the growth of a-C:H films for ETP conditions.

10.3.3.1 Hydrocarbon Reactions with a-C:H Surface Sites

Sticking and reflection coefficients have been determined by means of modeling by various researchers, many of which stem from the plasma fusion community.[111–114] Experimentally, obtaining hydrocarbon sticking coefficients is a difficult task, and only some data exist.[115] Because the a-C:H surface is essentially fully covered by C-H bonds, it is chemically passivated. Depending on their structure, different radicals will have different contributions. Diradicals, such as CH_2, can insert directly into C-C and C-H surface bonds. Hence, these species have sticking coefficients approaching 1. Closed shell neutrals, on the other hand, such as CH_4, have very low sticking coefficients and their effect is negligible. Monoradicals, such as CH_3, have a moderate effect. They can react with the film surface if dangling bonds are present, because they cannot insert directly into surface bonds. These dangling bonds can be created by removal of H atoms at the surface. The latter can occur either by an ion displacing the H atom, or by a H atom abstracting H from the C-H surface bond, or by an incoming radical such as CH_3. The latter mechanism is shown to be responsible for the synergistic effect of H on the sticking coefficient of CH_3.[115]

Neutral hydrocarbon radicals can only react at the surface, because they are too large to penetrate into the layer. Hydrogen atoms, on the other hand, can penetrate about 2 nm into the film,[116] where they can create subsurface dangling bonds, abstracting H from subsurface C-H bonds. In this way, H_2 is formed, which can desorb from the film or become trapped interstitially.

Using MD simulations, we investigated the reaction mechanisms of various hydrocarbon species on amorphous carbon. Reaction coefficients for hydrocarbons interacting with diamond surfaces will be presented in Section 10.3.4 on (U)NCD. In the following sections, examples are given on how the reaction behavior can be affected by certain parameters specific for the reacting species, such as chemical connectivity and structural stability, as well parameters specific for the a-C:H sites, such as steric hindrance.

To investigate the effect of both site-specific and species-specific factors, we applied the following simulation setup. In total, 11 different sites were created on a diamond {111} surface that represent various bonding sites on an a-C:H surface.[117] The impacting species were chosen based on threshold ionization mass spectrometry (TIMS) applied during the growth of a-C:H thin films in the ETP source. The radicals are positioned above the site and rotated randomly. Then, the radicals are allowed to impinge on the reaction site with a kinetic energy equal to the measured gas temperature of about 1500 K. Below, we describe a few examples demonstrating the influence of both site-specific and species-specific factors.

10.3.3.1.1 Site-Specific Factors: Steric Hindrance

Let us first consider a simple example: the impact of a linear C_3H radical on two a-C:H sites, shown in Figure 10.2. The first site is a passivated diamond {111} surface, in which one hydrogen atom is replaced by a carbon atom. The second site is the same substrate, but now a C_2 molecule is bound to the surface instead of a single C atom. The impinging radical is aimed at the topmost C-atom on the substrate. Hence, on the first site, the C-atom bearing the dangling bond and on which the radical will impinge, is surrounded by the H-atoms passivating the substrate, while on the second site the C-atom on which the radical impinges is fully exposed.

The calculated reflection coefficient of C_3H on the first site is 0.35, while on the second site it is 0.16 (i.e., less than half). The radical is reflected more often on the first site due to steric hindrance: the H-atoms on the surface prevent the bulky C_3H from easily binding to the dangling bond. This also results in different ratios in sticking configurations of the radical. On the first site, the radical sticks in 85% of its sticking events by the end-C to which the H-atom is not bound. It sticks to the surface with the other end-C in the remaining 15% of its sticking events. On the second site, however, C_3H binds with the free end-C in 55% of its sticking events, and with the other end-C atom in the remaining 45% of the cases.

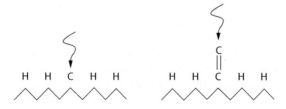

FIGURE 10.2
2D representation of two a-C:H sites. The wavy lines represent the impact location of the impacting radical.

In nearly all investigated cases, we observed that the site on which a specific radical sticks affects both the total sticking coefficient of that species as well as the number and structure of the various resulting configurations. This illustrates that the availability of surface sites strongly affects the reaction probability.[117]

10.3.3.1.2 *Species-Specific Factors: Chemical Reactivity*

Species-specific factors are those that influence the sticking behavior as determined by the properties of the impinging species, rather than due to the site on which the particle impinges. The examples below indicate how sticking affects the resulting structure of the deposited radical, and how the structural stability and the chemical connectivity in turn influence the sticking behaviour.[117–119]

As a first example, consider the reaction behavior of linear C_3 versus linear C_3H. Both radicals have a fully bound central C atom that does not bind to the surface. The two other C atoms, however, both have at least one unbound electron, and hence are available for direct sticking to a dangling bond. Structurally, the difference between both radicals is the presence of the H atom on the C_3H radical. This limits the availability for sticking of the C atom to which this H atom is connected, resulting in the lower reactivity of C_3H compared to C_3. As a result, the C_3 radical is on average found to be about 10% more reactive compared to the C_3H radical.

As a second example, consider the impact behavior of linear C_3H versus cyclic C_3H. These radicals are configurational isomers. Averaged over the total number of impacts on all investigated sites, the cyclic C_3H radical has a sticking coefficient of about 0.75, while the linear C_3H isomer has a sticking coefficient of about 0.40. There are two factors responsible for this difference.

First, the triangular shape of the cyclic radical renders this species structurally unstable. This can be easily seen by comparing cyclopropane to *n*-propane: in cyclopropane, the C-C bonds are about 32% weaker than in the linear isomer, due to a severe ring strain of 117 kJ/mol. In cyclic C_3H, the effect is even more pronounced: the C-C bonds in cyclic C_3H are about 50% weaker than in linear C_3H. Hence, the release of this ring strain is a driving force for the radical to break up, enhancing drastically its reactivity. As a result, the molecule breaks up in more than 70% of the sticking events.

Second, the bonding configuration is different in the two isomers. In cyclic C_3H, the sp bonds are not fully developed (because the molecule is not linear), and each of the C atoms bears electrons that do not participate in a bond. Hence, each of the three C atoms is very reactive. In the linear C_3H radical on the other hand, the three C atoms form a linear chain and the C atoms are sp-hybridized. Hence, the middle C atom does not bear a free electron and is not reactive.

The breaking up of a cyclic C_3H can occur through different mechanisms. Strikingly, while several mechanisms seem to occur on nearly all investigated sites, each site also leads to mechanisms not observed on the other sites. The main effect of a break-up event is the transformation of the cyclic structure into a linear one. The remaining bonds become stronger, depending on which bond is broken and which atom sticks to the surface.

Consider again the first site shown in Figure 10.2. On this site, the linear isomer never breaks up, and sticks to the surface with either the free end-C (85% of the cases) or the end-C to which the H atom is connected (15%). In the case of the cyclic isomer, however, eight different mechanisms were observed, in all of which the molecule breaks at least one bond. In fact, in almost half of these events, one or more atoms were sent into the gas phase again (i.e., the radical partially sticks and partially reflects). Furthermore, the radicals were observed to stick either with 1 or 2 C atoms to the surface. A few of the observed break-up patterns are shown in Figure 10.3.

The reactivity and breaking-up pattern of a radical may also be influenced by its internal energy (i.e., by its excitation state). For example, it was observed by MD simulations that the sticking coefficient of highly reactive species such as CH or C_2H decreases by nearly a factor of 2 upon increase of

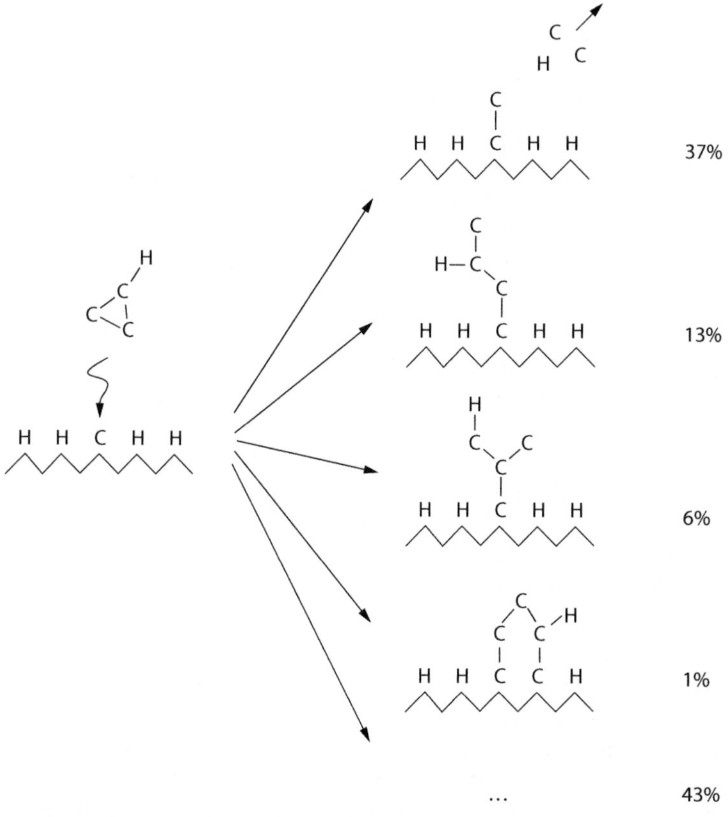

FIGURE 10.3
Observed break-up patterns of cyclic C_3H upon impact on a typical a-C:H surface site.

the internal energy from 0.026 eV to 2.6 eV. This strong decrease in reactivity was only observed for H-containing radicals. Also, the breaking-up pattern of the radical can be influenced. The CH radical was observed to break up in about half of its impacts when having an internal energy of 2.6 eV, while it never breaks up when it has an internal energy of 0.026 eV.[120]

Therefore, we can conclude that when a-C:H films are deposited using radicals as growth species, the connectivity of the film will be determined by (1) the available surface sites, (2) the structure of the radical, as well as (3) its excitation state.

10.3.3.2 Simulated a-C:H Growth

We now turn our attention to the simulated growth of thin a-C:H films. The formation of dense and hard films (typically by ion beam deposition) is usually understood in terms of the "subplantation model."[121] Energetic ions bombard the substrate and are implanted in the subsurface region. This creates a local density increase that leads to the formation of an interconnected sp^3-sp^3 network.

In a typical PECVD setup, however, the neutrals will also contribute significantly to the growth. Note that while subplantation is a physical process, chemisorption of neutrals is a chemical process. The contribution of each neutral species to the growth rate depends on its sticking coefficient, which is in turn determined by its chemical surface reactivity, as described above. Although many simulations have been performed for ion beam growth of a-C:H,[122–126] only few simulations deal with chemical growth. Here, we summarize some of our results on this topic, specifically for growth from an ETP plasma.

In the ETP setup, it was observed that the most important factor in determining the species fluxes (and hence the growth species) and the resulting film quality is the hydrocarbon-to-ion ratio F.[110] Specifically, when C_2H_2 is used as hydrocarbon source gas and Ar as carrier gas, it was found that high C_2H_2 flows lead to the best quality films in terms of density, hardness, and refractive index.[110] TIMS was used to measure the species fluxes.[127] When $F < 1$, the main growth species are small, highly reactive radicals such as C, CH, and C_2. For $F > 1$, mainly C_3 and C_3H were found to be important. Unfortunately, the H-flux toward the substrate could not be measured.

First, consider simulated growth under $F < 1$ conditions, with an arbitrary H-flux of 50% toward the substrate. Under these conditions, a film was formed containing a relatively high fraction of sp^3 carbon of 50%.[128] This is in reasonable agreement with the experimental value of 67% as measured using electron energy loss spectroscopy (EELS).[129] The hydrogen fraction was about 45% and the density was around 1.73 g.cm^{-3}. A graphical representation of the grown structure is shown in Figure 10.4. During the growth, the C_2 species was found to be the most efficient and most important growth species, contributing almost 70% to the total carbon content of the film. Furthermore, the C_2 radical was found to abstract hydrogen from the surface. Surprisingly, the role of the H atoms was

- 4-fold coordinated C-atom
- 3-fold coordinated C-atom
- 1-or 2-fold coordinated C-atom
- H-atom

FIGURE 10.4
Side view of a simulated a-C:H film grown by an expanding thermal plasma (ETP), under conditions of $F < 1$ (see text). (Reproduced from [128] with permission of Elsevier.)

found to be passivating dangling bonds only (i.e., etching by atomic H was not observed). The structure of the film was found to be polymeric, also in agreement with the experiment. Therefore, the simulated film is overall in agreement with the experiment, albeit for an arbitrarily chosen H-flux.

As the H-flux toward the substrate could not be measured, it is of interest to investigate how it influences the resulting film properties. Therefore, simulations were performed in which the H-flux was varied from 0% to 45%.[130] It was found that the H-fraction in the film is nearly linearly dependent on the H-flux toward the substrate. Even at the highest H-fluxes, nearly no chain-terminating CH_3 fragments were found, in agreement with the experiment. The increasing H-fraction in the film triggers a change in hybridization. At low H-fluxes, the sp carbon is found to hybridize to sp^2. This in turn leads to an increase in the mass density from about 1.7 g.cm^{-3} to about 1.8 g.cm^{-3} at ca. 12% of hydrogen in the film.[130] Increasing the H-flux further leads to a rehybridization from sp to sp^2 and from sp^2 to sp^3, such that the sp^2 fraction remains nearly constant. On the other hand, the increase in H-fraction leads to an increase in CH and especially bulky CH_2 fragments, resulting in a decrease in the mass density.[130] At the highest H-fluxes investigated, most sp carbons have been transformed into sp², inducing a strong increase in the sp³ fraction. Therefore, these simulations indicate that under the specified conditions, the density of the grown films can be controlled to a certain extent by controlling the exact H-flux toward the substrate.

As a second example, consider the growth of a-C:H under $F > 1$ conditions.[131] As determined by TIMS, C_3 and C_3H species are identified as the main growth species. Again, the H-flux toward the substrate was varied between 0% and 45%. When no H-flux toward the substrate is applied, the final H-content in the film is about 9%, all originating from the C_3H radical. Also in this case, an increase in the mass density was observed with increasing H-content in the film up to 22%, corresponding to a H-flux of 25% toward the substrate. The maximum mass density under these conditions was calculated to be about 1.34 g.cm⁻³ (see Figure 10.5). When increasing the H-flux toward the substrate even further, a decrease in mass density is observed. This simulation therefore indicates that under chemical growth conditions, densification of a-C:H films is possible by applying a suitable H-flux toward the substrate, independent from the exact growth species.

When no bias is applied to the substrate in the ETP setup, the growth proceeds entirely through chemical reactions (i.e., physical processes such as sputtering, implantation, or knock-on penetration do not play a role). This corresponds to the simulations outlined above. However, it is also possible to apply a substrate bias so that the ions will also contribute to the growth. Therefore, a series of simulations was also conducted to investigate the effect of the Ar⁺ ion bombardment.[132] In these simulations, the ion energy was varied (from 0.13 eV to 100 eV), as was the relative ion flux (from 0.1 to 0.47). The

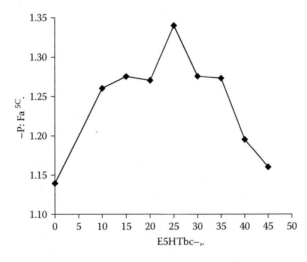

FIGURE 10.5
Calculated mass density of the simulated a-C:H films, grown by an ETP, as a function of the H-flux towards the substrate.

growth species were again chosen based on the experimentally measured fluxes for $F < 1$, $F = 1$, and $F > 1$ conditions, and included C, C_2, C_2H, linear C_3, linear C_3H, cyclic C_3, cyclic C_3H, C_5, and C_5H. Under these conditions, growth must still proceed from the selected hydrocarbon growth species, but knock-on penetration is now a possible mechanism.

The growth and especially the resulting properties were found to be strongly dependent on the Ar^+ ion energy and relative flux (see Figure 10.6). At low Ar^+ energies, the relative ion flux was found to be of little importance, and the differences in structure of the obtained films were a direct consequence of the growth species. As before, the films grown from the smaller, more reactive radicals, such as C and C_2, were found to be somewhat denser than the films grown from C_3 and C_3H, resulting in densities of 1.65 g.cm^{-1} and about 1.2 g.cm^{-1}, respectively. As the ion energy is increased, the probability per ion impact for a knock-on penetration event also increases. The total number of knock-on penetrations is also dependent on the total flux of the ions. As these knock-on penetration events cause an increase in the mass density of the film, the densest films were found at the highest ion energies investigated (100 eV) at the highest relative fluxes (47%). Under these conditions, a mass density of 2.8 g.cm^{-3} was obtained.[132]

The carbon coordination was also found to be strongly dependent on the ion energy and flux on the one hand, and on the growth species on the other hand. The carbon coordination was found to be always higher in the films grown from the smaller growth species compared to the films grown from the larger species. A higher ion energy and ion flux increase the carbon coordination in all films, again by the mechanism of knock-on penetration.[132]

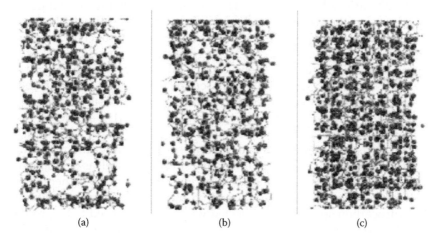

(a)　　　　　　　　(b)　　　　　　　　(c)

FIGURE 10.6
Representative parts of the simulated a-C:H films grown by an ETP, for conditions of $F = 1$ (see text), and an ion energy of (a) 0.13 eV; (b) 20 eV; and (c) 100 eV. Reproduced from [132] with permission of Wiley.

Finally, the H-content in the film was found to be determined by the growth species only. The ion beam does not sputter hydrogen from the film. Rather, reactive species such as C_2 were found to abstract hydrogen from the surface of the film, as was observed earlier (see above).

10.3.4 MD Simulations for Nanocrystalline diamond (NCD) and Ultrananocrystalline Diamond (UNCD) Deposition

In this section, we describe some of our modeling efforts related to (U)NCD growth. The goal is to identify which species may act as growth precursors for (U)NCD and to investigate how these growth species can pursue the diamond lattice—that is, how the (U)NCD films grow. Similar to our investigations related to amorphous carbon, we have therefore carried out various simulations, on the one hand to study the exact chemical processes that take place when hydrocarbons react with the diamond surface, and on the other hand to study the growth process.

10.3.4.1 Hydrocarbon Reactions at the Diamond Surface

Many of the MD simulations relevant for amorphous carbon are equally relevant for diamond growth. There are, however, a number of studies that were performed specifically on diamond surfaces. The interaction of CH_3 and C_2H_y species with diamond (100) and (111) surfaces was previously studied by several scientists.[112,114,133–134] The simulations by Träskelin et al. were performed to gain insight into the erosion of hydrogenated carbon-based films under ITER-relevant conditions[112,114,133] and consist of bombarding a diamond (111) surface with various hydrocarbons. The authors found that the sticking process depends on the angle of incidence and the local atomic neighborhood of the impacted reactive site.[133] Zhu et al. identified different chemisorption configurations, including the cross-linking between two neighboring reactive sites, when examining the behavior of C_2H_2 molecules impacting diamond (100) surfaces.[134] Cross-linking describes the sticking of two carbon atoms of the impacting C_xH_y species to two different surface sites. Analogous configurations were found for C_2H_2 impacting diamond (111) surfaces.[114]

In our group, we performed detailed MD simulations to calculate chemisorption probabilities and to determine sticking configurations of various C_xH_y species on diamond (100) and (111) surfaces, as well as on diamond step edges.[135–136] The results of these simulations are summarized here.

As (U)NCD films are often deposited by microwave-enhanced PECVD, those hydrocarbon species were selected which are supposed to be present close to the substrate in this deposition source.[35] These species include CH_x ($x = 0$–4), C_2H_x ($x = 0$–6), C_3H_x ($x = 0$–3), and C_4H_x ($x = 0$–2). Furthermore, as we wish to understand why different process conditions result in different film types, the simulations were carried out at two different substrate temperatures of 800 K and 1100 K, typical for UNCD and NCD, respectively.

First it was found that the sticking efficiency depends on both the number of free electrons and hydrogen atoms of these species,[135] in agreement with similar observations by Träskelin et al. who investigated the reaction behavior of C_2H_y species on diamond (111) surfaces.[114]

Moreover, a higher substrate temperature was found to promote a higher adatom coordination, which is especially important for the growth of diamond structures. Also, this result partially explains why larger diamond crystals, as typical for NCD, can be grown at higher temperature, whereas UNCD, with its high percentage of disordered phases, is grown at lower substrate temperature. Furthermore, the different bonding structure of the two investigated diamond surfaces (i.e., diamond (100)2 × 1 and diamond (111)1 × 1) was found to have a significant effect on the resulting sticking efficiency. Specifically, it was observed that diamond (111) growth is promoted by a higher temperature above diamond (100), in agreement with experiments.[136]

As the total contribution of a specific species to the growth of the film is dependent on its flux to the substrate, on the one hand, and its chemical reactivity, on the other hand, the obtained sticking data can be combined with data on species concentrations to determine the main growth precursors. Thus, using the plasma chemistry simulation data of May et al.[35] for the species concentrations above the growing diamond film, it was suggested that, within their series, C, C_2H_2, C_3, and C_4H_2 are the most important growth species for UNCD growth, whereas CH_3, C_2H_2, C_3H_2, and C_4H_2 are predicted as the major growth species for NCD growth. More information about these simulations can be found in the literature.[136]

10.3.4.2 Simulations Related to (U)NCD Film Growth

One of the most famous simulation papers concerning reaction mechanisms at diamond surfaces was written by Garrison et al. in 1992.[137] Applying the Brenner potential in an MD simulation, the reaction mechanism of dimer opening at the diamond (100)2 × 1 surface and subsequent insertion of CH_2 was discovered. Based on these findings, investigations at higher level of theory were carried out. Until now, this reaction mechanism is believed to be the essential part of the standard growth mechanism of diamond.

However, various processes that are relevant for the evolution of thin film growth, such as relaxation processes and diffusive events, take place on much longer time scales than can be handled by MD. In order to overcome this "time-scale problem" of MD simulations, different methodologies have been designed. One possibility is to couple the MD simulations to MC simulations. In contrast to the deterministic MD simulations, MC simulations are probabilistic and computationally (much) less demanding. One famous example is the MMC algorithm, which was developed in the 1950s.[138] In MMC, no activation barriers are taken into account. In an MMC simulation, the system evolves by random displacements ("moves") of the atoms and clusters of atoms. Depending on the energy change caused by this random displacement, the move is accepted or rejected. This sampling is performed by applying the Boltzmann distribution function—that is, the canonical ensemble (or "NVT ensemble") is sampled. Essentially, the Metropolis method is a Markov process in which a random walk is constructed.

By means of combined MD-MMC simulations, we demonstrated how the diamond structure is pursued at the atomic level, taking into account longer time scale events.[139–141] In the standard growth model of diamond, it is assumed that only the CH_x species contribute significantly to the growth of nanostructured diamond.[142] However, our simulations demonstrated that C_xH_y species with $x \geq 2$ also significantly affect the growth process. For instance, the adatom surface behavior of C and C_2H_2 at diamond $(111)1 \times 1$, C and C_4H_2 at diamond $(111)1 \times 1$, and C_3 at diamond $(100)2 \times 1$ has been investigated, exemplary for adatom arrangements during the growth of UNCD. For all species, the formation of diamond hexagons was observed during the MMC simulation, thereby pursuing the diamond crystal structure. In Figure 10.7, an example is shown of how a carbon atom and an acetylene molecule can pursue the diamond structure when stuck to a diamond (111) surface.

The same conclusions were drawn when the longer time scale behavior of C_xH_y species at step edges delimiting diamond terraces were investigated[141]: C_2H, C_2H_2, and C_3H_2 all have a high reactivity, implying a prominent role during the growth of (U)NCD. For example, C_2H has a high density above the growing NCD surface.[35,37] This confirms the recent doubts that the C_1Hy species are the only important growth species of (U)NCD.[143]

Furthermore, the MD-MMC simulations of C_xH_y species at diamond step edges have elucidated the different growth regimes of UNCD and NCD.[141] Crystal growth through the extension of the step edges, the "step flow growth mechanism," is believed to result in well-faceted, smooth diamond

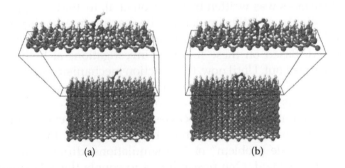

(a)   (b)

FIGURE 10.7
Side view of the input (a) and final (b) configuration of a MMC simulation. The white and gray spheres indicate H and C atoms, respectively. The input simulation is obtained by an MD simulation of an impacting C atom and an impacting C_2H_2 molecule; their C atoms are marked red. In the final configuration, one can observe the formation of a new C six ring (marked by red spheres), which proves that the diamond crystal structure is pursued. (Reproduced from [139] with permission of the Royal Society of Chemistry.)

films.[144] During the growth of NCD films, the species that are found to contribute to the step-flow growth mechanism are generally accepted to be important for NCD growth.[141] In other words, the species present above the growing NCD film enhance the step-flow growth mechanism, resulting in the well-faceted morphology that is characteristic for NCD films. For UNCD, however, the species that cause the formation of defects at the step edges, are generally considered to be important for UNCD growth.[141] The absence of the step-flow growth mechanism during UNCD growth therefore explains the nonfaceted morphology and the characteristic high fraction of noncrystalline phase within the UNCD films.[141]

10.3.5 MD Simulations for the Growth of Carbon Nanotubes (CNTs)

A general model for describing CNT growth on metals is the "vapor-liquid-solid" model.[145,146] In this model, a vapor (i.e., the hydrocarbon source) is in contact with a liquid (i.e., molten) catalyst particle. The hydrocarbon is assumed to decompose to its elements, and the C dissolves in the catalyst particle. When the latter becomes supersaturated, the excess C precipitates from the catalyst and forms a solid tube or fiber. On the other hand, a surface-mediated carbon transport model was put forward by Helveg et al. to describe low-temperature (PECVD) growth of CNTs.[147] Another model was proposed to describe growth on nonmetallic particles that do not form a eutectic alloy such as diamond, Si, or SiC.[148] Here, the CNT is believed to nucleate on a sp^2 carbon surface layer, covering the nanoparticle.

Various scientists have simulated the growth of CNTs by classical MD. The first of these simulations was performed by Maiti et al.,[149,150] albeit without taking into account the metal atoms explicitly. The first MD simulations taking into account all atoms (carbon and metal) were carried out by Shibuta et al.[151] A simplified Brenner-type potential was used for the C-C interactions, while parameters for C-metal and metal-metal interactions were fit to DFT energies of small clusters. Large, defect-rich tubes were formed from a random distribution of a large number of C atoms and a small number of Ni atoms. These simulations correspond to a laser ablation process. Shibuta et al. also investigated the catalytic CVD process by allowing C-atoms to impinge on small Ni clusters at 2500 K. In the total simulation time of 130 ns, a cap structure was formed that subsequently lifted off from the surface of the cluster.[152] The same authors also investigated the effect of the substrate on the catalytic particles during single-walled carbon nanotube (SWNT) growth.[153] A layered metal structure and a graphene layer parallel to the substrate were formed in the case of strong cluster–substrate interaction, while the metal did not adopt a specific orientation and the graphene sheet separated from the cluster in a random direction in the case of weak metal–substrate interaction.

Balbuena and coworkers updated the Shibuta potential and presented a step-by-step overview of the observed processes in the simulation during SWNT growth.[154,155] Ribas et al.[156] and Burgos et al.[157] applied this potential to investigate the effect of the adhesion strength of the graphitic cap to the catalyst, as well as the effect of temperature, on the SWNT growth. The catalyst encapsulation was found to depend on the work of adhesion at $T > 600$ K. At lower temperature, limited carbon diffusion hinders cap formation and cap lift off.

Numerous simulations were performed by Ding et al. to investigate the influence of various parameters and growth conditions on the growth mechanism. The growth mechanism was found to shift from bulk diffusion mediated to surface diffusion mediated at around 900 K to 1000 K.[158] Larger clusters resulted in an enhanced growth of SWNTs compared to smaller clusters (<20 atoms). Moreover, in agreement with the experiment, it was also found that clusters with diameters smaller than 0.5 nm yield tubes with slightly larger diameters of 0.6 to 0.7 nm.[159] Further, this group reported that while a temperature gradient may be important for larger particles, it is not required for SWNT growth from small particles.[160,161]

However, all of these simulations mentioned above invariably resulted in highly defected structures, due to the high C-addition rates used. Typically, a C atom is added to the cluster every 50 ps, or faster. Therefore, none of these simulations take into account relaxation effects which would enable defects to be healed out. Unfortunately, accelerated MD simulations (such as hyperdynamics, parallel replica, or temperature accelerated dynamics[162]) seem not to be directly applicable in the case of SWNT growth.[163] Alternatively, deterministic MD simulations can be coupled to stochastic MC simulations.[164–167] This is also the method we adopted to take into account relaxation effects in (U)NCD growth (see previous section). Recently, we successfully applied a hybrid MD/UFMC approach for the simulation of the melting mechanisms of nickel nanoclusters.[166]

Furthermore, it was demonstrated in the literature[167] that this methodology can also be successfully applied to the growth of SWNTs. In Figure 10.8, a carbon cap grown from a small Ni-cluster can be observed, together with some amorphous material.

The growth process was simulated as follows.[167] First, a small Ni_{32} nanocluster was prethermalized at 1200 K. A carbon atom was placed in the simulation box with thermal velocity. This C atom then travels through the box until it impinges on the cluster. Every 2 ps, another C atom is added to the box. Every 4 ps, the MD simulation is stopped, and the resulting MD configuration is used as input to the MC simulation. In the MC simulation, the maximum displacement of the atoms was limited to 0.1 Å. After 10^4 MC steps, the resulting configuration is used again as input to the MD simulation. During the MC cycle, no C atoms are added to the simulation volume.

Various stages during the growth were identified in the simulation. Initially, the C dissolves in the cluster, until (super)saturation is reached. Adding more C to the cluster results in the formation of surface and subsurface dimers and trimers. Additon of C to these entities forms short chains that can rearrange

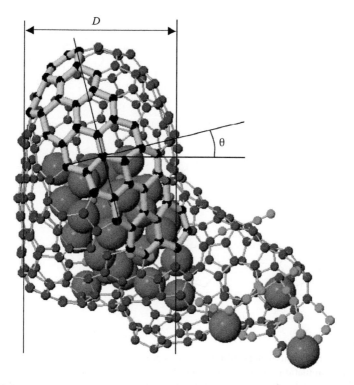

FIGURE 10.8
(See color insert.) SWNT cap grown by means of the hybrid MD/MC model. Determination of the chiral angle $\theta = 14°$ and diameter $D = 11.45$ Å allows to assign a (12,4) chirality to the cap. (Reproduced from [167] with permission of the American Chemical Society.)

to form the first rings. Longer chains are also formed, which eventually also rearrange, thereby also forming rings. In this stage, rings become connected to each other by surface chains. Addition of C to these systems allows the formation of graphitic patches, which subsequently coalesce into large graphitic islands. Finally, these islands grow until they detach from the metal surface and form a SWNT cap. More detailed information on these growth steps can be found in the literature.[167]

The effect of applying the MC relaxation stage is demonstrated by the observed healing of defects. A number of defect healing processes were identified during this relaxation stage, as shown in Figure 10.9. The simulations indicate that these processes occur only when the carbon network is still connected to the metal cluster (i.e., these processes are metal mediated). This was also observed in recent tight-binding simulations.[168] Due to the healing of these defects, a carbon cap was finally formed with a definable diameter and chiral angle, from which a (12,4) chirality could be assigned.[167]

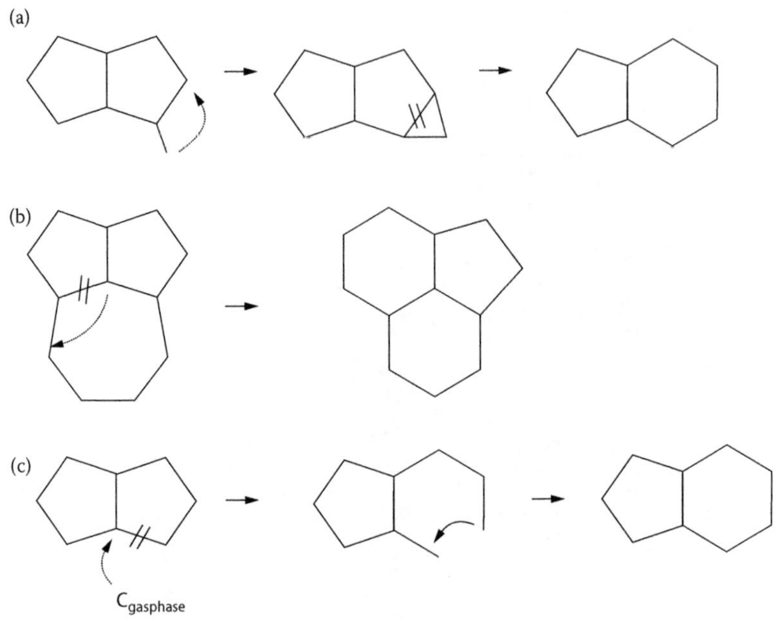

FIGURE 10.9
(a)–(c) Observed metal-mediated healing processes during the relaxation stage of the structure in the MD-MC simulation. All three processes involve the transformation of a pentagon into a hexagon. (Reproduced from [167] with permission of the American Chemical Society.)

10.4 Conclusions

In this chapter, we presented different modeling approaches that describe the plasma behavior and the plasma-based growth mechanisms during PECVD of thin films and nanostructured materials. The first section deals with plasma modeling, and more specifically the description of the plasma chemistry, to elucidate possible growth precursors. An overview is given of different modeling approaches. It is illustrated that the most suitable approaches for plasma chemistry modeling are 0D chemical kinetics modeling (also called global modeling), (1D or 2D) fluid modeling, and hybrid MC-fluid simulations. These approaches can take into account a rich plasma chemistry without significant computational effort. Moreover, in some hybrid models, the interaction with the substrate can be described, accounting for thin film deposition, albeit in a simplified way. Some calculation results from our own research group, for hydrocarbon-based plasmas used for CNT growth, are presented. Such plasma chemistry modeling can provide useful information on the precursors for thin film deposition or nanostructure formation.

The second part of the chapter deals with the simulation of the (plasma-based) growth mechanisms for thin films and nanostructured materials, where we have focused mainly on carbon-based materials. Several papers report on mechanistic modeling that gives an overall, albeit qualitative, picture of the growth mechanisms. Moreover, the mechanistic modeling can be combined with plasma chemistry modeling, as illustrated, for example, in the literature.[57,61,78,87] In this approach, results from the plasma simulations, such as the fluxes of species arriving at the substrate, are used as input for the mechanistic surface models, and vice versa, the plasma–surface interactions provide boundary conditions for the plasma simulations. In this way, an integrated picture of the PECVD of thin films or nanostructured materials is possible. However, it should be mentioned that these mechanistic models depend strongly on the availability of reaction rate coefficients. The latter need to be obtained from experiments, or they can be extracted from detailed atomistic simulations, such as DFT and classical MD simulations, which are the alternative approach for studying the growth mechanisms.

DFT simulations provide detailed and accurate information, but they are limited to very small systems and timescales (i.e., in the order of 100 atoms for a few picoseconds). Classical MD simulations can deal with larger systems (i.e., thousands to even millions of atoms) and somewhat longer timescales, especially when combined with MC simulations to treat the (longer timescale) surface relaxation processes. However, the quality of these MD simulations strongly depends on the reliability of the interatomic interaction potential used. We illustrated some examples of classical MD-MC simulations carried out in our research group, for different carbon-based materials, such as amorphous hydrogenated carbon (a-C:H) thin films, nanodiamond (UNCD and NCD) thin films, and CNTs. It is clear that these simulations can provide detailed insight into the growth process, without making any *a priori* assumptions, in contrast to the mechanistic models. However, the classical MD simulations suffer from long calculation times, so in practice, they are limited to certain aspects of the growth process, and they cannot yet be integrated in plasma chemistry simulations.

Therefore, in order to fully understand the growth process from a theoretical point of view, a multilevel simulation setup is needed, ranging from atomistic (i.e., quantum mechanical and classical MD) simulations to macroscale modeling. Even if MD simulations cannot be integrated into plasma modeling because of the different time and length scales, a more loosely coupled modeling approach should be possible, where the output of classical MD simulations (i.e., surface reaction probabilities but also detailed insights in the importance of certain growth mechanisms) serves as input for the other models. In our opinion, the ultimate approach for describing the PECVD process of thin films and nanostructured materials would be a combination of detailed MD simulations, mechanistic surface growth modeling, and plasma chemistry modeling.

Acknowledgments

The examples given in this chapter for the modeling work from our own research group have been realized with the financial support from the Fund for Scientific Research-Flanders (FWO), the Institute for Promotion of Innovation through Science and Technology in Flanders (IWT), and the Federal IAP-VI program. This work was carried out in part using the Turing HPC infrastructure at the CalcUA core facility of the Universiteit Antwerpen, a division of the Flemish Supercomputer Center VSC, funded by the Hercules Foundation, the Flemish Government (department EWI) and the Universiteit Antwerpen.

References

1. Bruno, G., P. Capezzuto, and Madan, A. eds. 1995. *Plasma Deposition of Amorphous Silicon-Based Materials (Plasma-Materials Interactions)*. New York: Academic Press.
2. Meyyappan, M. 2009. A review of plasma enhanced chemical vapour deposition of carbon nanotubes. *Journal of Physics D-Applied Physics*, 42: 213001.
3. Okada, K. 2007. Plasma-enhanced chemical vapor deposition of nanocrystalline diamond. *Science and Technology of Advanced Materials*, 8: 624–634.
4. Gordillo-Vazquez, F. J., and Albella, J. M. 2002. A quasianalytic kinetic model for nonequilibrium $C_2H_2(1\%)/H_2/Ar$ RF plasmas of interest in nanocrystalline diamond growth. *Plasma Sources Science and Technology*, 11: 498–512.
5. Gordillo-Vazquez, F. J., and Albella, J. M. 2003. Distinct nonequilibrium plasma chemistry of C_{-2} affecting the synthesis of nanodiamond thin films from C_2H_2 (1%)/H_2/Ar-rich plasmas. *Journal of Applied Physics*, 94: 6085–6090.
6. Gordillo-Vazquez, F. J., and Albella, J. M. 2004. Influence of the pressure and power on the non-equilibrium plasma chemistry of C_2, C_2H, C_2H_2,CH_3 and CH_4 affecting the synthesis of nanodiamond thin films from C_2H_2 (1%)/H_2/Ar-rich plasmas. *Plasma Sources Science and Technology*, 13: 50–57.
7. Dandy, D. S., and Coltrin, M. E. 1995. A simplified analytical model of diamond growth in direct-current arcjet reactors. *Journal of Materials Research*, 10: 1993–2010.
8. Lombardi, G., Hassouni, K., Bénédic, F., Mohasseb, F., Röpcke, J., and Gicquel, A. 2004. Spectroscopic diagnostics and modeling of $Ar/H_2/CH_4$ microwave discharges used for nanocrystalline diamond deposition. *Journal of Applied Physics*, 96: 6739–6751.
9. Hassouni, K., Mohasseb, F., Bénédic, F., Lombardi, G., and Gicquel, A. 2006. Formation of soot particles in $Ar/H_{-2}/CH_4$ microwave discharges during nanocrystalline diamond deposition: A modeling approach. *Pure and Applied Chemistry*, 78: 1127–1145.

10. Delzeit, L., McAninch, I., Cruden, B. A. et al. 2002. Growth of multiwall carbon nanotubes in an inductively coupled plasma reactor. *Journal of Applied Physics*, 91: 6027–6033.
11. Denysenko, I. B., Xu, S., Long, J. D., Rutkevych, P. P., Azarenkov, N. A., and Ostrikov, K. 2004. Inductively coupled $Ar/CH_4/CH_4/H_2$ plasmas for low-temperature deposition of ordered carbon nanostructures. *Journal of Applied Physics*, 95: 2713–2724.
12. Yuji, T., and Sung, Y. M. 2007. RF PECVD characteristics for the growth of carbon nanotubes in a CH_4-N_2 mixed gas. *IEEE Transactions on Plasma Science*, 35: 1027–1032.
13. Möller, W., 1993. Plasma and surface modeling of the deposition of hydrogenated carbon-films from low-pressure methane plasmas. *Applied Physics A—Materials Science and Processing*, 56: 527–546.
14. Dorai, R., Hassouni, K., and Kushner, M. J. 2000. Interaction between soot particles and NOx during dielectric barrier discharge plasma remediation of simulated diesel exhaust. *Journal of Applied Physics*, 88: 6060–6071.
15. Dorai, R., and Kushner, M. J. 2003. A model for plasma modification of polypropylene using atmospheric pressure discharges. *Journal of Physics D—Applied Physics*, 36: 666–685.
16. Hash, D., Bose, D., Govindan, T. R., and Meyyappan, M. 2003. Simulation of the dc plasma in carbon nanotube growth. *Journal of Applied Physics*, 93: 6284–6290.
17. Teo, K. B. K., Hash, D. B., Lacerda, R. G. et al. 2004. The significance of plasma heating in carbon nanotube and nanofiber growth. *Nano Letters*, 4: 921–926.
18. Hash, D. B., Bell, M. S., Teo, K. B. K., Cruden, B. A., Milne, W. I., and Meyyappan, M. 2005. An investigation of plasma chemistry for dc plasma enhanced chemical vapour deposition of carbon nanotubes and nanofibres. *Nanotechnology*, 16: 925–930.
19. Bell, M. S., Teo, K. B. K., Lacerda, R. G., Milne, W. I., Hash, D. B., and Meyyappan, M. 2006. Carbon nanotubes by plasma-enhanced chemical vapor deposition. *Pure and Applied Chemistry*, 78: 1117–1125.
20. Oda, A., Suda, Y., and Okita, A. 2008. Numerical analysis of pressure dependence on carbon nanotube growth in CH_4/H_2 plasmas. *Thin Solid Films*, 516: 6570–6574.
21. Okita, A., Suda, Y., Ozeki, A. et al. 2006. Predicting the amount of carbon in carbon nanotubes grown by CH_4 rf plasmas. *Journal of Applied Physics*, 99: 014302.
22. Herrebout, D., Bogaerts, A., Yan, M., Gijbels, R., Goedheer, W., and Dekempeneer, E. 2001. One-dimensional fluid model for an rf methane plasma of interest in deposition of diamond-like carbon layers. *Journal of Applied Physics*, 90: 570–579.
23. De Bleecker, K., Bogaerts, A., and Goedheer, W. 2006. Detailed modeling of hydrocarbon nanoparticle nucleation in acetylene discharges. *Physical Review E*, 73: 026405.
24. De Bleecker, K., Bogaerts, A., and Goedheer, W. 2006. Aromatic ring generation as a dust precursor in acetylene discharges. *Applied Physics Letters*, 88: 151501.
25. Mao, M., Benedikt, A. J., Consoli, A., and Bogaerts, A. 2008. New pathways for nanoparticle formation in acetylene dusty plasmas: A modelling investigation and comparison with experiments. *Journal of Physics D—Applied Physics*, 41: 225201.

26. Hassouni, K., Leroy, O., Farhat, S., and Gicquel, A. 1998. Modeling of H_2 and H_2/CH_4 moderate-pressure microwave plasma used for diamond deposition. *Plasma Chemistry and Plasma Processing*, 18: 325–362.

27. Herrebout, D., Bogaerts, A., Yan, M., Gijbels, R., Goedheer, W., and Vanhulsel, A. 2002. Modeling of a capacitively coupled radio-frequency methane plasma: Comparison between a one-dimensional and a two-dimensional fluid model. *Journal of Applied Physics*, 92: 2290–2295.

28. Bera, K., Farouk, B., and Vitello, P. 2001. Inductively coupled radio frequency methane plasma simulation. *Journal of Physics D—Applied Physics*, 34: 1479–1490.

29. Farouk, T., Farouk, B., Gudsol, A., and Fridman, A. 2008. Atmospheric pressure methane-hydrogen dc micro-glow discharge for thin film deposition. *Journal of Physics D—Applied Physics*, 41: 175202.

30. Ostrikov, K., Yoon, H. J., Rider, A. E., and Ligatchev, V. 2007. Reactive species in $Ar+H_2$ plasma-aided nanofabrication: Two-dimensional discharge modelling. *Physica Scripta*, 76: 187–195.

31. Ostrikov, K., Yoon, H. J., Rider, A. E., and Vladimirov, S. V. 2007. Two-dimensional simulation of nanoassembly precursor species in $Ar+H_2+C_2H_2$ reactive plasmas. *Plasma Processes and Polymers*, 4: 27–40.

32. Ivanov, V., Proshina, O., Rakhimova, T. V., Rakhimov, A., Herrebout, D., and Bogaerts, A. 2002. Comparison of a one-dimensional particle-in-cell-Monte Carlo model and a one-dimensional fluid model for a CH_4/H_2 capacitively coupled radio frequency discharge. *Journal of Applied Physics*, 91: 6296–6302.

33. Proshina, O.V., Rakhimova, T. V. and Rakhimov, A. T. 2006. A particle-in-cell Monte Carlo simulation of an rf discharge in methane: frequency and pressure features of the ion energy distribution function. *Plasma Sources Science and Technology*, 15: 402–09.

34. Alexandrov, A. L., and Schweigert, I. V. 2005. Two-dimensional PIC-MCC simulations of a capacitively coupled radio frequency discharge in methane. *Plasma Sources Science and Technology*, 14: 209–218.

35. May, P. W., Harvey, J. N., Smith, J. A., and Mankelevich, Yu A. 2006. Reevaluation of the mechanism for ultrananocrystalline diamond deposition from $Ar/CH_4/H_2$ gas mixtures. *Journal of Applied Physics*, 99: 104907.

36. Mankelevich, Y.A., Rakhimov, A. T., and Suetin, N. V. 1998. Three-dimensional simulation of a HFCVD reactor. *Diamond and Related Materials*, 7: 1133–1137.

37. May, P. W., Smith, J. A., and Mankelevich, Y. A. 2006. Deposition of NCD films using hot filament CVD and $Ar/CH_4/H_2$ gas mixtures. *Diamond and Related Materials*, 15: 345–352.

38. May, P. W., and Mankelevich, Y. A. 2006. Experiment and modeling of the deposition of ultrananocrystalline diamond films using hot filament chemical vapor deposition and $Ar/CH_4/H_2$ gas mixtures: A generalized mechanism for ultrananocrystalline diamond growth. *Journal of Applied Physics*, 100: 024301.

39. Mankelevich, Y. A., Ashfold, M. N. R., and Ma, J. 2008. Plasma-chemical processes in microwave plasma-enhanced chemical vapor deposition reactors operating with C/H/Ar gas mixtures. *Journal of Applied Physics*, 104: 113304.

40. Mankelevich, Y. A., Ashfold, M. N. R., and Orr-Ewing, A. J. 2007. Measurement and modeling of $Ar/H_2/CH_4$ arc jet discharge chemical vapor deposition reactors II: Modeling of the spatial dependence of expanded plasma parameters and species number densities. *Journal of Applied Physics*, 102: 063310.

41. Mao, M., and Bogaerts, A. 2010. Investigating the plasma chemistry for the synthesis of carbon nanotubes/nanofibres in an inductively coupled plasma enhanced CVD system: The effect of different gas mixtures. *Journal of Physics D—Applied Physics*, 43: 205201.
42. Mao, M., and Bogaerts, A. 2010. Investigating the plasma chemistry for the synthesis of carbon nanotubes/nanofibres in an inductively coupled plasma-enhanced CVD system: The effect of processing parameters. *Journal of Physics D—Applied Physics*, 43: 315203.
43. Amanatides, E., Stamou, S., and Mataras, D. 2001. Gas phase and surface kinetics in plasma enhanced chemical vapor deposition of microcrystalline silicon: The combined effect of rf power and hydrogen dilution. *Journal of Applied Physics*, 90: 5786–5798.
44. Kushner, M. J. 1988. A model for the discharge kinetics and plasma chemistry during plasma enhanced chemical vapor-deposition of amorphous-silicon. *Journal of Applied Physics*, 63: 2532–2551.
45. Strahm, B., Howling, A. A., Sansonnens, L., and Hollenstein, C. 2007. Plasma silane concentration as a determining factor for the transition from amorphous to microcrystalline silicon in SiH_4/H_2 discharges. *Plasma Sources Science and Technology*, 16: 80–89.
46. De Bleecker, K., Herrebout, D., Bogaerts, A., Gijbels, R., and Descamps, P. 2003. One-dimensional modelling of a capacitively coupled rf plasma in silane/helium, including small concentrations of O_2 and N_2. *Journal of Physics D—Applied Physics*, 36: 1826–1833.
47. Nienhuis, G. J., Goedheer, W. J., Hamers, E. A. G., vanSark, W. G. J. H. M., and Bezemer, J. 1997. A self-consistent fluid model for radio-frequency discharges in SiH_4-H_2 compared to experiments. *Journal of Applied Physics*, 82: 2060–2071.
48. Petrov, G. M., and Giuliani, J. L. 2001. Model of a two-stage rf plasma reactor for SiC deposition. *Journal of Applied Physics*, 90: 619–636.
49. Collins, D. J., Strojwas, A. J., and White, D. D. 1994. A CFD model for the PECVD of silicon-nitride. *IEEE Transactions on Semiconductor Manufacturing*, 7: 176–183.
50. Bavafa, M., Ilati, H., and Rashidian, B. 2008. Comprehensive simulation of the effects of process conditions on plasma enhanced chemical vapor deposition of silicon nitride. *Semiconductor Science and Technology*, 23: 095023.
51. Lyka, B., Amanatides, E., and Mataras, D. 2006. Simulation of the electrical properties of SiH_4/H_2 RF discharges. *Japanese Journal of Applied Physics Part 1*, 45: 8172–8176.
52. Yan, M., and Goedheer, W. J. 1999. A PIC-MC simulation of the effect of frequency on the characteristics of VHF SiH_4/H_2 discharges. *Plasma Sources Science and Technology*, 8: 349–354.
53. Kim, D. J., Kang, J. Y., Nasonva, A., Kim, K. S., and Choi, S. J. 2007. Numerical simulation on silane plasma chemistry in pulsed plasma process to prepare a-Si:H thin films. *Korean Journal of Chemical Engineering*, 24: 154–164.
54. Sato, N., and Tagashira, H. 1991. A hybrid Monte-Carlo fluid model of RF plasmas in a SiH_4/H_2 mixture. *IEEE Transactions on Plasma Science*, 19: 102–112.
55. Sugai, H., and Toyoda, H. 1992. Appearance mass-spectrometry of neutral radicals in radio-frequency plasmas. *Journal of Vacuum Science and Technology A*, 10: 1193–1200.

56. Ventzek, P. L. G., Sommerer, T. J., Hoekstra, R. J., and Kushner, M. J. 1993. 2-Dimensional hybrid model of inductively-coupled plasma sources for etching. *Applied Physics Letters*, 63: 605–607.
57. Zhang, D., and Kushner, M. J. 2000. Surface kinetics and plasma equipment model for Si etching by fluorocarbon plasmas. *Journal of Applied Physics*, 87: 1060–1069.
58. Arunachalam, V., Rauf, S., Coronell, D. G., and Ventzek, P. L. G. 2001. Integrated multi-scale model for ionized plasma physical vapor deposition. *Journal of Applied Physics*, 90: 64–73.
59. Grapperhaus, M. J., Krivokapic, Z., and Kushner, M. J. 1998. Design issues in ionized metal physical vapor deposition of copper. *Journal of Applied Physics*, 83: 35–43.
60. Arakoni, R., Stafford, D. S., Babaeva, N. J., and Kushner, M. J. 2005. $O_2((1)Delta)$ production in flowing He/O_2 plasmas. II. Two-dimensional modeling. *Journal of Applied Physics*, 98: 073304.
61. Bhoj, A. N., and Kushner, M. J. 2007. Continuous processing of polymers in repetitively pulsed atmospheric pressure discharges with moving surfaces and gas flow. *Journal of Physics D—Applied Physics*, 40: 6953–6968.
62. Hash, D. B., and Meyyappan, M. 2003. Model based comparison of thermal and plasma chemical vapor deposition of carbon nanotubes. *Journal of Applied Physics*, 93: 750–752.
63. Matthews, K., Cruden, B. A., Chen, B., Meyyappan, M., and Delzeit, J. 2002. Plasma-enhanced chemical vapor deposition of multiwalled carbon nanofibers. *Journal of Nanoscience and Nanotechnology*, 2: 475–480.
64. Chhowalla, M., Teo, K. B. K., and Ducati, C. 2001. Growth process conditions of vertically aligned carbon nanotubes using plasma enhanced chemical vapor deposition. *Journal of Applied Physics*, 90: 5308–5317.
65. Cruden, B. A., and Meyyappan, M. 2005. Characterization of a radio frequency carbon nanotube growth plasma by ultraviolet absorption and optical emission spectroscopy. *Journal of Applied Physics*, 97: 084311.
66. Sankaran, A., and Kushner, M. J. 2003. Fluorocarbon plasma etching and profile evolution of porous low-dielectric-constant silica. *Applied Physics Letters*, 82: 1824–1826.
67. von Keudell, A. 2000. Surface processes during thin-film growth. *Plasma Sources Science and Technology*, 9: 12.
68. Mantzaris, N. V., Gogolides, E., Boudouvis, A. G., Rhallabi, A., Turban, G., and Paal, J. 1996. Surface and plasma simulation of deposition processes: CH_4 plasmas for the growth of diamondlike carbon. *Journal of Applied Physics*, 79: 3718–3729.
69. May, P. W., Harvey, J. N., Allan, N. L., Richley, J. C., and Mankelevich, Y. A. 2010. Simulations of chemical vapor deposition diamond film growth using a kinetic Monte Carlo model. *Journal of Applied Physics*, 108: 014905.
70. Battaile, C. C., and Srolovitz, D. J. 2002. Kinetic Monte Carlo simulation of chemical vapor deposition. *Annual Review of Materials Research*, 32: 297–319.
71. Berg, S., and Nyberg, T. 2005. Fundamental understanding and modeling of reactive sputtering processes. *Thin Solid Films*, 476: 215.
72. Depla, D., Heirwegh, S., Mahieu, S., and De Gryse, R. 2007. Towards a more complete model for reactive magnetron sputtering. *Journal of Physics D—Applied Physics*, 40: 1957.

73. Puretzky, A. A., Geohegan, D. B., Jesse, S., Ivanov, I. N., and Eres, G. 2005. In situ measurements and modeling of carbon nanotube array growth kinetics during chemical vapor deposition. *Applied Physics A*, 81: 223–240.
74. Lee, D. H., Kim, S. O., and Lee, W. J. 2010. Growth kinetics of wall-number controlled carbon nanotube arrays. *Journal of Physical Chemistry C*, 114: 3454–3458.
75. Zhang, Y., and Smith, K. J. 2005. A kinetic model of CH_4 decomposition and filamentous carbon formation on supported Co catalysts. *Journal of Catalysis*, 231: 354–364.
76. Naha, S., Sen, S., De, A. K., and Puri, I. K. 2007. A detailed model for the flame synthesis of carbon nanotubes and nanofibers. *Proceedings of the Combustion Institute*, 31: 1821–1829.
77. Naha, S., and Puri, I. K. 2008. A model for catalytic growth of carbon nanotubes. *Journal of Physics D—Applied Physics*, 41: 065304.
78. Denysenko, I., and Ostrikov, K. 2007. Ion-assisted precursor dissociation and surface diffusion: Enabling rapid, low-temperature growth of carbon nanofibers. *Applied Physics Letters*, 90: 251501.
79. Denysenko, I., and Ostrikov, K. 2009. Plasma heating effects in catalyzed growth of carbon nanofibres. *Journal of Physics D—Applied Physics*, 42: 015208.
80. Latorre, N., Romeo, E., Cazana, F. et al. 2010. Carbon nanotube growth by catalytic chemical vapor deposition: A phenomenological kinetic model. *Journal of Physical Chemistry C*, 114: 4773–4782.
81. Grujicic, M., Cao, G., and Gersten, B. 2002. Optimization of the chemical vapor deposition process for carbon nanotubes fabrication. *Applied Surface Science*, 199: 90–106.
82. Lysaght, A. C., and Chiu, W. K. S. 2008. Modeling of the carbon nanotube chemical vapor deposition process using methane and acetylene precursor gases. *Nanotechnology*, 19: 165607.
83. Lysaght, A. C., and Chiu, W. K. S. 2009. The role of surface species in chemical vapor deposited carbon nanotubes. *Nanotechnology*, 20: 115605.
84. Hosseini, M. R., Jalili, N., and Bruce, D. A. 2009. A time-dependent multiphysics, multiphase modeling framework for carbon nanotube synthesis using chemical vapor deposition. *Aiche Journal*, 55: 3152–3167.
85. Levchenko, I., and Ostrikov, K. 2008. Plasma/ion-controlled metal catalyst saturation: Enabling simultaneous growth of carbon nanotube/nanocone arrays. *Applied Physics Letters*, 92: 063108.
86. Levchenko, I., Ostrikov, K., Mariotti, D., and Murphey, A. B. 2008. Plasma-controlled metal catalyst saturation and the initial stage of carbon nanostructure array growth. *Journal of Applied Physics*, 104: 073308.
87. Levchenko, I., Ostrikov, K., Khachan, J., and Vladimirov, S. V. 2008. Growth of carbon nanocone arrays on a metal catalyst: The effect of carbon flux ionization. *Physics of Plasmas*, 15: 103501.
88. Tam, E., and Ostrikov, K. 2009. Catalyst size effects on the growth of single-walled nanotubes in neutral and plasma systems. *Nanotechnology*, 20: 375603.
89. May, P. W., Allan, N. L., Ashfold, M. N. R., Richley, J. C., and Mankelevich, Y. A. 2009. Simplified Monte Carlo simulations of chemical vapour deposition diamond growth. *Journal of Physics—Condensed Matter*, 21: 364203.
90. Gruen, D. M. 1999. Nanocrystalline diamond films. *Annual Review of Materials Science*, 29: 211–259.

91. Agacino, E., and de la Mora, P. 2003. Theoretical study of intradimer mechanism for diamond growth over diamond (100). *Structural Chemistry*, 14: 541–550.

92. Kang, J. K., and Musgrave, C. B. 2000. A theoretical study of the chemical vapor deposition of (100) diamond: An explanation for the slow growth of the (100) surface. *Journal of Chemical Physics*, 113: 7582–7587.

93. Larsson, K. 1997. Adsorption of hydrocarbon species on a stepped diamond (111) surface. *Physical Review B*, 56: 15452–15458.

94. Tamura, H., Zhou, H., Hirano, Y., et al., 2000. Periodic density-functional study on oxidation of diamond (100) surfaces. *Physical Review B*, 61: 11025–11033.

95. Cheesman, A., Harvey, J. N., and Ashfold, M. N. R. 2008. Studies of carbon incorporation on the diamond {100} surface during chemical vapor deposition using density functional theory. *Journal of Physical Chemistry A*, 112: 11436–11448.

96. Tamura, H., and Gordon, M. S. 2005. Ab initio study of nucleation on the diamond(100) surface during chemical vapor deposition with methyl and H radicals. *Chemical Physics Letters*, 406: 197–201.

97. Garrison, B. J., Kodali, P. B. S., and Srivastava, D. 1996. Modeling of surface processes as exemplified by hydrocarbon reactions. *Chemical Reviews*, 96: 1327–1341.

98. van Beest, B. W. H., Kramer, G. J., and van Santen, R. A. 1990. Force fields for silicas and aluminophosphates based on ab initio calculations. *Physical Review Letters*, 64: 1955.

99. Stillinger, F. H., and Weber, T. A. 1985. Computer-simulation of local order in condensed phases of silicon. *Physical Review B*, 31: 5262–5271.

100. Vashishta, P., Kalia, R. K., and Rino, J. P. 1990. Interaction potential for SiO_2—A molecular-dynamics study of structural correlations. *Physical Review B*, 41: 12197–12209.

101. Tersoff, J. 1988. New empirical-approach for the structure and energy of covalent systems. *Physical Review B*, 37: 6991–7000.

102. Brenner, D. W. 1990. Empirical potential for hydrocarbons for use in simulating the chemical vapor-deposition of diamond films. *Physical Review B*, 42: 9458–9471.

103. Daw, M. S., and Baskes, M. I. 1983. Semiempirical, quantum-mechanical calculation of hydrogen embrittlement in metals. *Physical Review Letters*, 50: 1285–1288.

104. van Duin, A. C. T., Dasgupta, S., Lorant, F. et al. 2001. ReaxFF: A reactive force field for hydrocarbons. *Journal of Physical Chemistry A*, 105: 9396–9409.

105. Han, S. S., Kang, J. K., Lee, H. M., van Duin, A. C. T., and Goddard, W. A. 2005. Liquefaction of H_2 molecules upon exterior surfaces of carbon nanotube bundles. *Applied Physics Letters*, 86: 203108.

106. Lewis, G. V., Catlow, C. R. A., and Cormack, A. N. 1985. Defect structure and migration in Fe_3O_4. *Journal of Physics and Chemistry of Solids*, 46: 1227–1233.

107. Georgieva, V., Saraiva, N., Jehanathan, N., Lebelev, O. I., Depla, D., and Bogaerts, A. 2009. Sputter-deposited Mg-Al-O thin films: Linking molecular dynamics simulations to experiments. *Journal of Physics D—Applied Physics*, 42: 065107.

108. Baguer, N., Georgieva, V., Calderin, L., Todorov, I. T., Van Gils, S., and Bogaerts, A. 2009. Study of the nucleation and growth of TiO_2 and ZnO thin films by means of molecular dynamics simulations. *Journal of Crystal Growth*, 311: 4034–4043.

109. Meziani, T., Colpo, P., and Rossi, F. 2007. PECVD of diamond-like carbon (a-C:H) from the decomposition of methane in a high-density inductively coupled discharge. *Journal of Superhard Materials*, 29: 153–157.
110. Benedikt, J., Eijkman, D. J., Vandamme, W., Agarwal, S., and van de Sanden, M. C. M. 2005. Threshold ionization mass spectrometry study of hydrogenated amorphous carbon films growth precursors. *Chemical Physics Letters*, 402: 37–42.
111. de Rooij, E. D., Kleyn, A. W., and Goedheer, W. J. 2010. Sticking of hydrocarbon radicals on different amorphous hydrogenated carbon surfaces: A molecular dynamics study. *Physical Chemistry Chemical Physics*, 12: 14067–14075.
112. Träskelin, P., Salonen, E., Krasheninnikov, A. V., and Keinonen, J. 2003. Molecular dynamics simulations of CH_3 sticking on carbon surfaces. *Journal of Applied Physics*, 93: 1826–1831.
113. Sharma, A. R., Schneider, R., Toussaint, U., and Nordlund, K. 2007. Hydrocarbon radicals interaction with amorphous carbon surfaces. *Journal of Nuclear Materials*, 363–365: 1283–1288.
114. Träskelin, P., Saresoja, O., and Nordlund, K. 2008. Molecular dynamics simulations of C_2, C_2H, C_2H_2, C_2H_3, C_2H_4, C_2H_5, and C_2H_6 bombardment of diamond (111) surfaces. *Journal of Nuclear Materials*, 375: 270–274.
115. von Keudell, A., and Jacob, W. 2004. Elementary processes in plasma-surface interaction: H-atom and ion-induced chemisorption of methyl on hydrocarbon film surfaces. *Progress in Surface Science*, 76: 21–54.
116. Ugolini, D., Eitle, J., and Oelhafen, P. 1990. Influence of process gas and deposition energy on the atomic and electronic-structure of diamond-like (a-C:H) films. *Vacuum* 41: 1374–1377.
117. Neyts, E., Tacq, M., and Bogaerts, A. 2006. Reaction mechanisms of low-kinetic energy hydrocarbon radicals on typical hydrogenated amorphous carbon (a-C:H) sites: A molecular dynamics study. *Diamond and Related Materials*, 15: 1663–1676.
118. Neyts, E., Bogaerts, A., and van de Sanden, M. C. M. 2006. Unraveling the deposition mechanism in a-C:H thin-film growth: A molecular-dynamics study for the reaction behavior of C_3 and C_3H radicals with a-C:H surfaces. *Journal of Applied Physics*, 99: 014902.
119. Neyts, E., Bogaerts, A., Gijbels, R., Benedikt, J., and van de Sanden, M. C. M. 2005. Molecular dynamics simulation of the impact behaviour of various hydrocarbon species on DLC. *Nuclear Instruments and Methods in Physics Research Section B—Beam Interactions with Materials and Atoms*, 228: 315–318.
120. Neyts, E., and Bogaerts, A. 2006. Influence of internal energy and impact angle on the sticking behaviour of reactive radicals in thin a-C:H film growth: A molecular dynamics study. *Physical Chemistry Chemical Physics*, 8: 2066–2071.
121. Robertson, J. 2002. Diamond-like amorphous carbon. *Materials Science and Engineering R-Reports*, 37: 129–281.
122. Jäger, H. U., and Belov, A. 2003. Ta-C deposition simulations: Film properties and times-resolved dynamics of film formation. *Physical Review B*, 68: 1–13.
123. Belov, A., and Jäger, H. U. 2003. Atomistic study of ion-beam deposition conditions for hard amorphous carbon. *Computational Materials Science*, 24: 16–22.
124. Marks, N. 2002. Modeling diamond-like carbon with the environment dependent interaction potential. *Journal of Physics: Condensed Matter*, 14: 2901–2927.

125. Quan, W. L., Li, H. X., Zhao, F. et al. 2010. Molecular dynamical simulations on a-C:H film growth from atomic flux of C and H: Effect of H fraction. *Physics Letters A*, 374: 2150–2155.

126. Quan, W. L., Li, H. X., Zhao, F. et al. 2010. Molecular dynamical simulations on a-C:H film growth from C and H atomic flux: Effect of incident energy. *Chinese Physics Letters*, 27: 088102.

127. Benedikt, J., Agarwal, S., Eijkman, D., Vandamme, W., Creatore, M., van de Sanden, M. C. M. 2005. Threshold ionization mass spectrometry of reactive species in remote Ar/C_2H_2 expanding thermal plasma. *Journal of Vacuum Science and Technology A*, 23: 1400–1412.

128. Neyts, E., Bogaerts, A., Gijbels, R., Benedikt, J., and van de Sanden, M. C. M. 2004. Molecular dynamics simulations for the growth of diamond-like carbon films from low kinetic energy species. *Diamond and Related Materials*, 13: 1873–1881.

129. Hamon, A. -L. 2004. ELNES study of carbon K-edge spectra of plasma deposited carbon films. *Journal of Materials Chemistry*, 14: 2030–2035.

130. Neyts, E., Bogaerts, A., and van de Sanden, M. C. M. 2006. Effect of hydrogen on the growth of thin hydrogenated amorphous carbon films from thermal energy radicals. *Applied Physics Letters*, 88: 141922.

131. Neyts, E., Bogaerts, A., and van de Sanden, M. C. M. 2006. Densification of thin a-C:H films grown from low-kinetic energy hydrocarbon radicals under the influence of H and C particle fluxes: A molecular dynamics study. *Journal of Physics D: Applied Physics*, 39: 1948–1953.

132. Neyts, E., Eckert, M., and Bogaerts, A. 2007. Molecular dynamics simulations of the growth of thin a-C:H films under additional ion bombardment: Influence of the growth species and the Ar^+ ion kinetic energy. *Chemical Vapor Deposition*, 13: 312–318.

133. Träskelin, P., Salonen, E., Nordlund, K. Keionen, J. And Wu, C. H. 2004. Molecular dynamics simulations of CH_3 sticking on carbon surfaces, angular and energy dependence. *Journal of Nuclear Materials*, 334: 65–70.

134. Zhu, W. J., Pan, Z. Y., Ho, Y. K., and Man, Z. Y. 1999. Molecular dynamics simulation of C_2H_2 deposition on diamond (001)-(2 × 1) surface. *European Physical Journal D*, 5: 83–88.

135. Eckert, M., Neyts, E., and Bogaerts, A., 2008. On the reaction behaviour of hydrocarbon species at diamond (100) and (111) surfaces: A molecular dynamics investigation. *Journal of Physics D: Applied Physics*, 41: 032006.

136. Eckert, M., Neyts, E., and Bogaerts, A. 2008. Molecular dynamics simulations of the sticking and etch behavior of various growth species of (ultra)nanocrystalline diamond films. *Chemical Vapor Deposition*, 14: 213–223.

137. Garrison, B. J., Dawnkaski, E. J., Srivastava, D., and Brenner, D. W. 1992. Molecular-dynamics simulations of dimer opening on a diamond (001)(2 × 1) surface. *Science*, 255: 835–838.

138. Metropolis, N., Rosenbluth, A. W., Rosenbluth, M. N., Teller, A. H., and Teller, E. 1953. Equation of state calculation by fast computing machines. *Journal of Chemical Physics*, 21: 1087–1092.

139. Eckert, M., Neyts, E., and Bogaerts, A. 2009. Modeling adatom surface processes during crystal growth: A new implementation of the Metropolis Monte Carlo algorithm. *CrystEngComm*, 11: 1597–1608.

140. Eckert, M., Neyts, E., and Bogaerts, A. 2010. Insights into the growth of (ultra) nanocrystalline diamond by combined molecular dynamics and Monte Carlo simulations. *Crystal Growth and Design*, 10: 3005–3021.
141. Eckert, M., Neyts, E., and Bogaerts, A. 2010. Differences between ultrananocrystalline and nanocrystalline diamond growth: Theoretical investigation of C_xH_y species at diamond step edges. *Crystal Growth and Design*, 10: 4123–4134.
142. Butler, J. E., Mankelevich, Y. A., Cheesman, A., Ma, J., and Ashfold, M. N. R. 2009. Understanding the chemical vapor deposition of diamond: Recent progress. *Journal of Physics: Condensed Matter*, 21: 364201.
143. May, P. W., and Mankelevich, Y. A. 2008. From ultrananocrystalline diamond to single crystal diamond growth in hot filament and microwave plasma-enhanced CVD reactors: A unified model for growth rates and grain sizes. *Journal of Physical Chemistry C*, 112: 12432–12441.
144. Netto, A., and Frenklach, M. 2005. Kinetic Monte Carlo simulations of CVD diamond growth-interlay among growth, etching, and migration. *Diamond Related Materials*, 14: 1630–1646.
145. Wagner, R. S., and Ellis, W. C. 1964. Vapor-liquid-solid mechanism of single crystal growth. *Applied Physics Letters*, 4: 89–90.
146. Baker, R. T. K., Barber, M. A., Waite, R. J., Harris, P. S., and Feates, F. S. 1972. Nucleation and growth of carbon deposits from nickel catalyzed decomposition of acetylene. *Journal of Catalysis*, 26: 51–62.
147. Helveg, S., Lopez-Cartes, C., Sehested, J. et al. 2004. Atomic-scale imaging of carbon nanofibre growth. *Nature* 427: 426–429.
148. Homma, Y., Liu, H. P., Takagi, D., and Kobayashi, Y. 2009. Single-walled carbon nanotube growth with non-iron-group "catalysts" by chemical vapor deposition. *Nano Research*, 2: 793–799.
149. Maiti, A., Brabec, C. J., Roland, C., and Bernholc, J. 1995. Theory of carbon nanotube growth. *Physical Review B*, 52: 14850–14858.
150. Maiti, A., Brabec, C. J., and Bernholc, J. 1997. Kinetics of metal-catalyzed growth of single-walled carbon nanotubes. *Physical Review B* 55: R6097–R6100.
151. Shibuta, Y., and Maruyama, S. 2002. Molecular dynamics simulation of generation process of SWNTs. *Physica B*, 323 187–199.
152. Shibuta, Y., and Maruyama, S. 2003. Molecular dynamics simulation of formation process of single-walled carbon nanotubes by CCVD method. *Chemical Physics Letters*, 382: 381–386.
153. Shibuta, Y., and Maruyama, S. 2007. A molecular dynamics study of the effect of a substrate on catalytic metal clusters in nucleation process of single-walled carbon nanotubes. *Chemical Physics Letters*, 437: 218–223.
154. Martinez-Limia, A., Zhao, J., and Balbuena, P. B. 2007. Molecular dynamics study of the initial stages of catalyzed single-wall carbon nanotubes growth: Force field development. *Journal of Molecular Modeling*, 13: 595–600.
155. Zhao, J., Martinez-Limia, A., Balbuena, P. B. 2005. Understanding catalysed growth of single-wall carbon nanotubes. *Nanotechnology*, 16: S575–S581.
156. Ribas, M. A., Ding, F., Balbuena, P. B., and Yakobson, B. I. 2009. Nanotube nucleation versus carbon-catalyst adhesion—Probed by molecular dynamics simulations. *Journal of Chemical Physics*, 131: 224501.

157. Burgos, J. C., Reyna, H., Yakobson, B. I., and Balbuena, P. B. 2010. Interplay of catalyst size and metal-carbon interactions on the growth of single-walled carbon nanotubes. *Journal of Physical Chemistry C*, 114: 6952–6958.

158. Ding, F., Rosén, A., and Bolton, K. 2005. Dependence of SWNT growth mechanism on temperature and catalyst particle size: Bulk versus surface diffusion. *Carbon*, 43: 2215–2217.

159. Ding, F., Rosén, A., and Bolton, K. 2004. Molecular dynamics study of the catalyst particle size dependence on carbon nanotube growth. *Journal of Chemical Physics*, 121: 2775–2779.

160. Ding, F., Bolton, K., and Rosén, A. 2006. Molecular dynamics study of SWNT growth on catalyst particles without temperature gradients. *Computational Materials Science*, 35: 243–246.

161. Ding, F., Rosén, A., and Bolton, K. 2004. The role of the catalytic particle temperature gradient for SWNT growth from small particles. *Chemical Physics Letters*, 393: 309–313.

162. Voter, A., Montalenti, F., and Germann, T. 2002. Extending the time scale in atomistic simulation of materials. *Annual Review of Materials Research*, 32: 321–346.

163. Neyts, E., Shibuta, Y., and Bogaerts, A. 2010. Bond switching regimes in nickel and nickel-carbon nanoclusters. *Chemical Physics Letters*, 488: 202–205.

164. Dereli, G. 1992. Stillinger-Weber type potentials in Monte-Carlo simulation of amorphous-silicon. *Molecular Simulation*, 8: 351–360.

165. Timonova, M., Groenewegen, J., and Thijsse, B. J. 2010. Modeling diffusion and phase transitions by a uniform-acceptance force-bias Monte Carlo method. *Physical Review B*, 81: 144107.

166. Neyts, E. C., and Bogaerts, A. 2009. Numerical study of the size-dependent melting mechanisms of nickel nanoclusters. *Journal of Physical Chemistry C*, 113: 2771–2776.

167. Neyts, E. C., Shibuta, Y., van Duin, A. C. T., and Bogaerts, A. 2010. Catalyzed growth of carbon nanotube with definable chirality by hybrid molecular dynamics-force biased Monte Carlo simulations. *ACS Nano*, 4: 6665–6672.

168. Page, A. J., Ohta, Y., Okamoto, Y., Irle, S., and Morokuma, K. 2009. Defect healing during single-walled carbon nanotube growth: A density-functional tight-binding molecular dynamics investigation. *Journal of Physical Chemistry C*, 113: 20198–20207.

11

Modeling Catalytic Growth of One-Dimensional Nanostructures

Eugene Tam
Kostya (Ken) Ostrikov
Tony Murphy

CONTENTS

11.1 Introduction

The unique properties of one-dimensional (1D) nanostructures, such as nanotubes and nanowires, have stirred the interest of researchers trying to utilize them in advanced new devices. These nanostructures have many potential uses such as drug and gene delivery, for hydrogen storage, as electron field emitters, and several others.[1-8] Traditionally in materials science, new structures are first fabricated, their properties characterized, and attempts to model their properties are performed last to produce some sort of understanding of how these structures form and how they can be tuned. With the improvements of computational resources and techniques, it is now possible to model the growth and properties of new materials and predict how they will be fabricated before any experiment is performed.[9] This allows for the tailoring of nanostructures for specific functions rather than trial-and-error experiments to determine which nanostructure fits the desired purpose.

The catalytic fabrication of 1D nanostructures, such as nanowires and nanotubes, can be split into two distinct stages: nucleation and growth. Nucleation occurs before the growth and, in some cases, fully determines the physical properties of the nanostructures.

This chapter will be divided into two main sections. The first will address the actual models, what they need to take into account, their limitations, and where can they be used. The second section will address computational techniques that may be applied to solve these problems.

11.2 Modeling

When modeling the growth of nanowires and nanotubes, one must consider many factors, including conditions for nucleation and growth, delivery of building units to the structures, and the role of different plasma species in the nanostructure nucleation and growth.

Growth and nucleation models can be broadly divided into three categories (not including hybrids): models that only consider growth in thermal equilibrium, those that model kinetic growth, and ab initio models.

Pure thermal equilibrium models (such as Wulff constructions) are the simplest types of model that consider only the most energetically favorable outcomes.[10] They are typically used to determine the equilibrium structure of a crystal after thermal annealing. In terms of modeling the growth of nanowalls and nanotubes, these models are generally restricted to determining whether nucleation occurs or not and are therefore usually used in conjunction with a kinetic model.

Kinetic models determine motion of the particles (atoms, ions, radicals, as well as nonreactive species) in the system. These models can generally provide temporal information and can produce structures characterized by a global minimum of the total energy.

Ab initio models refer to any model that uses the first principles of quantum dynamics to model the system. These models are extremely computationally taxing and are typically used only for the nucleation stages or to determine the interatomic interaction potentials for molecular dynamic simulation or again, equilibrium states (using density functional theory, for example) of small atomic stacks.

11.2.1 Structure

As simple as this may sound, it is important to first model the structure before modeling its growth. Choosing how to represent the structure can determine how best to model its growth and what information can be obtained from the models. For example, a nanowire can be represented as a collection of a large number of atoms, as a polygon with many facets, or even as a cylinder (Figure 11.1). The cylinder is the simplest representation and will only allow for simulations to determine the growth rates and aspect ratios

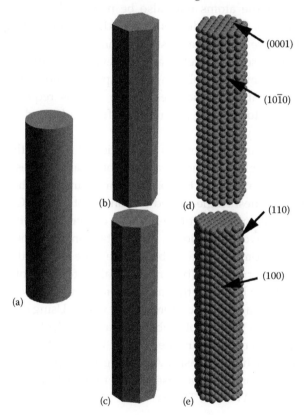

FIGURE 11.1
Different representations of the same one-dimensional (1D) nanostructure. (a) A cylinder is the simplest way a 1D nanostructure can be represented. A cylinder can be used to represent the structure if one is only interested in growth rates or aspect ratios. (b,c) Polygonal representation of nanowires with a hexagonal and an octagonal cross section, respectively. This representation is only slightly more complicated and can be used in models that require the determination of the equilibrium shapes and reactivity of different facets. (d,e) Atomic representation of nanowires with different crystal structures. This type of representation is required when more details about the nanostructure are needed for certain models (e.g., crystallographic structure). A nanowire with a hexagonal lattice is shown in (d) with the {10$\bar{1}$0} and the {0001} planes exposed. A face centered cubic (FCC) lattice is shown in (e) with the {100} and {110} planes exposed.

of the structures. This is a truly 1D model, which is therefore very simple to solve, requiring very little computational time. In some cases analytical solutions are possible. Using a polygon representation of the nanowire allows for growth simulations to account for the differences (such as surface area, surface free energy, etc.) between each crystal facet.

However, for a model that describes the full details of growth and nucleation, one has to represent the nanowire as a collection of atoms, and depending on what details are required, reliable information on the electronic states of the atoms may also be needed. By representing the nanostructure in this fashion, details such as defects, branching, and even whether the structure is a single crystal or amorphous can be determined. Models that require so much detail are, however, generally very complex because they have fewer simplifying assumptions and require massive amounts of computational resources. These models require significant computational time to produce any meaningful results and are generally restricted to smaller systems.

Representing the nanostructures as a collection of atoms rather than simply as a cylinder or a group of facets can have other advantages, even before any real calculations are made. For example, an estimate of the diameter is required for the thinnest single-walled carbon nanotube (SWCNT) that can be formed. The walls of a defect-free SWNCT can be described as a roll of graphene (that is a two-dimensional, 2D, crystal consisting of a hexagonal network of carbon atoms, all in a sp^2 bond state). The ends of the tube, however, must either be terminated (by a metal catalyst, for example) or capped. To form a SWCNT cap however, the carbon atoms in the network must form shapes other than hexagons to be energetically viable. Using energetic arguments, the only shapes that can plausibly form a network of sp^2 carbon (other than hexagons) are pentagons and heptagons. Of these two shapes, pentagons are formed much more easily. Assuming the caps are made only of hexagons and pentagons, one can use its Euler characteristic to determine the minimum size of SWCNTs. That is

$$f + v - e = \chi, \tag{11.1}$$

where f is the number of faces in a polyhedron, v the number of vertices, e the number of edges, and χ the characteristic of the polyhedron, which is 2 for convex shapes.

Using Equation (11.1), one can show that every fullerene has 12 pentagons (assuming that the fullerenes consists of only hexagons and pentagons) by introducing two new variables h and p representing the number of hexagons and pentagons in the polyhedron, respectively. The relationships

$$f = h + p,$$

$$2e = 5p + 6h,$$

$$3v = 5p + 6h, \tag{11.2}$$

can be obtained because every pentagon has five sides and every hexagon six, every edge adjoins two shapes, every pentagon has five vertices and every hexagon six, and every vertex is at the intersection of three shapes. Substituting Equation (11.2) back into Equation (11.1) will yield $p = 12$, meaning that every fullerene has 12 pentagons regardless of size. Thus, the smallest fullerene, C_{20}, has 12 faces in total (or 20 carbon atoms). Assuming the cap of the SWCNT is at least half a fullerene, we can then determine that the smallest-diameter SWCNT is ~0.39 nm, which is approximately the size of the smallest-diameter SWCNT found in experiment (~0.42 nm for free standing and ~0.39 nm when extracted from a multiwalled tube).[11,12] This prediction obviously could not have been made by representing the nanotube as a capped cylinder.

11.2.2 Free Energy

Free-energy models are generally more applicable to nanowire rather than nanotube growth. However, these models can still be used in determining whether conditions are suitable for the nucleation of the caps in SWCNTs. The aim of these models is to minimize the free energy in the system. This is typically done by considering the sum of the free energies of each of the components in the system, that is

$$\Delta G = G_{bulk} + G_{surface} + G_{edges} + G_{corners}, \tag{11.3}$$

where G_x is the free energy due to either the bulk form (crystal or amorphous) or to various interfaces with the catalyst, vacuum, or gas.[13] In the case of substrate-bound nanostructures, there are additional interface energies for the surface, edge, and corner terms.

In general, each term is a function of the temperature, and the bulk term is also proportional to the volume of the structure. However, the nonbulk terms can also be represented as

$$G_{surface} = \sum_i A_i \gamma_i,$$

$$G_{edges} = \sum_j L_j \lambda_j,$$

$$G_{corners} = \sum_k W_k \varepsilon_k, \tag{11.4}$$

where A_i and γ_i are, respectively, the surface area and the surface free energy of facet i. Furthermore, L_j and λ_j are the lengths and edge free energy of the edges of the nanostructure, respectively. In Equation (11.4) W_k and ε_k are the number and corner free energy of the corners, respectively. In larger crystals, the bulk term determines the most favorable morphology at a given temperature. However, it is well known that smaller structures have a much larger surface area to volume ratio, as well as the edge length to surface area ratio. Therefore, as the structure size decreases, the significance of the nonbulk term increases, which leads to either minimization of the surface area or minimization of the number of facets (which increases the surface area but minimizes the edges), depending on the relative free-energy values.

Another factor that affects the total free energy is the surface tension. In liquids, the phenomena of surface tension and the surface free energy are the same. However, in solids these effects are distinct. The extra contribution to the total energy is, however, usually considered to be insignificant. Therefore, the Laplace–Young approximation (in which surface tension is equated to the surface free energy) is typically used.[13]

When a structure is represented as a collection of atoms, the total free energy can be calculated from the summation of bond energies using ab initio models. This generally only applies to the nanostructure nucleation stage. This approach has typically been used to determine the conditions for the formation of SWCNT caps and to predict the probability of nucleation of SWNCTs of certain chiralities.[14,15]

11.2.3 Kinetic Considerations

By only considering the free energy, one can consider the growth of only the most stable nanostructures. However, the structure at thermal equilibrium may not be the one of interest, or the growth environment maybe highly nonequilibrium. In these cases it is necessary to consider kinetic processes that can dominate the growth. Kinetic processes also need to be considered when a time-resolved model is required.

11.2.3.1 The Role of the Gas and Plasma

A comprehensive treatment of the gas or plasma is rarely included in a model where the main focus is on the growth of 1D nanostructures, as large amounts of computational resources are required to solve a complete gas or plasma model on its own. However, atoms, ions, and other particles in the gas phase also play a vital role in the production of nanostructures, particularly in plasma-based systems for the growth of surface-bound nanostructures. In this case, some control of the energy and trajectories of the particles can be obtained by applying a bias to the substrate. The gas phase can play a number of roles ranging from milling of the surface by ions to simply delivery of the building units to the surface. A large range of particle species from radicals to relatively large molecules and nanostructures can be produced in the gas phase. Additionally, these ions and electrons can heat the surface, removing or reducing the need for an external substrate heater and potentially heating the surface only where it is needed.[16–21] Finally, some plasma systems can be magnetically enhanced, allowing for the growth of longer SWCNTs in the plasma.[22]

Typically, when the effects of the plasma are required to be included in the model, simplifications are usually made so that the solutions can be obtained in a reasonable amount of computational time. For this purpose, the key plasma-related aspects should be chosen. For example, plasma calculations can be restricted to specie generation rates, when modeling the effects of different species, or the spatial location of some atoms/ions in a very small region in a vacuum system to determine where species influx occurs. Simplifications can also be made in the growth model, for example, an extensive plasma model can be coupled with a growth model simplified by assuming certain types of growth can only occur under certain conditions.[20,23,24]

11.2.4 Surface-Supported Nanostructures

Other considerations are required when modeling surface-supported nanostructures. In particular, mobile adatoms on the surface become a major source of building material to arrive at the nanostructures, especially with 1D nanostructures grown via the root-growth mechanism. The growth rates of nanostructures formed with this mechanism, in which building units must find their way to the interface of the nanostructure with the substrate (or to a catalyst bound to the substrate or buffer), typically decrease as the nanostructure grows larger.[20,21,25,26]

This phenomenon is due to a number of processes, the main ones being catalyst poisoning and where the nanostructure prevents building units from reaching the growth regions. Catalyst poisoning occurs regardless of whether the nanostructure grows by either the root- or tip-growth mechanism, and even for growth in the vapor phase. Building units deposit onto the surface of the catalyst, forming an amorphous layer that encapsulates it, and prevents further incorporation of building units.

Treatment of catalyst poisoning usually comes in the form of physical or chemical etching.[27–29] For example, in the case of carbon nanotubes on a substrate in a plasma-assisted deposition system (such as in inductively coupled plasma-enhanced chemical vapor deposition), transition metals such as iron, nickel, or cobalt are typically used as the catalyst. Over time, a thin layer of carbon forms over the catalyst, reducing its activity. To remove the carbon layer, one can bias the substrate surface in a plasma system such that atoms and ions of the carrier gas (typically nitrogen, argon, hydrogen, or some mixture in these systems) impinge onto the surface with high enough energy to remove the amorphous carbon. However, care should be taken so as not to increase the bias so much that it destroys the nanostructures being fabricated. Alternatively, one can use a water plasma by adding water into the gas mixture. Water is a weak oxidizer that selectively reacts with the amorphous carbon formed on the catalyst surface and not with the carbon in the nanotube structure (assuming there is only a small number of defects), thus promoting the activity of the catalyst.

The second process that causes the reduction in the growth rates of the nanostructures that develop via the root growth mechanism is the redistribution of building units (BUs) as the nanostructure grows. As the nanostructure height increases, the surface area on which the BUs can deposit also increases.[20,21] In the case of a root growth mechanism, the average distance that BUs have to travel to the catalyst nanoparticles increases as the nanotube becomes longer. Specifically, in neutral gas and narrow-sheath plasma systems, BUs predominantly land closer to the tips of the nanotubes. Consequently, as the SWCNTs grow, the precursor species have to travel much longer distances before reaching the nanotube base. This in turn leads to a significantly larger fraction of the precursor species desorbing from the surface before they can reach the catalyst nanoparticle.

To illustrate this point, Figure 11.2 shows the relative number of precursors that land on the surface and then reach the base of the SWCNTs, and the relative heights of the SWCNTs at different times. This figure also illustrates that a plasma system can deliver more BUs to the base of SWCNTs in an array than is possible in a neutral gas system.

11.2.4.1 Surface Processes

Once a particle in the gas phase comes into contact with a surface, there are many possibilities. If the particle was highly energetic (>10 kV) then implantation can occur. With sufficient energy, part of the surface may be milled away. At lower energies, the particle has a finite probability of being adsorbed on to the surface. After being adsorbed, the particle follows one of three courses. The particle can move along the surface, it may leave the surface, or it may react with other particles or features. Some of these processes are shown schematically in Figure 11.3.

Let us now consider adsorbed particles moving on the surface. In most vacuum-based physical vapor deposition systems, particles can move along the surface via a mechanism called adatom diffusion. This is caused by atoms spontaneously moving in a random walk motion caused by thermal "kicks." The rate at which the adatoms move is determined by the lattice vibrational frequency (the rate of the thermal adatom kicks), the temperature of the surface, and the energy barrier of the surface. The diffusion coefficient is generally represented as[30]

$$D_i = v \frac{\lambda^2}{4} \exp\left(-\frac{\varepsilon_i^{ads}}{k_B T}\right),$$

(11.5)

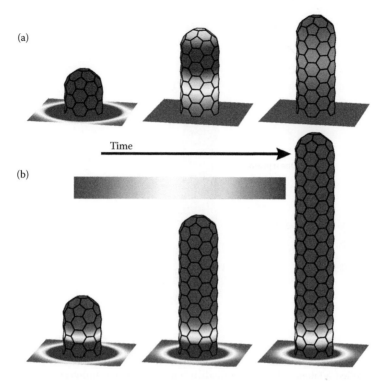

FIGURE 11.2
(See color insert.) The progression of the growth of the single-walled carbon nanotubes
(SWCNTs) in (a) neutral and (b) plasma (wide sheath) systems. The color represents the flux
incident on the specific region from which a building unit (BU) will reach the base of the
SWCNT; this quantity is proportional to the SWCNT growth rate. In neutral gas systems, BUs
primarily deposit on the tips of the SWCNTs, and as the SWCNTs get longer, fewer BUs reach
the base of the SWCNT, leading to slower, stifled growth. However, in a wide plasma sheath
system, BU trajectories are directed toward the base of the SWCNTs, and a higher flux of BUs
reaches the base of the SWCNT, leading to the rapid growth of long SWCNTs. (Figure and
caption reproduced from Tam, E., and Ostrikov, K. 2008. Plasma-controlled adatom delivery
and (re)distribution—Enabling uninterrupted, low-temperature growth of ultralong vertically
aligned single walled carbon nanotubes. *Appl. Phys. Lett.*, 93(26): 261504. With permission.)

where v is the vibrational frequency, λ is the lattice constant, ε_i^{ads} is the
energy barrier for species i, k_B is the Boltzmann constant, and T is the surface
temperature. This equation is used to model adatom diffusion in fluid and
kinetic Monte Carlo (KMC) models and requires some literature research
to determine material-specific constants. The exponential term in the equation appears in every spontaneous surface process for which the surface is
at thermal equilibrium, with the only difference being the energy barrier
term for each specific process. In addition, the energy barrier term can be
influenced by the presence of an electric field which, in general, reduces the
energy barrier and thus increases the D_i.[31]

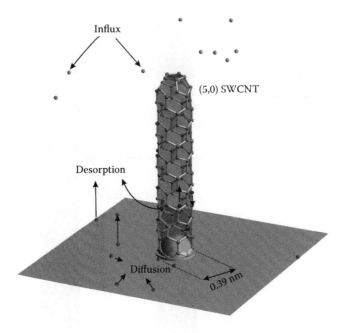

FIGURE 11.3
Some of the processes to be considered in simulations of the growth of a surface-bound one-dimensional (1D) nanostructure. (Figure reproduced from Tam, E., and Ostrikov, K. 2008. Plasma-controlled adatom delivery and (re)distribution—Enabling uninterrupted, low-temperature growth of ultralong vertically aligned single walled carbon nanotubes. *Appl. Phys. Lett.*, 93(26): 261504. With permission.)

Processes that can remove the adatoms from the surface include spontaneous desorption and induced desorption. Similar to adatom diffusion, spontaneous desorption is caused by thermal "kicks," though the energy barrier for desorption is generally significantly higher.

The rate of spontaneous desorption of particle i from a surface (μ_i^s) is[30]

$$\mu_i^e = n_i v \exp\left(-\frac{e_i^{des}}{k_B T}\right), \tag{11.6}$$

where n_i is the number of adatoms on the surface, and ε_i^{des} is the energy barrier that an individual adatom must overcome to leave the surface.

However, one must remember that there are generally high-energy particles in the gas phase that may also induce desorption by either imparting some of their energy to the adatom upon physical collision (sputtering) or by chemically bonding (etching) with the particle and pulling it off the surface. The rate at which vapor-phase-particle-induced desorption occurs is equal to the product of the cross section of the reaction between the adatoms and gas phase particles (σ_{im}) and the gas phase particle current density to the surface (j), or[24]

$$\mu_i^g = n_i \sum_m \sigma_{im} j_m \qquad (11.7)$$

Finally, desorption can also be induced by other adatoms on the surface through the formation of other molecules that have a significantly lower desorption energy than the original components.

By considering all the processes by which an adatom can be removed from the surface [Equations (11.6) and (11.7) as well as the rate due to surface reactions], one can find that the average time an adatom resides on the surface is

$$\tau_i = \frac{n_i}{\mu_i^e + \mu_i^g + \mu_i^s} \qquad (11.8)$$

where μ_i^s is the rate at which desorption of species i occurs due to surface reactions. By combining Equations (11.6) and (11.8), one can determine the average distance that an adatom will move on a surface before returning to the gas phase:

$$\left\langle x_i^2 \right\rangle = D_i \tau_i \qquad (11.9)$$

where D_i is given by Equation (11.5). Finally, adatoms can move onto sites on the surface at which they become bonded. These sites include those with dangling bonds from the crystal structure, specific molecules purposely introduced onto the surface to allow building units to bond to specific areas (such as is the case in atomic layer deposition), or catalysts that allow the adatom to bond with other adatoms and form the nanostructure.

11.2.5 The Role of the Catalyst

By far the most cited mechanism for catalyst-assisted growth of 1D nanostructures is the vapor-liquid-solid (VLS) mechanism. This model, proposed by Wagner in 1964,[32] involves vapor precursors, liquid catalysts, and a solid precipitate. Essentially, the vapor precursors are dissolved into the liquid catalyst. At some point the liquid catalyst become saturated with the precursor. At this point it still continues to dissolve precursors from gas (and the surface if applicable). However, in order to do so, some of the dissolved precursors must first precipitate out of the catalyst. At this stage, the catalyst is supersaturated, and depending on the fabrication conditions, a 1D nanostructure may nucleate. This mechanism is, however, restricted to compatible materials; that is, the metal catalyst must be capable of dissolving the gas-phase precursors.

Binary phase diagrams show the phases of an alloy at thermal equilibrium for a given temperature, pressure, and mixture. However, in the vast majority of such phase diagrams in the literature, only the bulk case is considered and no account is taken of the size or charge of an alloy particle, which may in fact change the state of the particle.[33] Even with this limitation, binary phase diagrams represent an invaluable tool to help determine the conditions under which certain nanostructures will form. These diagrams provide a rough guideline to the conditions under which the catalyst is a liquid, the liquid catalyst will absorb the vapor building units, and solid precipitates can be expected. Figure 11.4 shows a Au-Si binary phase diagram, which is of the simple eutectic type.[34]

By choosing a mechanism by which the catalyst acts, one can determine rates of extrusion of the building units, allowing the prediction of nanowire growth rates or energy calculations to be performed to determine the nanowire diameter.[35]

Although not common, there is evidence of growth of nanowires and nanotubes on a variety of different solid catalysts.[36,37] The vapor-solid-solid (VSS) mechanism had been proposed relatively more recently than the VLS mechanism that has been studied quite extensively. Although not common, there is evidence of growth of nanowires and nanotubes on a variety of different solid catalysts.[36,37] The vapor-solid-solid (VSS) mechanism was proposed more recently than the VLS mechanism. It is still disputed whether nanowires and nanotubes can be nucleated and grown via the VSS mechanism, with various anomalies being cited, including the low diffusion rates through the solids not accounting for the nanowire growth rates, and difficulty in explaining increases in the nanowire diameter, which can be attributed to the increase of volume of a liquid catalyst after dissolving the

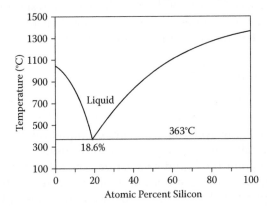

FIGURE 11.4
A typical Au–Si binary phase diagram. Note that there is a eutectic point at 363°C with ~18.6% (at) Si. The phase diagram is widely available in numerous publications such as Sunkara, M. K., and Meyyappan, M. 2010. *Inorganic Nanowires: Applications, Properties, and Characterization.* Boca Raton, FL: CRC Press.

solutes.[38] However, recent ab initio and molecular dynamics simulations address some of these concerns and provide crucial theoretical evidence for the existence of the VSS mechanism.[39] Likewise, nanotexturing of the solid catalyst is a requirement, for nanowire nucleation, that is commonly cited in solid catalyst growth, something which one would not expect to be a prerequisite in liquid catalyst growth. Nanotexturing the catalyst surface is typically done using an O_2 or H_2 plasma.[40,41]

11.3 Computational Techniques

Depending on the complexity of the model used to describe a nanostructure, computational techniques may be required to produce quantitative results from the model. At one extreme, simple models can in many cases be solved analytically; however, only a relatively small amount of information is generally produced by such solutions. In addition, simple models typically have many assumptions, based on experiment or phenomenological reasoning, to simplify the calculations; further, they are limited to only a small set of environments and process conditions.

At the other extreme, complex models generally involve the solution of many complicated equations and in most cases cannot be solved analytically. Even with the aid of large supercomputing clusters, it can take months to produce numerical approximations of the solutions to the equations required for very small systems. However, these models produce the most information, including many physical properties that may not relate to the growth of the structure directly. In addition, ab initio models and simulations typically have very few physical assumptions, and most of the approximations are in the form of mathematical tricks to simplify calculations; they are therefore generally more robust. Below, an outline of some of the most commonly used computational techniques in modeling the growth of nanostructures will be provided.

11.3.1 Phenomenological Models

The generation of a model begins with the listing of all the processes observed in experiments. The dominant processes in the system are included in the model, and processes that have "little effect" on the system are generally discarded. A balance must be struck between the calculation time and accuracy when deciding which processes should be discarded. The model can be further simplified by assuming some symmetry in the growth system (e.g., a simplified geometry). A series of partial differential equations representing the group behavior of the particles for each process is then formulated. These equations can then be numerically solved using various techniques. The phenomenological models are popular as they are generally easy to formulate, require relatively little computational effort, and can be used to check some of the general trends predicted by more sophisticated models.

These models, however, require experimental values for certain parameters (e.g., the diffusion and desorption rates when considering adatom diffusion along the surface). Furthermore, there may be processes that may be difficult to represent as partial differential equations, and complex geometries can make the model difficult to formulate. Thus, the phenomenological models are generally restricted to simple cases as first approximations, or used for larger systems for longer time periods where other methods are not viable for treating the growth of 1D nanostructures. These models are generally used to determine the growth rates of the 1D nanostructures, the aspect ratios, and sometimes the preferential morphology under certain growth conditions.[24,26]

11.3.2 Kinetic Monte Carlo Technique

A different approach to solving phenomenological models is the kinetic Monte Carlo (KMC) technique. KMC differs from other methods in that the motion of each individual particle in the system is simulated, rather than the group behavior being modeled. The main advantage of this is that the 1D nanostructure does not have to be represented by a cylinder or a long polygon, but rather as a collection of atoms, which allows for the formation of defects, and the bending or even the branching of the nanostructures to be treated if desired.

These models work by calculating the probability that an adatom will undergo a particular process at a given time. For every particle in the system, the probability of each process is recalculated at every time step and a pseudo random number is used to determine which process the particle undergoes. This is repeated throughout the rest of the simulation.

KMC codes can be used to determine the growth rates of 1D nanostructures, their aspect ratios, and the preferential morphology, and are typically the method of choice for larger systems with more complex geometries and longer timescales.[20,21]

11.3.3 Molecular Dynamics

Molecular dynamics (MD) is a form of computer simulation in which the motion of atoms and molecules in the system is determined by the input force fields or potentials. That is, every atom and molecule in the system (including the substrate and catalyst atoms) induces an interaction force, typically with only its nearest neighbors (a common assumption to reduce computation time), which determines the motion of every other particle. Because of this, MD is much more computationally taxing than the previously mentioned methods; however, the approach is not without advantages.

The only assumption in MD models is the choice of the potentials that are used. These potentials may be empirical or calculated ab initio potentials. Either way, the system is not restricted to predefined events, as is the case for the phenomenological models. The MD models allow the nanostructure to "wiggle" and move, and can allow for the 1D nanostructure to bend out of shape, which is not possible (or at least not easy to implement) in most phenomenological models. MD simulations tend to be the most computationally taxing method used to simulate the growth stages of a nanostructure and are generally restricted to smaller systems and their dynamics over shorter timescales. Typical uses include determining the role of the catalyst in the growth and the effects of dopants on the growth, as well as the formation of specific SWCNT caps.[14,42,43]

11.3.4 Ab Initio Models

There are many approximation techniques that can be used in the construction of ab initio models. The most popular, in the context of structural properties of nanostructures, is the density functional theory (DFT). Ab initio techniques are, however, generally far too computationally taxing to model the entire growth of nanowires or nanotubes. These models are usually restricted to the nucleation stages, where one can determine under what conditions these structures will nucleate. For instance, in the case of carbon nanotubes, one can find the conditions that favor the formation of the most energetically viable caps.[15]

11.4 Closing Remarks

As computational resources improve, it is possible to use more complex models, and the occurrence of new structures can be predicted even before they are produced by experiments. In this chapter, the considerations required to model catalytic growth of nanowires and nanotubes and the computational techniques by which these models can be solved were considered. To model the growth of 1D nanostructures, one must consider what information is required from the model, and from this select how best to represent the structure. A suitable model must then be developed and the right technique chosen to solve the problem.

Comprehensive modeling and simulation of plasma-based nanostructure growth require multiscale (in both space and time) models that bridge processes in the plasma bulk (reactor size) to atomic processes during nucleation. The different approaches, both physical and computational, discussed in this chapter need to be used for each spatial and timescale, and have to be combined in a tractable yet reliable manner. The development of such multiscale models represents a significant challenge for the coming years.

References

1. Cai, D., Mataraza, J. M., Qin, Z. -H. et al. 2005. Highly efficient molecular delivery into mammalian cells using carbon nanotube spearing. *Nat. Methods*, 2(6): 449–454.
2. Goldberg, M., Langer, R., and Jia, X. 2007. Nanostructured materials for applications in drug delivery and tissue engineering. *J. Biomat. Sci.-Polym. E.*, 18(3): 241–268.
3. Ha, B., and Lee, C. J. 2007. Electronic structure and field emission properties of in situ potassium-doped single-walled carbon nanotubes. *Appl. Phys. Lett.*, 90: 23108.
4. Liu, Y. -C., and Wang, Q. 2000. Dynamic behaviors on zadaxin getting into carbon nanotubes. *J. Chem. Phys.*, 126: 124901.
5. McKnight, T. E., Melechko, A. V., Hensley, D. K., Mann, D. G. J., Griffin, G. D., and Simpson, M. L. 2004. Tracking gene expression after DNA delivery using spatially indexed nanofiber arrays. *Nano Lett.*, 4(7): 1213–1219.
6. Miller, A. J., Hatton, R. A., Chen, G. Y., and Silva, S. R. P. 2007. Carbon nanotubes grown on In_2O_3-Sn glass as large area electrodes for organic photovoltaics. *Appl. Phys. Lett.*, 90: 23105.
7. Pantarotto, D., Briand, J. -P., Prato, M., and Bianco, A. 2004. Translocation of bioactive peptides across cell membranes by carbon nanotubes. *Chem. Comm.*, 1: 16–17.
8. Sidorenko, A., Krupenkin, T., Taylor, A., Fratzl, P., and Aizenberg, J. 2007. Reversible switching of hydrogel-actuated nanostructures into complex micropatterns. *Science*, 315: 487–490.
9. Yang, H. G., Sun, C. H., Qiao, S. Z. et al. 2008. Anatase TiO_2 single crystals with a large percentage of reactive facets. *Nature*, 453(7195): 638–641.
10. Kitayama, M., Narushima, T., Carter, W. C., Cannon, R. M., and Glaeser, A. M. 2000. The Wulff shape of alumina-I, modeling the kinetics of morphological evolution. *J. Am. Ceram. Soc.*, 83(10): 2561–2571.
11. Hayashi, T., Kim, Y. A., Matoba, T. et al. 2003. Smallest freestanding single-walled carbon nanotube. *Nano Lett.*, 3(7): 887–889.
12. Peng, H. Y., Wang, N., Zheng, Y. F. et al. 2000. Smallest diameter carbon nanotubes. *Appl. Phys. Lett.*, 77(18): 2831–2833.
13. Barnard, A. S., and Zapol, P. 2004. A model for the phase stability of arbitrary nanoparticles as a function of size and shape. *J. Chem. Phys.*, 121(9): 4276–4283.
14. Gómez-Gualdrón, D. A., and Balbuena, P. B. 2009. Effect of metal cluster-cap interactions on the catalyzed growth of single-wall carbon nanotubes. *J. Phys. Chem. C*, 113(2): 698–709.
15. Reich, S., Li, L., and Robertson, J. 2005. Structure and formation energy of carbon nanotube caps. *Phys. Rev. B*, 72: 165423.

16. Levchenko, I., Ostrikov, K., Keidar, M., and Xu, S. 2006. Deterministic nanoassembly—Neutral or plasma route? *Appl. Phys. Lett.*, 89: 033109.
17. Ostrikov, K., and Murphy, A. B. 2007. Plasma-aided nanofabrication—where is the cutting edge? *J. Phys. D*, 40: 2223–2241.
18. Ostrikov, K. 2005. Reactive plasma as a versatile nanofabrication tool. *Rev. Mod. Phys.*, 77(2): 489–511.
19. Ostrikov, K. 2008. Surface science of plasma exposed surfaces—A challenge for applied plasma science. *Vacuum*, 83(1): 4–10.
20. Tam, E., and Ostrikov, K. 2008. Plasma-controlled adatom delivery and (re)distribution—Enabling uninterrupted, low-temperature growth of ultralong vertically aligned single walled carbon nanotubes. *Appl. Phys. Lett.*, 93(26): 261504.
21. Tam, E., and Ostrikov, K. 2009. Catalyst size effects on the growth of single-walled nanotubes in neutral and plasma systems. *Nanotechnology*, 20(37): 375603.
22. Keidar, M., Levchenko, I., Arbel, T., Alexander, M., Waas, A. M., and Ostrikov, K. 2008. Increasing the length of single-wall carbon nanotubes in a magnetically enhanced arc discharge. *Appl. Phys. Lett.*, 92(4): 043129.
23. Kim, K. S., Cota-Sanchez, G., Kingston, C. T., Imris, M., Simard, B., and Soucy, G. 2007. Large-scale production of single-walled carbon nanotubes by induction thermal plasma. *J. Phys. D*, 40(8): 2375–2387.
24. Denysenko, I., Ostrikov, K., Yu, M. Y., and Azarenkov, N. A. 2007. Effects of ions and atomic hydrogen in plasma-assisted growth of single-walled carbon nanotubes. *J. Appl. Phys.*, 102(7): 074308.
25. Louchev, O. A., Laude, T., Sato, Y., and Kanda, H. 2003. Diffusion-controlled kinetics of carbon nanotube forest growth by chemical vapor deposition. *J. Chem. Phys.*, 118(16): 7622–7624.
26. Louchev, O. A., Kanda, H., Rosen, A., and Bolton, K. 2004. Thermal physics in carbon nanotube growth kinetics. *J. Chem. Phys.*, 121(1): 446–456.
27. Hata, K., Futaba, D. N., Mizuno, K., Namai, T., Yumura, M., and Iijima, S. 2004. Water-assisted highly efficient synthesis of impurity-free single-walled carbon nanotubes. *Science*, 306(5700): 1362–1364.
28. Yun, Y., Shanov, V., Tu, Y., Subramaniam, S., and Schulz, M. J. 2006. Growth mechanism of long aligned multiwall carbon nanotube arrays by water-assisted chemical vapor deposition. *J. Phys. Chem. B*, 110: 23920–23925.
29. Zhong, G., Iwasaki, T., Robertson, J., and Kawarada, H. 2007. Growth kinetics of 0.5 cm vertically aligned single-walled carbon nanotubes. *J. Phys. Chem. B*, 111(8): 1907–1910.
30. Krasheninnikov, A. V., Nordlund, K., Lehtinen, P. O., Foster, A. S., Ayuela, A., and Nieminen, R. M. 2004. Adsorption and migration of carbon adatoms on carbon nanotubes—Density-functional ab initio and tight-binding studies. *Phys. Rev. B*, 69(7): 073402.
31. Ostrikov, K., Levchenko, I., and Xu, S. 2008. Self-oganised nanoarrays—Plasma-related controls. *Pure Appl. Chem.*, 80(9): 1909–1918.
32. Wagner, R. S., and Ellis, W. C. 1964. Vapor-liquid-solid mechanism of single crystal growth. *Appl. Phys. Lett.*, 4(5): 89–90.
33. Schmidt, V., Wittemann, J. V., and Gösele, U. 2010. Growth, thermodynamics, and electrical properties of silicon nanowires. *Chem. Rev.*, 110(1): 361–388.
34. Sunkara, M. K., and Meyyappan, M. 2010. *Inorganic Nanowires: Applications, Properties, and Characterization*. Boca Raton, FL: CRC Press.

35. Wang, C. -X., Wang, B., Yang, Y. -H., and Yang, G.-W. 2005. Thermodynamic and kinetic size limit of nanowire growth. *J. Phys. Chem. B*, 109(20): 9966–9969.
36. Homma, Y., Liu, H., Takagi, D., and Kobayashi, Y. 2009. Single-walled carbon nanotube growth with non-iron-group "catalysts" by chemical vapor deposition. 2: 793–799.
37. Tuan, H. -Y., Ghezelbash, A., and Korgel, B. A. 2008. Silicon nanowires and silica nanotubes seeded by copper nanoparticles in an organic solvent. *Chem. Mater.*, 20(6): 2306–2313.
38. Mohammad, S. N. 2009. For nanowire growth, vapor-solid-solid (vapor-solid) mechanism is actually vapor-quasisolid-solid (vapor-quasiliquid-solid) mechanism. *J. Chem. Phys.*, 131(22): 224702.
39. Page, A. J., Chandrakumar, K. R. S., Irle, S., and Morokuma, K. 2011. SWNT nucleation from carbon-coated SiO_2 nanoparticles via a vapor-solid-solid mechanism. *J. Am. Chem. Soc.*, 133(3): 621–628.
40. Ostrikov, K., Levchenko, I., Cvelbar, U., Sunkara, M., and Mozetic, M. 2010. From nucleation to nanowires: A single-step process in reactive plasmas. 2: 2012–2027.
41. Nordmark, H., Nagayoshi, H., Matsumoto, N. et al. 2009. Si substrates texturing and vapor-solid-solid Si nanowhiskers growth using pure hydrogen as source gas. *J. Appl. Phys.*, 105(4): 043507.
42. Neyts, E. C., Shibuta, Y., van Duin, A. C. T., and Bogaerts, A. 2010. Catalyzed growth of carbon nanotube with definable chirality by hybrid molecular dynamics—Force biased Monte Carlo simulations. 4(11): 6665–6672.
43. Shibuta, Y., and Maruyama, S. 2003. Molecular dynamics simulation of formation process of single-walled carbon nanotubes by CCVD method. *Chem. Phys. Lett.*, 382: 381–386.

12

Diagnostics of Energy Fluxes in Dusty Plasmas

Horst R. Maurer
Holger Kersten

CONTENTS

12.1 Introduction: Diagnostics of Reactive Dusty Plasmas

Today, plasma technology is a key feature in many emerging industrial sectors like optics, nanotechnology, microelectronics, medicine, and many others dealing with surface modification.

Besides preparing macroscopic wafers in plasmas, plasma-based synthesis or modification of nano- and microparticles with specific properties offers a variety of new applications [1–9]. Some of them are concerned with the improvement of optical or mechanical coatings [1,2], sintering processes [3], disperse composite catalysts [6], polymorphous solar cells [7,8], or optical devices using plasmon resonances [10].

Particle growth in reactive plasmas, which is the result of a series of chemical reactions and physical processes, can be described by a three-step model [11]. The chain of processes is determined by the precursor gas, the discharge type, and the discharge parameters. In the first nucleation phase, primary clusters are formed consisting of particles smaller than 1 nm. Due to their small size, charge fluctuations are relatively large, and the clusters can be positively or negatively charged. In a second step, agglomeration processes lead to the formation of particle clusters. During the agglomeration phase, the particle size quickly increases up to some hundred nanometers, while the particle density is dramatically reduced. When the charge fluctuations become less important due to a large number of elementary charges, the floating potential of the clusters remains negative and further growth can only occur due to accretion processes (e.g., by deposition of atoms or radicals). The different stages of the particle growth in the gas phase can be observed, for example, by fixed-angle [2,12] and angle resolved [13] Rayleigh-Mie ellipsometry, laser extinction photometry [14,15], or laser-induced evaporation [16]. The related processes in the plasma or the gas phase can also be monitored via microwave cavity measurements [17] combined with photodetachment [17,18], absorption spectroscopy [19], mass spectrometry [20], or self-excited electron resonance spectroscopy (SEERS) [15]. Common diagnostics of particle formation also use the observation and analysis of electrical characteristics [15], harmonics [21], and other discharge characteristics.

In the case of small particles in low-pressure plasmas, their electronic and energetic surface conditions are affected by their size and geometry [22–25], and their surface temperature is determined by the energy balance with the environment.

The energetic conditions at the surface of substrates in processes like thin film deposition, plasma etching or sputtering, as well as for nanoparticle formation are crucial for their improvement with respect to process rates, morphology, and stoichiometry [26–30]. Thus, monitoring and controlling the constitutional parameters like gas composition, gas pressure, or substrate temperature is essential, and understanding the plasma–surface interactions is critical to the design of process conditions. For example, a classical tool for the quantification of energy fluxes toward substrates are calorimetric probes, invented by Thornton [31]. Different energy contributions can be separated by biasing the probe substrate (e.g., kinetic energies of electrons and ions as well as the released recombination energy) [32–35].

Several studies, and references therein, have addressed the grain temperature of micro- and nanoparticles in plasmas, both theoretically [23,25,36–38] and experimentally [39–41,42]. Daugherty and Graves measured the temperature-dependent decay time of the fluorescence of phosphor microparticles in a pulsed argon discharge during the plasma-off phase [39]. Swinkels et al. utilized Rhodamine-B dyed microparticles and compared their temperature-dependent emission spectrum measured in the plasma to spectra from a calibration furnace [40]. Oliver and Enikov measured incandescent radiation from particles in a plasma jet [41], which of course is only possible at rather high particle temperatures above $T_\mu \approx 1000$ K. Pyrometric techniques for temperature measurements of micro- and nanoparticles in plasmas are also reviewed by Magunov [42].

Knowledge of the temperature of micro- and nanoparticles is not solely beneficial for the improvement of processing plasmas. The interaction of small particles with plasmas is still not fully understood. A new, semi-invasive diagnostic tool for plasma–surface interaction could offer the possibility to obtain valuable supplemental information and, thus, to improve the understanding of related basic phenomena. This could, in turn, be of benefit for theoretical and astrophysical questions, as well as for plasma physics and nanotechnology in general.

12.2 Methods for the Determination of Particulate Temperature in Plasmas

The temperature of nano- and microparticles in a gas or plasma can, in principle, be measured using one of the following approaches.

A solid body always emits thermal radiation, whose wavelength and intensity is determined by the temperature [42] and, if the body is very small, also by its size and shape [43,44]. The absolute intensity of this radiation increases with temperature, while the wavelength of maximum emission decreases. Incandescent emission provides the possibility to easily determine temperatures of nano- and microparticles above 800 K. Spectral pyrometry has been utilized in situ to determine the temperature of clusters and particles [45–47] in vacuum and gases as well as in plasmas [41,42].

In addition to thermal emission, indirect temperature indicators also exist. For example, different temperature-dependent features in the visible luminescent emission of some phosphors allow for the determination of surface temperatures. As luminescence usually is quickly quenched by phonons, this approach is especially feasible close to room temperature and below where the luminescence is still bright and not governed by thermal emission. However, some phosphors exist that allow phosphor thermometric measurements up to more than 2000 K [48]. Microparticles in a low-pressure discharge typically reach temperatures between 300 K and 400 K. Thus, temperature-sensitive luminescence of specific dyes and phosphors offers the opportunity to measure the in situ temperature of microparticles in low-pressure radio-frequency (RF) discharges [39,40,49].

In addition to the optical techniques mentioned above, the crystallinity of carbon nanoparticles grown in a process plasma provides access to the grain temperature during the growth process [24,50].

In the following, a promising technique of luminescence thermometry is described and experimental results for the temperature of microparticles in an RF discharge are presented.

12.3 Experiments: How to Use Special Luminescent Particles to Access the Energetic Conditions in a Dusty Plasma

As already mentioned, higher temperature causes quenching of luminescent optical transitions. Usually, with increasing temperature an optical transition is more and more dominated by nonradiative deexcitation processes. This also has an influence on its relative emission intensity and linewidth as well as on the decay time of its luminescence. However, if the excitation of the optical transition is fed through a different state separated by an energy barrier, an increase in temperature can also amplify the optical emission. This is also the case for some rare-earth-activated phosphors if they are excited in the ultraviolet (UV) region [48,51,52], enabling very high luminescence yields. These materials have, therefore, been optimized for fluorescent lamps and cathode ray tubes [53]. The *4f* states of the activator ions, involved in the optical transitions, are shielded from the crystal field of the surrounding host lattice by *5s* and *5p* electrons, yielding in a line-type emission spectrum. Due to their strong spectral intensity, these lines can easily be detected and separated from a plasma background spectrum.

To determine the particle temperature, YVO_4:Eu grains are used, and the spectral distribution of their temperature-dependent luminescence is evaluated by comparison to calibration measurements. The spectral region of this phosphor, containing the most dominant emission lines, is shown in Figure 12.1a for different particle temperatures. The ratio of the two brightest emission lines shows a reproducible temperature dependence, as demonstrated in Figure 12.1b. Similar features have been observed for other phosphors, too. As YVO_4:Eu is also well suited for measurements in a harsh plasma environment without noticeable bleaching effects, it has been chosen to be the best candidate for experiments among a large phosphor arsenal. The lamp industry usually adds small amounts of quenchers to the phosphors to suppress unwanted optical transitions. Hence, individual temperature calibration is necessary for different particle charges.

Figure 12.2a shows a sketch of the experimental setup (PULVA–INP). This experiment was particularly designed for the confinement and manipulation of microparticles in plasmas [54], and for the excitation and observation of luminescence of confined phosphor grains. The temperature diagnostics

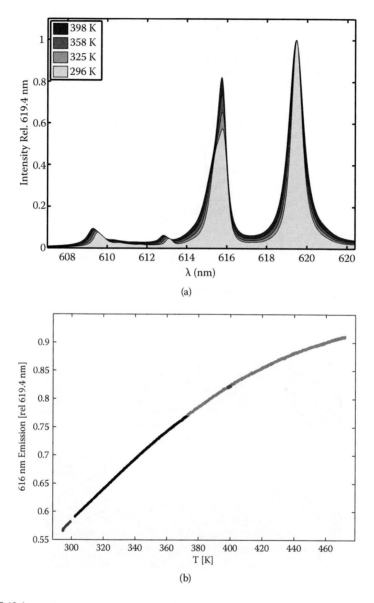

(a)

(b)

FIGURE 12.1

Temperature dependence of the luminescent YVO$_4$:Eu emission. (a) Comparison of different emission spectra of YVO$_4$:Eu, recorded in a calibration furnace at different temperatures. The wavelength of excitation is $\lambda_c = 313$ nm. For comparison, all spectra are normalized to the emission maximum. When the temperature is changed, different features like line shifting, line broadening, or changes in the relative intensity can be observed. (b) Relative intensity of the YVO$_4$:Eu emission lines at 616 nm and 619.4 nm as a function of temperature. The wavelength of excitation is $\lambda_c = 313$ nm. Different colors mark different successive measurement days. Although the absolute intensity of the exciting radiation is not exactly constant, the relative intensity of both lines is highly reproducible.

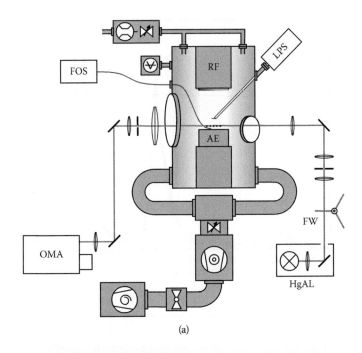

(a)

(b)

FIGURE 12.2

(a) Sketch of the experimental setup PULVA–INP: RF, driven RF-electrode; AE, adaptive (seg-mented) electrode; LPS, Langmuir probe system; FOS, fiber optical sensor; HgAL, mercury arc lamp; FW, filter wheel; OMA, optical multichannel analyzer. The vessel is evacuated by a turbomolecular pump and a scroll pump in series. Gas flow is adjusted by mass flow con-trollers for each gas species, respectively. Gas pressure is arranged by adjusting the effective pump exhaustion rate in an automatic feedback control. (b) Melamine-formaldehyde particles of approximately 10 µm in diameter, confined in front of the AE by setting a negative bias voltage to certain pixels. The pixel bias can be changed individually during the experiment, allowing for the manipulation of the particles in real time.

shown in Figure 12.2a were used in situ, while the Langmuir probe (LP) system was used under reproduced discharge conditions. The device consists of a discharge chamber of approximately 70 liters volume, which is evacuated by a turbomolecular pump and a scroll pump in series. Between pumps and chamber, a butterfly valve is used to adjust the effective exhaustion rate in a closed-loop control. The gas feed can be adjusted by digital mass flow controllers. The gas pressure is monitored by a Baratron gauge that measures the pressure independently of the gas species. All pumps, controllers, and valves are controlled by a personal computer.

The plasma is generated between the upper capacitively coupled RF-electrode, which is driven at 13.56 MHz, and a segmented bottom electrode. The segmented adaptive electrode (AE) [54,55] is the most prominent component of PULVA-INP, it consists of more than 100 square pixels with an area of $7 \cdot 7$ mm^2 that can be biased individually in real time [56]. Injected or grown particles can be confined in front of the AE using the biasing options of the segments. A photograph of the AE is shown in Figure 12.2b, where microparticles are confined in a Φ-shaped configuration. Generally, bias voltages for confining purposes are between $V_{bias} = -5$ V and -20 V.

Typical discharge parameters are gas pressures between $p_{gas} = 1$ Pa and 100 Pa and RF-powers from $P_{rf} = 5$ W to 100 W with an amplitude up to $V_{rf} = 1$ kV. Electron densities in the plasma bulk are in the range of $n_e = 10^9$ cm^{-3} to 10^{11} cm^{-3} at temperatures from $k_B T_e = 0.8$ eV to 2.8 eV (argon). The vertical equilibrium position of the particles varies with discharge power and remains almost independent of gas pressure. Typically the particles are confined one millimeter above the AE.

Langmuir probe measurements are performed through a tilted side flange of the chamber as shown in Figure 12.2a using a SmartProbe™ system (Scientific Systems) and a personal computer. The probe tip consists of a piece of tungsten wire, several millimeters long and approximately a hundred microns in diameter. It is fixed on a linear translation stage that holds electrical connection and filter electronics. This allows vertically resolved measurements to be performed, and the probe can also be moved back and forth to avoid unnecessary contact with the plasma. For data acquisition, the commercial software SmartSoft™ is used [57]. The current-voltage characteristics are evaluated from a self-made analysis code [58].

After the phosphor particles of approximately 11 μm in diameter are confined above the center of the AE, their luminescence can be excited by means of ultraviolet light emitted by a mercury arc lamp. The lamp housing is equipped with a cooled dichroic mirror, reflecting only the ultraviolet emission of the lamp. A second mirror is used to image the mercury arc onto the particles, with the image rotated into a horizontal orientation. This mirror is also used for fine tuning when the vertical equilibrium position of the particles is changed. To turn the excitation on and off, a filter wheel is placed between the lamp and plasma (see Figure 12.2a). This wheel is equipped with a broad band interference filter ($\lambda_c = 313$ nm), a cover to interrupt illumination and record reference spectra, and an empty slot that is used for aligning purposes and as a benchmark. Particle emission is observed at a right angle to the incident excitation. The light is collected by a large, partially shaded lens, passing an aperture and collimated. To allow for vertical tracking of the particles, a periscope-like construction is implied. Particle emission is finally detected by an 500 mm imaging spectrograph (ACTON SpectraPro 2560 i), using a 1200 mm^{-1} grating blazed at 500 nm, and a back-illuminated charge coupled device (CCD) camera (Princeton Instruments PIXIS 400 B).

Especially when hydrogen is added, a lot of particles fall down if the discharge power is increased. As the particles are confined very close to the AE (typically 1 mm), the measurement of their luminescence can in principle be influenced by stray light from lost particles on the electrode surface. To avoid this, the features of the AE are very useful because the particles can be moved to a different position before the discharge parameters are changed.

The emission, related to the particles under UV excitation is separated from the plasma background by subtracting the reference spectrum. An example is shown in Figure 12.3. The occurrence of stray light from the Hg lamp in the spectrum is suppressed by the dichroic mirror in the lamp housing and by the interference filter. Several measurements are averaged, which allows for the rejection of spikes and noisy plasma emission lines. For comparison, calibration spectra are measured inside a special calibration furnace at known temperatures. The carefully recorded spectra provide a temperature resolution of 0.1 K. To determine the particle temperature in the plasma, their

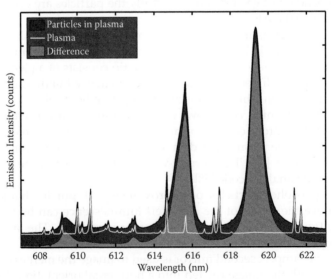

FIGURE 12.3
The emission of the YVO$_4$:Eu particles (gray area) is obtained from the emission spectrum of excited phosphor particles in the plasma (black area) by subtracting the plasma background (white line).

background corrected spectrum is compared to the calibration measurements using a least-square fit.

Unfortunately, heating of the electrodes and the chamber walls due to the plasma cannot be avoided, and because of very large heat capacities the thermal conditions of the environment are barely reproducible in practice. To benchmark the environmental conditions, the temperature of the adaptive electrode T_{AE} has to be continuously recorded during the experiments [59]. A fiber-optic temperature sensor [60] is fixed at the AE as an in situ reference, as shown in Figure 12.2a. It is covered with a thin metal sheet to avoid direct plasma interaction. This sensor delivers the absolute temperature T_{AE} at the surface of the AE. The knowledge of the temperature T_{AE} provides an estimate for the environmental temperature as well as for the gas temperature.

12.4 Experimental Determination of Particle Temperatures under Dusty Plasma Conditions

Particle temperature measurements have been performed in argon as well as in an argon-hydrogen plasma environment.

In argon, the sequence of measurements was P_{rf} = 10 W, 20 W, 35 W, 50 W, 70 W, and 80 W. At each discharge power, the argon pressure was either increased or decreased step by step between p_{Ar} = 10 Pa and 50 Pa. This procedure has been chosen to minimize effects due to heating of plasma-exposed surfaces between subsequent measurements. Typical results are shown in Figures 12.4a to 12.4d. When the discharge conditions were changed, the next measurement was performed after waiting for approximately 30 min. In Figures 12.4a to 12.4d, with increasing gas pressure, a decrease in T_μ is observable. This observation can be explained by increasing efficiency of heat conduction, which is proportional to gas pressure within the Knudsen regime. A comparable behavior in T_{AE} seems to also be visible, but here one has to keep in mind the sequence of measurements and the large heat capacity of the electrode. Above p_{Ar} = 40 Pa, the particle temperature increases but not T_{AE}. This could be caused by a transition of the gas kinetics into the collisional regime where the conductive heat flux is independent of gas pressure. Furthermore, with increasing ion collisionality, the ion current toward the particles is increased [61,62], causing enhanced heat release to the particles due to electron-ion recombination at the grain surface. Thus, the behavior of T_μ qualitatively shows the expected behavior.

In Figure 12.5a, the measurements at p_{Ar} = 10 Pa, shown in Figure 12.4, are displayed as a function of discharge power. Here, steps are visible in T_{AE}, caused by the sequence of measurements and the large heat capacity of

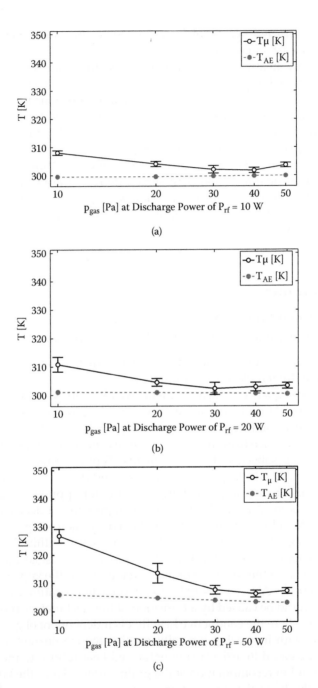

FIGURE 12.4
Particle temperature T_μ and the temperature of the Adaptive Electrode T_{AE} as a function of gas pressure at different discharge powers. Error bars show the standard deviation for T_μ of 10 subsequent measurements. The error for T_{AE} is assumed to be negligible. (a) $P_{rf} = 10$ W, p_{Ar} stepwise decreased. (b) $P_{rf} = 20$ W, p_{Ar} stepwise increased. (c) $P_{rf} = 50$ W, p_{Ar} stepwise decreased.

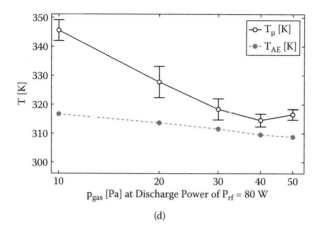

(d)

FIGURE 12.4 (continued)
Particle temperature T_μ and the temperature of the Adaptive Electrode T_{AE} as a function of gas pressure at different discharge powers. Error bars show the standard deviation for T_μ of 10 subsequent measurements. The error for T_{AE} is assumed to be negligible. (d) $P_{rf} = 80$ W, p_{Ar} stepwise decreased.

the AE. If the temperature was measured in a cycle with stepwise reduced pressure, T_{AE} is systematically lower than in a cycle with stepwise increased pressure. The same behavior is also clearly visible in T_μ, and it becomes very obvious why the in situ observation of T_{AE} as a reference value is important.

Additionally, measurements for 9 Pa argon under admixture of 1 Pa hydrogen have been performed. Here the gas pressure remained constant and the discharge power was stepwise increased. The results are plotted in Figure 12.5b. In the argon-hydrogen discharge, the distance between T_μ and T_{AE} is comparable to that in argon, but the absolute temperatures are systematically higher. The increase in the error for T_μ at higher discharge power is caused by particle loss and in return by a decrease in the luminescence signal.

The plasma parameters (electron density n_e and electron temperature $k_B T_e$) measured by the Langmuir probe system, corresponding to the measurements at 10 Pa total gas pressure (argon, see Figure 12.5a and gas mixture, see Figure 12.5b) are given in Figures 12.5c and 12.5d. The results in argon agree with those measured under similar discharge conditions by Tatanova et al. [63]. When molecular hydrogen is added, the electron temperature is dramatically decreased due to inelastic electron-neutral collisions, which now become very efficient even at low energies. This also allows for stepwise excitation and ionization of hydrogen molecules, resulting in an increased electron density. Obviously, even if the particle temperatures in argon are similar to those in the mixture, the energetic conditions at the particle surface are substantially changed under the admixture of hydrogen.

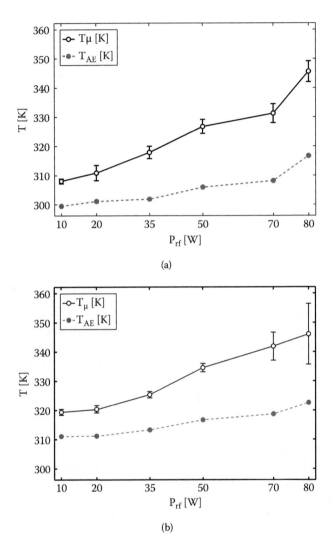

FIGURE 12.5
(a) Particle temperatures T_μ and electrode temperature T_{AE} for 10 Pa argon. (b) Particle temperatures T_μ and electrode temperature T_{AE} for 9 Pa argon and 1 Pa hydrogen.

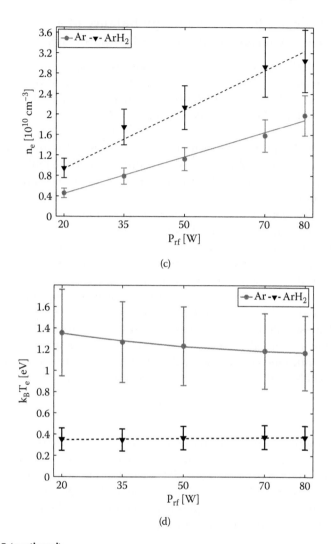

FIGURE 12.5 (continued)
(c) Electron density n_e as a function of discharge power P_{rf}. The lines show a linear fit of the data, the error is estimated to be within $\pm20\%$. (d) Electron temperature $k_B T_e$ as a function of discharge power P_{rf}. Lines show a polynomial fit of second order. The error is estimated to be within $\pm30\%$.

12.5 A Simple Model for the Energy Balance of a Spherical Microparticle in the Plasma Sheath

In the presented model, an idealized plasma is assumed where electrons show a Maxwellian velocity distribution with streaming ions at the sheath edge and where collisions are negligible. The model describes fluxes between plasma and the particle surface and was originally used in probe theory [22]. In this concept, the flux of positive ions toward the negatively charged dust grains is limited due to their orbital trajectories, and it is therefore called orbital motion limited (OML) theory. Of course, the collisionless assumption of ion kinetics is only applicable for the low-pressure conditions of our experiments (about $p_{gas} \approx 10$ Pa).

Microparticles, confined in the sheath of a plasma, are exposed to multispecies bombardment by neutrals, radicals, electrons, and ions as well as to electromagnetic plasma irradiation. The kinetic energy of the impinging electrons and ions as well as their recombination energy contributes to the heating of the particle surface. Furthermore, depending on the plasma environment, other processes like latent heat release of deposited material, exothermic chemical reaction processes, or association energy from recombination of dissociated molecules can account for the heating of a particle. Typically, due to its small heat capacity, a microparticle reaches a stable temperature T_μ within tens of milliseconds [39]. Then, the integral energy influx density J_{in} is balanced by integral energy loss density J_{out}:

$$J_{in} = J_{out}. \tag{12.1}$$

The contribution from plasma irradiation can be assumed to be very small in typical RF discharges [40], and the influence of the external excitation source is negligible in our experiment, too [58]. Also the role of metastables, which has been benchmarked by Do et al. [64] to be in the order of some μWcm^{-2}, is insignificant. In the case of argon-hydrogen plasmas, the integral energy influx density J_{in}, can thus be described by

$$J_{in} = J_e + J_i + J_{rec} + J_{ass}, \tag{12.2}$$

where J_e, J_i, J_{rec}, and J_{ass} denote the kinetic energy release by electrons and ions and the energy influx densities due to recombination of charge carriers and dissociated molecules, respectively. These terms are balanced by energy loss J_{out} due to radiation and conduction to the environment [36,40]. Other processes like thermionic electron emission or evaporation of the particle are neglected for the present conditions close to room temperature.

$$J_{out} = J_{rad} + J_{cond}. \tag{12.3}$$

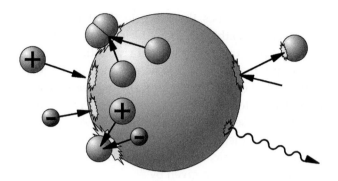

FIGURE 12.6
Relevant energy fluxes between a microparticle and the surrounding plasma. Energy influx (left) like the kinetic energy of electrons J_e and ions J_i, their recombination J_{rec}, and the release of energy due to association processes J_{ass} are balanced by energy loss (right) due to radiation J_{rad} and conduction J_{cond}.

A schematic of the mentioned energy flux densities is presented in Figure 12.6, where the incoming fluxes are located on the left and the losses on the right side. The electron flux density j_e towards a retarding surface at a yet unknown floating potential V_f is described by

$$j_e = \alpha \frac{1}{4} n_{e,0} \exp\left\{\frac{V_f}{V_e}\right\} \sqrt{\frac{8e_0 V_e}{\pi m_e}} \tag{12.4}$$

where $n_{e,0}$ is the electron density in the undisturbed (bulk) plasma, m_e the electron mass, $V_e = k_B T_e / e_0$ the electron temperature in Volts, e_0 the elementary charge, T_e the electron temperature in K, and k_B the Boltzmann constant. The exponential term describes the reduction of n_e due to repulsion from the negatively charged surface, and the square root describes the mean electron thermal velocity. The factor α is an approximation for the time-averaged electron density at the particle position in the RF-sheath during an RF-cycle. It also includes the electron sticking coefficient. The energy flux density due to kinetic energy of the electrons is then

$$J_e = j_e \cdot 2e_0 V_e. \tag{12.5}$$

The factor $2e_0 V_e = 2k_B T_e$ is the mean kinetic energy of the electrons arriving at the particle surface, which is obtained from integration over the Maxwellian electron energy distribution function (EEDF) [65].

By contrast, the dust particle is attractive for ions, and the ion flux density j_i toward the particle is

$$j_i = \frac{1}{4} n_{e,0} \exp\left\{-\frac{1}{2}\right\} \sqrt{\frac{e_0 V_e}{m_i}} \cdot \left(1 - 2\frac{V_f}{V_e}\right), \tag{12.6}$$

where the square root describes the ion velocity that approaches sound (Bohm) velocity v_B at the sheath edge. Here the ion density at the sheath edge is described by the bulk electron density times a reduction factor that accounts for the attenuation of ion density due to acceleration of the ions to v_B, and m_i denotes the ion mass. For impinging ions, the particles are assumed to be perfect absorbers. The factor 1/4 accounts for the area of a two-dimensional projection of the object, as seen by the streaming ions. Again, this equation is adopted from OML probe theory [22], and the geometric correction factor for ion collection by a small spherical object is given in the brackets. The kinetic energy flux density of ions is then given by

$$J_i = -j_i \cdot e_0 V_f = -j_e \cdot e_0 V_f. \tag{12.7}$$

However, if the plasma consists of more than one kind of positive ions (e.g., in an argon-hydrogen mixture), the behavior of the different ion species has to be taken into account. In a multiple-ion low-pressure plasma with comparable ion densities, each species enters the sheath with the bulk ion sound velocity [66] depending on the ion composition and the ion masses.

After hitting the particle, the ion can recombine at the particle surface. Because the particle is not connected to an external electrical circuit, the charge carriers have been collected from the surrounding plasma and any network function is zero. Assuming that the recombination energy E_{ion} is released to the particle, the energy influx density from recombination is

$$J_{rec} = j_e E_{ion} = j_i E_{ion}. \tag{12.8}$$

If the ions are molecular, such as H^+_3 or ArH^+, some energy might be required for their dissociation into stable atoms or molecules that should then be considered in E_{ion}. Just as in Equation (12.6), properly accounting for different ion species would result in a sum of partial fluxes where the single ion densities replace the uniform electron density. A more detailed description may be found elsewhere [67,68].

Dissociation of hydrogen molecules in the plasma mainly occurs by electron-neutral collisions [69]. In low-pressure plasmas, the association process where dissociated molecules recombine most probably occurs on surfaces [69]. The released energy from this process often plays an important role for the energy balance. Assuming a Maxwellian velocity distribution, the resulting energy flux density toward the particle can be estimated analogous to Equation (12.4) [40]:

$$J_{ass} = \frac{1}{2}\Gamma_k \frac{1}{4} n_k \sqrt{\frac{8k_B T_{gas}}{\pi m_k}} E_{diss,k}. \tag{12.9}$$

In this equation, n_k is the number density of the dissociated gas species, 1/2 is the stoichiometric factor, and the factor 1/4 accounts for the geometric consideration as for the impinging electrons or ions. Γ_k is the association probability at the particle surface, and $E_{diss,k}$ is the related dissociation energy.

The integral energy influx density toward the particles described by Equation (12.2) is thus determined by the plasma parameters, the electron duty cycle α, the gas mixture, and occasionally by the release of energy at the surface due to association processes.

In thermal equilibrium, J_{in} is balanced by energy loss and Equation (12.3) is valid. The radiative energy loss density can be estimated by Stefan Boltzmann's law:

$$J_{rad} = \sigma\varepsilon_\mu \left(T_\mu^4 - T_{env}^4\right),\tag{12.10}$$

where T_μ and T_{env} are the temperatures of the particles and the walls of the plasma chamber, respectively; ε_μ is the emissivity of the particles; and σ is the Stefan Boltzmann constant. Because the particles are confined very close to the adaptive electrode, the environmental temperature T_{env} can be approximated by T_{AE}.

In low-pressure conditions the behavior of the gas molecules is described by the Knudsen regime [36,40] where the energy loss density is proportional to the gas pressure:

$$J_{cond} = p_{gas} \frac{\gamma+1}{16(\gamma-1)} \alpha_\mu \sqrt{\frac{8k_B T_{gas}}{\pi m_{gas}}} \left(\frac{T_\mu - T_{gas}}{T_{gas}}\right).\tag{12.11}$$

The adiabatic coefficient $\gamma = c_p/c_v$ is 5/3 for argon and 7/5 for molecular hydrogen. The accommodation coefficient α_μ accounts for how efficient thermal energy is transferred between particle surface and the gas species.

The floating potential used in Equations (12.4) through (12.7) is self-consistently computed from the equilibrium condition (Equation 12.1). Elsewhere the floating potential for similar calculations has often been obtained by Langmuir probe measurements directly [33,40,49]. However, the floating potential of a surface within the quasi-neutral plasma bulk strongly differs from that within the RF sheath, and in the OML regime the geometrical shape of the ion collector is also important.

The model described above in principle is also valid for nanoparticles, as long as thermionic emission, evaporation, and so forth, can be neglected. However, due to their very small mass the location of nanoparticles in a plasma is not affected by gravity, and they are distributed in the whole quasi-neutral plasma volume. The ion kinetics have to be described in a different way. Moreover, due to their small size, their emissivity is very small, and energy loss due to radiation becomes inefficient [24,25,47] resulting in a remarkably higher particle temperature.

12.6 Use of Microparticles as Tiny Calorimetric Probes in Process Plasmas

The size of the phosphor microparticles in our experiments is comparable to the wavelength of thermal radiation emitted close to room temperature, and their effective emissivity ε_μ should be smaller than the value of bulk material. We assume $\varepsilon_\mu = 0.5$, which is slightly below the emissivity of the bulk material [70]. The accommodation coefficient of the particles is estimated to be $\alpha_\mu \approx 0.86$ [40,49]. The value of α is empirically expected to be in the order of $\alpha \approx 0.1$ [71,72]. In the calculations of the energy balance, α has been obtained by treating it as a free parameter in the energy balance for argon plasma.

In addition to the plasma parameters and the temperature of the environment, the knowledge of the gas temperature T_{gas} is required. Spatially resolved in situ measurements of the gas temperature with an accuracy of some Kelvins are not easy to perform in a plasma because any solid object would instantly interact with the plasma. Moreover, optical techniques usually do not offer the desired spatial resolution. We can expect the gas temperature T_{gas} to be slightly higher than the temperature of the walls, but smaller than T_μ. The most practical possibility is to use the microparticles as a gas temperature probe. Hence, we consider the gas temperature at the position of the particles in the Knudsen layer to be approximately $T_{gas} = 1/2(T_\mu + T_{AE})$.

Using these approximations in an argon plasma, agreement between J_{in} and J_{out} is found for a value of $\alpha = 0.12$ as shown in Figure 12.7a. This value for α is of the expected magnitude [71,72]. The resulting total power densities are comparable to results published elsewhere [33,40,49]. It should be mentioned that our experiments were performed with the particles confined above the passive electrode, whereas Swinkels measured in front of the driven RF-electrode where the plasma density is higher and the particles become hotter.

The calculated contributions to the integral power densities are shown in Figures 12.7b and 12.7c. The contribution to particle heating due to kinetic electrons and ions is exceeded by the release of energy due to the recombination of electrons and ions by a factor of approximately 3 as shown in Figure 12.7b. This ratio is determined by the floating potential V_f that in the OML regime is directly proportional to the electron temperature $k_B T_e$ for a fixed ion mass. This also implies that the term $\exp\{V_f/V_e\}$ is constant and the particle flux densities $j_e, j_i \sim n_{e,0} T_e^{1/2}$ are commensurate with the electron density. Energy is "lost" from the particle surface by radiation and heat conduction in nearly equal quantities as depicted in Figure 12.7c.

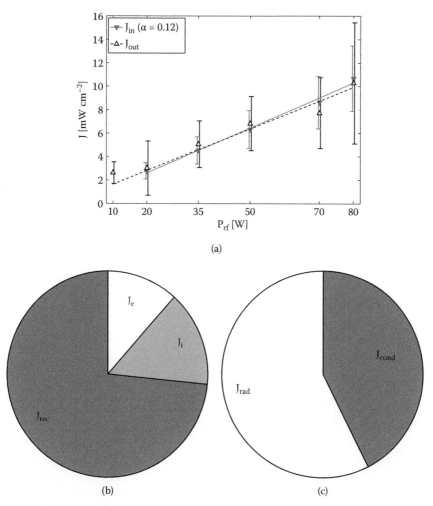

(a)

(b) (c)

FIGURE 12.7
Calculated incoming and outgoing energy flux densities, assuming $\alpha = 0.12$, and their composition. (a) The integral energy flux density toward the microparticles in argon, balanced by the integral energy loss. (b) Contributions to the integral energy influx are dominated by energy release due to electron-ion recombination. (c) The energy influx is balanced by energy losses due to radiation and conduction in comparable quantities.

In an argon-hydrogen plasma, different ion species are expected. Monte Carlo PIC simulations for discharge parameters similar to our experiments have been performed by Bogaerts et al. [73,74]. In agreement with mass spectrometric measurements, they reported comparable densities of ArH^+, Ar^+, and H_3^+, which were used for our calculations. Moreover, the density of atomic hydrogen n_H that is mainly produced by electron impact dissociation [69,75] and its association probability constant Γ_H are unknown. As the electron temperature is not changing much within the measured parameter range, n_H should be directly proportional to n_e. If we assume α to remain unchanged when hydrogen is added to the discharge, we can treat $J_{ass} \sim n_e$ as a free parameter. Considering the ion mixture published by Bogaerts et al. [73] for the calculation of the Bohm velocity v_B, the ion flux density j_i, and the floating potential V_f, agreement between J_{in} and J_{out} is found for $\Gamma_H \, n_H = 3n_e$ as shown in Figure 12.8a.

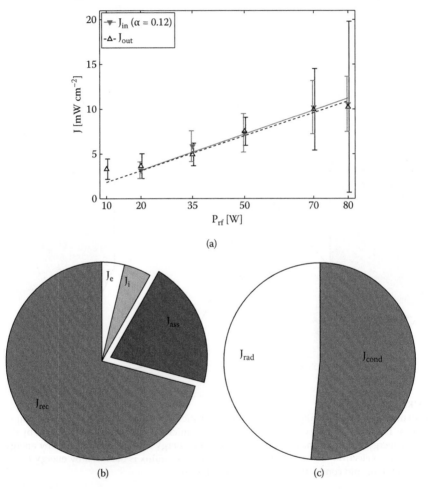

(a)

(b)   (c)

FIGURE 12.8

Calculated incoming and outgoing energy flux densities in argon-hydrogen mixture plasma, assuming an unchanged $\alpha = 0.12$ as in the argon case, and their composition. Now the energy flux by molecular association J_{ass} is treated as a free parameter to match J_{in} and J_{out}. (a) The integral energy flux density toward the microparticles in argon-hydrogen, balanced by the integral energy loss. (b) Contributions to the integral energy influx are dominated by energy release due to electron-ion recombination. Another important energy source is now the recombination of hydrogen atoms at the particle surface. (c) As in the argon case, the integral energy influx is balanced by energy loss due to radiative and conductive processes in comparable quantities.

The recombination of electrons and ions at the particle surface is still dominant for the energy balance, but the contribution due to association processes shown in Figure 12.8b is approximately 25%. Due to the low electron temperature and the resulting weak floating potential, the kinetic contribution by electrons and ions is smaller than in argon. Energy losses from the particle again occur due to radiation and heat conduction. The particles are confined very close to the metallic surface of the AE which is an effective sink for atomic hydrogen. Dissociation due to electron impact occurs in the plasma volume, and a gradient in the dissociation ratio toward metallic surfaces could be observed in comparable experiments [69].

Additionally, the contribution of J_{ass} to the total energy influx is about 25%, which is less than one may expect in the plasma volume. We assumed that the factor α is unchanged under the addition of hydrogen. However, the electron temperature is dramatically reduced under the admixture of hydrogen. Hence, the floating potential of the particles is much weaker than in argon. Thus, the confinement in the RF-sheath is modified, and the equilibrium position of the particles could be shifted. Confining the particles in the mixture was more difficult and a lot of particles got lost. If the particle position is farther away from the sheath edge, the electron duty cycle is also smaller, which would in return result in smaller contributions for J_e, J_i, and J_{rec} and thus in a larger power density J_{ass}.

The dissociation ratio, resulting from J_{in}

$$\frac{n_H}{n_H + 2n_{H2}} \approx \frac{3}{\Gamma_H} \cdot 10^{-4},$$

is in the order of 10^{-3}, if Γ_H is close to 1. This is comparable to results published by Pipa [69] when extrapolated to a position close to the electrode. Especially for nanoparticles, levitating in the plasma volume where the degree of dissociation is expected to be higher than at the sheath edge, even a small addition of H_2 to low-pressure Ar plasmas causes a dramatic increase in the energy deposition through H recombination at the surface. The heat release can be quenched by sequential addition of a hydrocarbon precursor. Such selective control offers an effective mechanism for deterministic control of growth shape, crystallinity, and density of nanostructures in plasma-aided nanofabrication. This opportunity to independently control the surface fluxes of energy and hydrogen-containing radicals, enabling selective control of the nanostructure heating and passivation, was demonstrated in a previous paper [76].

12.7 Summary and Conclusion

Characterization and diagnostics of process plasmas that contain nano- and microparticles (either grown in the plasma or externally injected) are absolutely necessary for the optimization of particle synthesis and their incorporation into nanocomposite films. Among the various diagnostic methods that have been briefly reviewed, the determination of the energy fluxes to and from the dust particles is essential. The energy fluxes determine the equilibrium temperature and, finally, the properties of the nano- and microparticles including crystallinity and stoichiometry.

The equilibrium temperature of microparticles confined in the RF-sheath of a plasma was measured in argon as well as in an argon-hydrogen mixture using temperature-dependent features of the luminescence of certain phosphor particles. The temperature of the microparticles appeared to be between room temperature and approximately 100°C. The accuracy of the temperature measurements is on the order of a few K.

Based on OML theory, a simple model was applied describing the contributions of the relevant energy fluxes to the thermal balance of the particles. The results for the energy fluxes are in good accordance with literature and show reasonable behavior when discharge power is increased or molecular gas is added. The main contribution to particle heating is recombination of electrons and ions at the particle surface. Additionally—depending on the plasma composition—for example, heating due to association processes is an important energy source. The energy losses due to radiation and heat conduction are of comparable quantity at the investigated gas pressures.

The combination of electrical and calorimetric diagnostics with the same particle could be valuable for an improvement in the understanding of plasma–particle interactions. The use of well-defined particles as well as of cooled electrodes would improve the reproducibility and reliability of such measurements. Furthermore, probing the RF sheath in front of a dielectric barrier as well as the combination with mass spectroscopy could be another future direction.

Acknowledgment

This work has been supported by the Deutsche Forschungsgemeinschaft (German Research Foundation) under SFB TR-24 "Fundamentals of Complex Plasmas," Project B4.

References

1. P. R. i Cabarrocas, P. Gay, and A. Hadjadj. Experimental evidence for nanoparticle deposition in continuous argon-silane plasmas: Effects of silicon nanoparticles on film properties. Volume 14, pp. 655–659. New York: AVS (1996).
2. E. Stoffels, W. W. Stoffels, G. Ceccone, R. Hasnaoui, H. Keune, G. Wahl, and F. Rossi. MoS 2 nanoparticle formation in a low pressure environment. *Journal of Applied Physics* 86 (6), 3442–3451 (1999).
3. T. Ishigaki, T. Sato, Y. Moriyoshi, and M. I. Boulos. Influence of plasma modification of titanium carbide powder on its sintering properties. *Journal of Materials Science Letters* 14 (23), 1694–1697 (1995).
4. H. Kersten, H. Deutsch, E. Stoffels, W. W. Stoffels, G. M. W. Kroesen, and R. Hippler. Micro-disperse particles in plasmas: From disturbing side effects to new applications. *Contributions to Plasma Physics* 41 (6), 589–609 (2001).
5. H. Kersten, H. Deutsch, E. Stoffels, W. W. Stoffels, and G. M. W. Kroesen. Plasma-powder interaction: Trends in applications and diagnostics. *International Journal of Mass Spectrometry* 223–224, 313–325 (2003).
6. A. V. Gavrikov, A. S. Ivanov, A. F. Pal, O. F. Petrov, A. N. Ryabinkin, A. O. Serov, Y. M. Shulga, A. N. Starostin, and V. E. Fortov. Dusty plasma technology of DCM with nanostructure surface layer production. Volume 1041, pp. 237–238. College Park, MD: AIP (2008).
7. P. R. i Cabarrocas, N. Chaâbane, A. V. Kharchenko, and S. Tchakarov. Polymorphous silicon thin films produced in dusty plasmas: Application to solar cells. *Plasma Physics and Controlled Fusion* 46 (12B), B235 (2004).
8. P. R. i Cabarrocas, Y. Djeridane, T. Nguyen-Tran, E. V. Johnson, A. Abramov, and Q. Zhang. Low temperature plasma synthesis of silicon nanocrystals: A strategy for high deposition rate and efficient polymorphous and microcrystalline solar cells. *Plasma Physics and Controlled Fusion* 50 (12), 124037 (2008).
9. U. Kortshagen. Nonthermal plasma synthesis of semiconductor nanocrystals. *Journal of Physics D: Applied Physics* 42 (11), 113001 (2009).
10. X. Hoa, A. Kirk, and M. Tabrizian. Towards integrated and sensitive surface plasmon resonance biosensors: A review of recent progress. *Biosensors and Bioelectronics* 23 (2), 151–160 (2007).
11. C. Hollenstein. The physics and chemistry of dusty plasmas. *Plasma Physics and Controlled Fusion* 42 (10), R93 (2000).
12. A. Bouchoule, A. Plain, L. Boufendi, J. P. Blondeau, and C. Laure. Particle generation and behavior in a silane-argon low-pressure discharge under continuous or pulsed radio-frequency excitation. *Journal of Applied Physics* 70 (4), 1991–2000 (1991).
13. W. W. Stoffels, E. Stoffels, G. H. P. M. Swinkels, M. Boufnichel, and G. M. W. Kroesen. Etching a single micrometer-size particle in a plasma. *Physical Review E* 59 (2), 2302–2304 (1999).
14. Y. Nakamura, and H. Bailung. A dusty double plasma device. *Review of Scientific Instruments* 70 (5), 2345–2348 (1999).
15. J. C. Schauer, S. Hong, and J. Winter. Electrical measurements in dusty plasmas as a detection method for the early phase of particle formation. *Plasma Sources Science and Technology* 13 (4), 636 (2004).

16. L. Boufendi, J. Hermann, A. Bouchoule, B. Dubreuil, E. Stoffels, W. W. Stoffels, and M. L. de Giorgi. Study of initial dust formation in an Ar-SiH 4 discharge by laser induced particle explosive evaporation. *Journal of Applied Physics* 76 (1), 148–153 (1994).

17. E. Stoffels, W. W. Stoffels, G. M. W. Kroesen, and F. J. de Hoog. Dust formation and charging in an Ar/SiH4 radio-frequency discharge. Volume 14, pp. 556–561. New York: AVS (1996).

18. T. Fukuzawa, K. Obata, H. Kawasaki, M. Shiratani, and Y. Watanabe. Detection of particles in rf silane plasmas using photoemission method. *Journal of Applied Physics* 80 (6), 3202–3207 (1996).

19. J. Röpcke, G. Lombardi, A. Rousseau, and P. B. Davies. Application of mid-infrared tuneable diode laser absorption spectroscopy to plasma diagnostics: A review. *Plasma Sources Science and Technology* 15 (4), S148 (2006).

20. C. Hollenstein, J. L. Dorier, J. Dutta, L. Sansonnens, and A. A. Howling. Diagnostics of particle genesis and growth in RF silane plasmas by ion mass spectrometry and light scattering. *Plasma Sources Science and Technology* 3 (3), 278 (1994).

21. S. Hong, J. Berndt, and J. Winter. Growth precursors and dynamics of dust particle formation in the Ar/CH$_4$ and Ar/C$_2$H$_2$ plasmas. *Plasma Sources Science and Technology* 12 (1), 46 (2003).

22. J. E. Allen. Probe theory—The orbital motion approach. *Physica Scripta* 45 (5), 497–503 (1992).

23. R. Piejak, V. Godyak, B. Alexandrovich, and N. Tishchenko. Surface temperature and thermal balance of probes immersed in high density plasma. *Plasma Sources Science and Technology* 7 (4), 590–598 (1998).

24. C. Arnas, and A. A. Mouberi. Thermal balance of carbon nanoparticles in sputtering discharges. *Journal of Applied Physics* 105 (6), 063301 (2009).

25. F. Galli, and U. Kortshagen. Charging, coagulation, and heating model of nanoparticles in a low-pressure plasma accounting for ion-neutral collisions. *Plasma Science, IEEE Transactions on* 38 (4), 803–809 (2010).

26. B. Window. Recent advances in sputter deposition. *Surface and Coatings Technology* 71 (2), 93–97 (1995).

27. S. D. Bernstein, T. Y. Wong, and R. W. Tustison. Comparison of the temperature dependence of the properties of ion beam and magnetron sputtered Fe films on (100) GaAs. *Journal of Vacuum Science and Technology A* 17 (2), 571–576 (1999).

28. C. Cardinaud, M. Peignon, and P. Tessier. Plasma etching: Principles, mechanisms, application to micro- and nanotechnologies. *Applied Surface Science* 164 (1–4), 72–83 (2000).

29. J. G. Han. Recent progress in thin film processing by magnetron sputtering with plasma diagnostics. *Journal of Physics D: Applied Physics* 42 (4), 043001 (2009).

30. A. von Keudell. Surface processes during thin-film growth. *Plasma Sources Science and Technology* 9 (4), 455 (2000).

31. J. A. Thornton. Substrate heating in cylindrical magnetron sputtering sources. *Thin Solid Films* 54 (1), 23–31 (1978).

32. H. Kersten, D. Rohde, H. Steffen, H. Deutsch, R. Hippler, G. Swinkels, and G. Kroesen. On the determination of energy fluxes at plasma-surface processes. *Applied Physics A: Materials Science and Processing* 72 (5), 531–540 (2001).

33. A. L. Thomann, N. Semmar, R. Dussart, J. Mathias, and V. Lang. Diagnostic system for plasma/surface energy transfer characterization. *Review of Scientific Instruments* 77 (3), 033501 (2006).

34. D. Lundin, M. Stahl, H. Kersten, and U. Helmersson. Energy flux measurements in high power impulse magnetron sputtering. *Journal of Physics D: Applied Physics* 42 (18), 185202 (2009).

35. M. Stahl, T. Trottenberg, and H. Kersten. A calorimetric probe for plasma diagnostics. *Review of Scientific Instruments* 81 (2), 023504 (2010).

36. E. Stoffels, W. W. Stoffels, H. Kersten, G. H. P. M. Swinkels, and G. M. W. Kroesen. Surface processes of dust particles in low pressure plasmas. *Physica Scripta* 2001 (T89), 168 (2001).

37. S. A. Khrapak and G. E. Morfill. Grain surface temperature in noble gas discharges: Refined analytical model. *Physics of Plasmas* 13 (10), 104506 (2006).

38. F. X. Bronold, H. Fehske, H. Kersten, and H. Deutsch. Towards a microscopic theory of particle charging. *Contributions to Plasma Physics* 49 (4–5, Sp. Iss. SI), 303–315 (2009).

39. J. E. Daugherty, and D. B. Graves. Particulate temperature in radio frequency glow discharges. *Journal of Vacuum Science and Technology A* 11 (4), 1126–1131 (1993).

40. G. Swinkels, H. Kersten, H. Deutsch, and G. M. W. Kroesen. Microcalorimetry of dust particles in a radio-frequency plasma. *Journal of Applied Physics* 88 (4), 1747–1755 (2000).

41. D. Oliver, and R. Enikov. Micro-particles temperature measurements in a plasma jet. *Vacuum* 58 (2–3), 244–249 (2000).

42. A. Magunov. Spectral pyrometry (review). *Instruments and Experimental Techniques* 52 (4), 451–472 (2009).

43. C. Bohren and D. Hoffman. *Absorption and Scattering of Light by Small Particles.* New York: Wiley (1998).

44. H. Odashima, M. Tachikawa, and K. Takehiro. Mode-selective thermal radiation from a microparticle. *Physical Review A* 80 (4), 041806 (2009).

45. U. Frenzel, U. Hammer, H. Westje, and D. Kreisle. Radiative cooling of free metal clusters. *Zeitschrift für Physik D: Atoms, Molecules and Clusters* 40 (1–4), 108–110 (1997).

46. L. Landström, J. Lu, and P. Heszler. Size-distribution and emission spectroscopy of W nanoparticles generated by laser-assisted CVD for different WF6/H2/Ar mixtures. *Journal of Physical Chemistry B* 107 (42), 11615–11621 (2003). Incandescence.

47. L. Landström, K. Elihn, M. Boman, C. Granqvist, and P. Heszler. Analysis of thermal radiation from laser-heated nanoparticles formed by laser-induced decomposition of ferrocene. *Applied Physics A: Materials Science and Processing* 81 (4), 827–833 (2005).

48. S. W. Allison, and G. T. Gillies. Remote thermometry with thermographic phosphors: Instrumentation and applications. *Review of Scientific Instruments* 68 (7), 2615–2650 (1997).

49. G. Swinkels. Optical studies of micron-sized particles immersed in a plasma. Ph.D. thesis (1999).

50. K. Elihn, L. Landström, O. Alm, M. Boman, and P. Heszler. Size and structure of nanoparticles formed via ultraviolet photolysis of ferrocene. *Journal of Applied Physics* 101 (3), 034311 (2007).

51. M. Yu, J. Lin, and J. Fang. Silica spheres coated with YVO4:Eu3+ layers via sol-gel process: A simple method to obtain spherical core-shell phosphors. *Chemistry of Materials* 17 (7), 1783–1791 (2005).

52. R. Ningthoujam, L. R. Singh, V. Sudarsan, and S. D. Singh. Energy transfer process and optimum emission studies in luminescence of core-shell nanoparticles: YVO_4:Eu-YVO_4 and surface state analysis. *Journal of Alloys and Compounds* 484 (1–2), 782–789 (2009).

53. S. Shionoya, and W. M. Yen (eds.). *Phosphor Handbook*. Boca Raton, FL: CRC Press (1999).

54. R. Basner, H. Fehske, H. Kersten, S. Kosse, and G. Schubert. Manipulation of micro-disperse particles in a process plasma. *Vakuum in Forschung und Praxis* 17 (5), 259–261 (2005).

55. R. Basner, F. Sigeneger, D. Loffhagen, G. Schubert, H. Fehske, and H. Kersten. Particles as probes for complex plasmas in front of biased surfaces. *New Journal of Physics* 11 (1), 013041 (2009).

56. B. M. Annaratone, M. Glier, T. Stuffler, M. Raif, H. M. Thomas, and G. E. Morfill. The plasma-sheath boundary near the adaptive electrode as traced by particles. *New Journal of Physics* 5 (1), 92 (2003).

57. Scientific Systems, Dublin, Ireland: Smart Probe Product Manual. http://www.scisys.com/langmuir.cfm.

58. H. R. Maurer, R. Basner, and H. Kersten. Temperature of particulates in low-pressure rf-plasmas in Ar, Ar/H_2 and Ar/N_2 Mixtures. *Contributions to Plasma Physics* 50 (10), 1521–3986 (2010).

59. H. R. Maurer, M. Hannemann, R. Basner, and H. Kersten. Measurement of plasma-surface energy fluxes in an argon rf-discharge by means of calorimetric probes and fluorescent microparticles. *Physics of Plasmas* 17 (11), 113707 (2010).

60. U. Roland, C. Renschen, D. Lippik, F. Stallmach, and F. Holzer. A new fiber optical thermometer and its application for process control in strong electric, magnetic, and electromagnetic fields. *Sensor Letters* 1 (1), 93–98 (2003).

61. M. Tichý, M. Šicha, P. David, and T. David. A collisional model of the positive ion collection by a cylindrical Langmuir probe. *Contributions to Plasma Physics* 34 (1), 59–68 (1994).

62. I. H. Hutchinson, and L. Patacchini. Computation of the effect of neutral collisions on ion current to a floating sphere in a stationary plasma. *Physics of Plasmas* 14 (1), 013505 (2007).

63. M. Tatanova, G. Thieme, R. Basner, M. Hannemann, Y. B. Golubovskii, and H. Kersten. About the EDF formation in a capacitively coupled argon plasma. *Plasma Sources Science and Technology* 15 (3), 507 (2006).

64. H. T. Do, H. Kersten, and R. Hippler. Interaction of injected dust particles with metastable neon atoms in a radio frequency plasma. *New Journal of Physics* 10 (5), 053010 (2008).

65. J. Reece Roth. *Industrial Plasma Engineering*, volume 1. Philadelphia: IOP (1995).

66. D. Lee, L. Oksuz, and N. Hershkowitz. Exact solution for the generalized Bohm criterion in a two-ion-species plasma. *Physical Review Letters* 99 (15), 155004 (2007).

67. H. R. Maurer, V. Schneider, M. Wolter, R. Basner, T. Trottenberg, and H. Kersten. *Contributions to Plasma Physics* 51, 218–227. (2011).

68. H. Maurer, and H. Kersten. On the heating of nano- and microparticles in process plasmas. *Journal of Physics D: Applied Physics* 44, 174029 (2011).

69. A. V. Pipa. On determination of the degree of dissociation of hydrogen in non-equilibrium plasmas by means of emission spectroscopy. Logos, Berlin (2004). PhD thesis, EMAU Greifswald (1999).

70. H. J. Hoffman, and R. K. Shori (eds.). High-resolution absolute temperature mapping of laser crystals in diode-end-pumped configuration, volume 5707. Bellingham, WA: SPIE (2005).

71. T. Trottenberg, A. Melzer, and A. Piel. Measurement of the electric charge on particulates forming Coulomb crystals in the sheath of a radio-frequency plasma. *Plasma Sources Science and Technology* 4 (3), 450 (1995).

72. J. Carstensen, F. Greiner, and A. Piel. Determination of dust grain charge and screening lengths in the plasma sheath by means of a controlled cluster rotation. *Physics of Plasmas* 17 (8), 083703 (2010).

73. E. Neyts, M. Yan, A. Bogaerts, and R. Gijbels. Particle-in-cell/Monte Carlo simulations of a low-pressure capacitively coupled radio-frequency discharge: Effect of adding H_2 to an Ar discharge. *Journal of Applied Physics* 93 (9), 5025–5033 (2003).

74. E. Neyts, M. Yan, A. Bogaerts, and R. Gijbels. PIC-MC simulation of an RF capacitively coupled Ar/H_2 discharge. Nuclear Instruments and Methods in Physics Research Section B: Beam Interactions with Materials and Atoms 202, pp. 300–304 (2003). Sixth International Conference on Computer Simulation of Radiation.

75. B. P. Lavrov, N. Lang, A. V. Pipa, and J. Röpcke. On determination of the degree of dissociation of hydrogen in non-equilibrium plasmas by means of emission spectroscopy: II. Experimental verification. *Plasma Sources Science and Technology* 15 (1), 147 (2006).

76. M. Wolter, I. Levchenko, H. Kersten, S. Kumar, and K. Ostrikov. Disentangling fluxes of energy and matter in plasma-surface interactions: Effect of process parameters. *Journal of Applied Physics* 108 (5), 053302 (2010).

68. A. V. Pipa. On determination of the degree of dissociation of hydrogen from equilibrium plasmas by means of emission spectroscopy. Lat. u. Berlin (2004). PhD thesis, EMAU Greifswald (1999).

69. H. J. Hoffmann and P. R. Shen (eds.), High-resolution absolute temperature mapping of laser crystals in diode-end-pumped configuration. Volume 3631, Bellingham, WA: SPIE (1999).

70. J. Trottenberg, A. Melzer and A. Piel. Measurement of the electric charge on particles forming Coulomb crystals in the sheath of a radio frequency plasma. Plasma Sources Science and Technology 4 (3), 450 (1995).

71. F. Castellanos, F. Greiner, and A. Piel. Determination of dust grain charge and screening lengths in the plasma sheath by means of a controlled cluster rotation. Physics of Plasmas 17 (4), 083701 (2010).

72. L. Marke, M. Tan, A. Bogaerts, and E. Neyts. Particle-in-cell/Monte Carlo simulations of a low-pressure capacitively coupled radio-frequency discharge: Effect of adding H₂ to an Ar discharge. Journal of Applied Physics 93 (3), 3078–3075 (2003).

73. E. Neyts, M. Yan, A. Bogaerts, and E. Neyts. PIC-MC simulation of an RF capacitively coupled Ar/H₂ discharge. Nuclear Instruments and Methods in Physics Research Section B: Beam Interactions with Materials and Atoms 202, pp. 300–304 (2003). Sixth International Conference on Computer Simulation of Radiation.

74. H. F. Lavrov, N. Lang, A. V. Pipa, and J. Röpcke. On determination of the degree of dissociation of hydrogen in nonequilibrium plasmas by means of emission spectroscopy. II. Experimental verification. Plasma Sources Science and Technology 18 (1), 112 (2004).

75. M. Weber, L. Levrunker, H. Kersten, S. Kumar, and K. Ostrikov. Mechanisms of energy and matter in plasma-surface interactions: Flux of process parameters. Journal of Applied Physics 108 (5), 053302 (2010).

13

Selective Functionalization and Modification of Carbon Nanomaterials by Plasma Techniques

Yuhua Xue
Liming Dai

CONTENTS

13.1 Introduction

Radio-frequency (RF) glow discharges are formed when an electric field at frequencies of kHz to GHz is applied to a gas at low pressure (<10 Torr).[1–3] In a glow discharge, excited electrons, ions, and free radicals are generated with high reactivity toward a surface, allowing surface modification (plasma treatment) or polymer deposition (plasma polymerization). Plasma polymerization produces cohesive, adhesive, and pinhole-free thin films with potential for a wide range of applications. Because of the highly cross-linked structure intrinsically associated with plasma deposition, these polymer films often show superb environmental stability toward high temperature, intensive light, and strong electric fields.[2-5] Apart from the preparation of polymer films, plasma techniques have been widely used for functionalization of a large variety of materials, including nanomaterials, by introducing various functional surface groups (via plasma treatment) or smooth and pinhole-free functional polymer layers (via plasma polymerization).[1,4] Along with plasma polymerization and surface functionalization, plasma techniques have been applied to pattern formation and selective etching by introducing functional surface groups or removing surface species.[3] In this case, the spatial resolution is limited mainly by the structure and resolution of the physical mask used because the plasma process involves submolecular species. Plasma techniques are highly generic, which ensures that the methodology developed in a particular case can be readily transferred to many other systems.

The aim of this chapter is to summarize our recent studies on the functionalization and growth of carbon nanomaterials by plasma techniques, though reference is also made to other complementary work as appropriate. Because focus is given to our own work and not necessarily a comprehensive literature survey of the subject, the examples presented here do not exhaust all significant work reported in the literature. Therefore, we apologize in advance to the authors of papers not cited.

13.2 Plasma Techniques for Selective Functionalization of Carbon Nanomaterials

Plasma techniques have played an important role in surface activation of various materials, ranging from organic polymers to inorganic ceramics and to nanomaterials. In this context, we developed an approach for chemical modification of carbon nanotubes (CNTs) via plasma activation followed by highly selective chemical reactions with functional groups generated by the plasma.[6,7] More specifically, amino-dextran chains have been immobilized onto acetaldehyde treated aligned carbon nanotubes through the formation of Schiff-base linkages, which were further stabilized by reduction with sodium cyanoborohydride (Figure 13.1).[6] Using the same approach, we also chemically grafted periodate-oxidized dextran chains prelabeled with fluorescein onto ethylenediamine-treated carbon nanotubes. The resulting polysaccharide-grafted carbon nanotubes are very hydrophilic and potentially useful for many biological applications.[6]

Figure 13.2a shows a typical scanning electron micrographs (SEMs) for the as-synthesized aligned carbon nanotubes before plasma activation. By plasma polymerization (e.g., acetaldehyde), a concentric layer of (acetaldehyde)

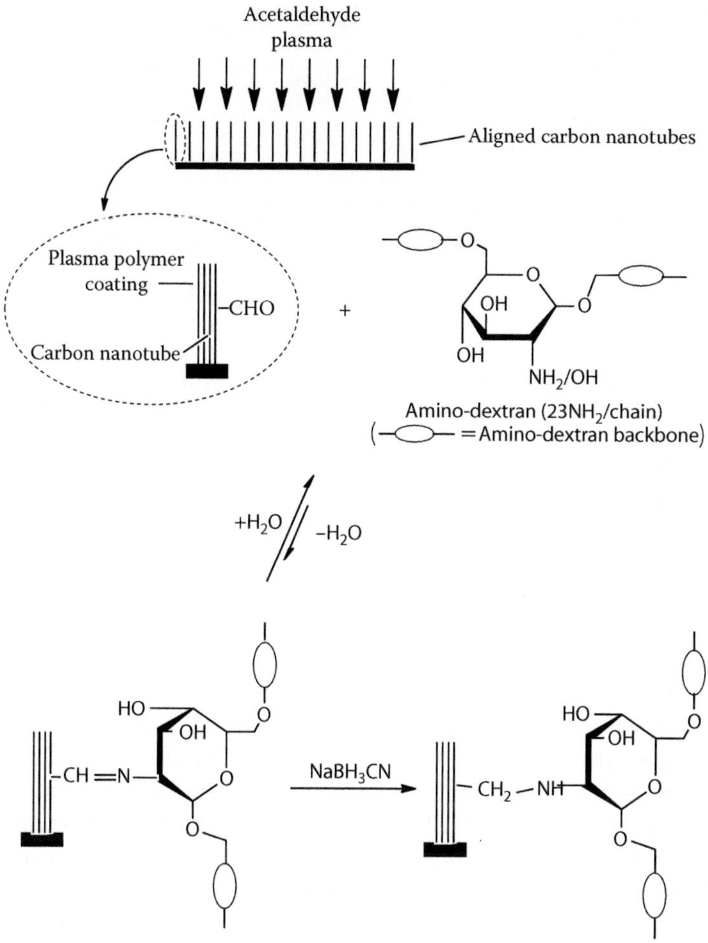

FIGURE 13.1
Reaction scheme for the covalent immobilization of amino-dextran chains onto plasma-activated carbon nanotubes. For reasons of clarity, only one of the many plasma-induced aldehyde surface groups is shown for an individual nanotube (Adapted from Chen, Q., Dai, L., Gao, M., Huang, S., and Mau, A. Plasma activation of carbon nanotubes for chemical modification. *J. Phys. Chem. B* 105, 618–622, 2001.)

polymer film was homogeneously deposited onto each of the aligned carbon nanotubes (Figure 13.2b). The SEM image for the polymer-coated nanotubes given in Figure 13.2b shows similar features as the aligned nanotube array of Figure 13.2a, but with a larger tubular diameter and smaller inter-tube distance due to the presence of the coating.

In a somewhat related but independent experiment, we also carried out acetic acid treatment on gold-supported aligned carbon nanotubes generated from pyrolysis of iron(II) phthalocyanine,[8] followed by

(a)   (b)

FIGURE 13.2

Scanning electron micrographs (SEMs) of the aligned carbon nanotubes (a) before and (b) after the plasma polymerization of acetaldehyde. The insets show transmission electron micrograph (TEM) images of an individual nanotube (a) before and (b) after being coated with a layer of the acetaldehyde-polymer. Note that the micrographs shown in (a) and (b) were not taken from the same spot due to technical difficulties. (Adapted from Chen, Q., Dai, L., Gao, M., Huang, S., and Mau, A. Plasma activation of carbon nanotubes for chemical modification. *J. Phys. Chem. B* 105, 618–622, 2001.)

grafting single-strand DNA (ssDNA) chains with an amino group at the 5A-phosphate end (i.e., [AmC6]TTGACACCAGACCAACTGGT-3A, **I**) onto the –COOH group through the amide formation in the presence of EDC [1-(3-dimethylaminopropyl)-3-ethylcarbodiimide hydrochloride] coupling reagent.[9] Complementary DNA (cDNA) chains prelabeled with ferrocenecarboxaldehyde, FCA, (designated as [FCA-C6]ACCAGTTGGTCTGGTGTCAA-3A, **II**) were then used for hybridizing the surface-immobilized oligonucleotides to form double-strand DNA (dsDNA) helices on the aligned carbon nanotube electrodes (Figure 13.3), providing the basis for DNA sequence sensing.[9]

The performance of aligned carbon nanotube-DNA sensors for sequence-specific DNA diagnoses was demonstrated in Figure 13.4. The strong oxidation peak seen at 0.29 V in curve (a) of Figure 13.4 is attributed to ferrocene and indicates the occurrence of hybridization of FCA-labeled cDNA (II) chains with the nanotube-supported ssDNA (I) chains, leading to a long-range electron transfer from the FCA probe to the nanotube electrode through the DNA duplex. In contrast, the addition of FCA-labeled, noncomplementary DNA chains (i.e., [FCA-C6]CTCCAGGAGTCGTCGCCACC-3′, **III**) under the same conditions did not show any redox response of FCA (curve (b) of Figure 13.4). Subsequent addition of target DNA chains (i.e., 5′-GAGGTCCTCAGCAGCG GTGGACCAGTTGGTCTGGTGTCAA-3′, **IV**) into the above solution, however, led to a strong redox response from the FCA-labeled DNA (III) chains (curve (c) of Figure 13.4) because the target DNA(IV) contains complementary sequences for both DNA (I) and DNA (III) chains. These results clearly indicate that the modified, ssDNA functionalized, aligned carbon nanotubes have applications as highly sensitive DNA sensors because of the large surface area associated with the aligned nanotube structure.

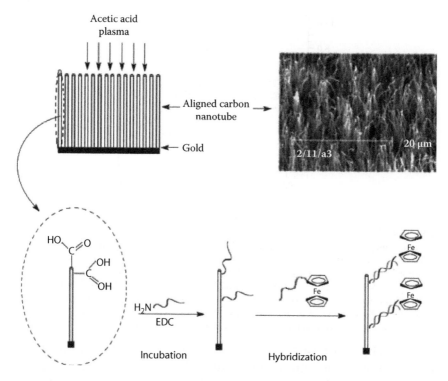

FIGURE 13.3
The aligned nanotube–DNA electrochemical sensor. The upper right scanning electron micrograph (SEM) shows the aligned carbon nanotubes after having been transferred onto a gold foil. For reasons of clarity, only one of the many carboxyl groups is shown at the nanotube tip and wall, respectively. (Adapted from He, P., and Dai, L. Aligned carbon nanotube–DNA electrochemical sensors. *Chemical Communications*, 348–349, 2004.)

The successful demonstration of surface functionalization of vertically aligned carbon nanotubes by plasma activation while largely retaining their structural integrity prompted us to *asymmetrically* functionalize individual CNTs with different molecular species/nanoparticles at the two end tips or along the nanotube length. In particular, we demonstrated a simple, but effective, method for asymmetric modification of the sidewall of CNTs with oppositely charged moieties by plasma treatment and π-π stacking interaction.[10] The as-prepared asymmetrically sidewall functionalized CNTs can be used as a platform for bottom-up self-assembly of complex structures or can be selectively self-assembled onto or between electrodes under an appropriate applied voltage for potential device applications.[10] As schematically shown in Figure 13.5a, we used a polymer-masking technique for the asymmetric functionalization of nanotube sidewalls by sequentially masking vertically aligned carbon nanotubes (VA-CNTs) twice with only half of the nanotube length being modified each time. In that particular case, the polymer-free nanotube surface for each of the constituent nanotubes in a poly-methyl methacrylate

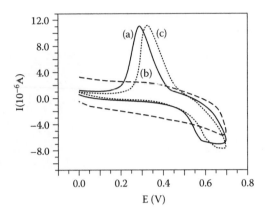

FIGURE 13.4
Cyclic voltammograms of the ssDNA (I)-immobilized aligned carbon nanotube electrode after hybridization with FCA-labeled complementary DNA (II) chains (a), in the presence of FCA-labeled noncomplementary DNA(III) chains (b), and after hybridization with target DNA (IV) chains in the presence of the FCA-labeled noncomplementary DNA (III) chains (c). All the cyclic voltammograms were recorded in 0.1 M H_2SO_4 solution with a scan rate of 0.1 $V \cdot s^{-1}$. The concentration of the FCA-labeled DNA probes is 0.05 $mg \cdot m^{-1}$. (Adapted from He, P., and Dai, L. Aligned carbon nanotube–DNA electrochemical sensors. *Chemical Communications*, 348–349, 2004.)

(PMMA)–embedded VA-CNT array was first negatively charged by acetic-acid polymerization to introduce the carboxylic surface functionalities. The plasma-treated VA-CNT array was then subjected to physical adsorption of positively charged gold nanocubes (Au-NCs) via electrostatic interaction to neutralize the nanotube surface charge (step 1 of Figure 13.5a). Subsequent removal of the PMMA supporting layer by ultrasonication in chloroform (typically, ~5 min) caused the release of the asymmetrically Au-NC-attached CNTs (step 2 of Figure 13.5a) and allowed modification of the gold nanopar-ticle (Au-NP) free surface of the same nanotubes with neutral R_3N-grafted pyrene-dimethylaminoethyl methacrylate (DMAEMA) copolymer chains through the specific pyrene-nanotube interaction (step 3 of Figure 13.5a). As the areas between the attached Au-NCs along the other half-tube length have been "premasked" by the acetic-acid polymer coating, its surface interaction with pyrene moieties along the DMAEMA polymer chains is minimal. After quat-ernization treatment with 1-bromohexane to convert the R_3N groups into posi-tively charged R_4N^+ moieties, negatively charged gold nanospheres (Au-NSs) were finally adsorbed onto the positively charged nanotube surface (step 4 of Figure 13.5a) *via* electrostatic interaction. Although the aforementioned asym-metric functionalization can in principle be applied to a large variety of charged moieties, the positively charged Au-NCs and negatively charged Au-NSs were used in the present study for easy characterization by SEM.

As expected, the SEM image of the Au-NC-attached CNT array in Figure 13.5b clearly shows many gold nanocubes deposited onto the PMMA-free region along the nanotube length. In comparison, Figure 13.5c unambiguously

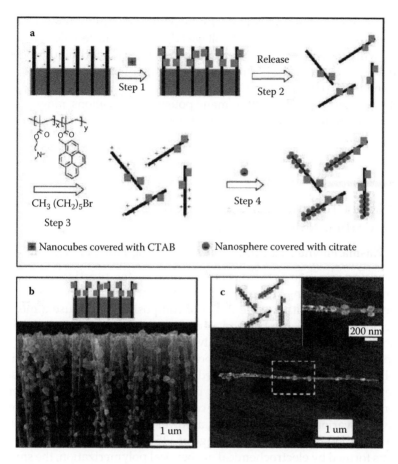

FIGURE 13.5

Asymmetric functionalization of carbon nanotubes (CNTs) with opposite charges: (a) procedures for asymmetric functionalization of CNTs with opposite charges, followed by tube-length-specific deposition of gold nanoparticles via electrostatic interactions. (b) A schematic representation and scanning electron micrograph (SEM) image of the CNT array partially functionalized with cubic gold nanoparticles; and (c) a schematic representation and SEM image of the sidewall-functionalized CNTs with half of the nanotubes' length covered by gold nanocubes and the other half covered by spherical gold nanoparticles through electrostatic assembly. (Inset shows a higher-magnification SEM image for the squared area. (Adapted from Peng, Q., Qu, L., Da, L., Park, K., and Vaia, R. A. Asymmetrically charged carbon nanotubes by controlled functionalization. *ACS Nano* 2, 1833–1840 (2008).)

reveals that the sidewall modified CNTs are coated over half the tube length by Au-NCs and the other half by Au-NSs. The relatively low number of Au-NCs along individual CNTs seen in Figure 13.5c with respect to Figure 13.5b indicates, most probably, that some of the adsorbed Au-NCs have been removed from the nanotube surface during the process to dissolve the PMMA protective layer and the subsequent solution steps shown in Figure 13.5a. Nevertheless, a direct current (DC) electrical field was used with the asymmetrically sidewall-charged CNTs across two parallel electrodes and successfully demonstrated to generate a pseudolinear current (I)-voltage (V) curve characteristic of the nanotube. Owing to the highly generic nature of the plasma technique, together with the versatile π-π stacking interaction between the pyrene-grafted molecules and CNT surface, the methodology developed in this study could be regarded as a general approach for many potential applications, ranging from nanotube electronic devices to biomedical systems.

13.3 Plasma Techniques for Fabrication of Carbon Nanomaterials

13.3.1 Plasma Polymerization of Semiconducting Polymer Nanofilms

Plasma polymerization is a solvent-free process for functionalizing thin films in a clean working environment, which enables the fabrication of thin polymeric coatings with a wide range of compositions, because almost all volatile organic vapors can be used as monomers. Although plasma polymerization typically produces electrically insulating organic films, it can also be employed to produce conducting polymer films. Among many conducting polymers, polyaniline has attracted a great deal of attention because of its exceptional electronic and photonic properties combined with good environment stability. Therefore, we have carried out plasma polymerization of aniline onto various substrates using a power input between 10 and 50 W and an excitation frequency in the range of 125 to 325 kHz.[5] Unlike polyaniline films formed by electrochemical or chemical polymerization, the smooth polyaniline films produced by plasma processes are free from oxidant and solvent, showing improved physicochemical characteristics (e.g., mechanical and environmental stability). Conductivity of the polyaniline films can be enhanced by three orders of magnitude in a controllable fashion through HCl treatment, implying that there is considerable room for tailoring the physical properties of the plasma-produced film (e.g., different conductivity values can be obtained by controlling the level of HCl treatment).

13.3.2 Plasma Patterning

13.3.2.1 Plasma Patterning of Conducting Polymers

For many optoelectronic and other applications involving semiconducting polymer films and carbon nanomaterials, it is highly desirable to produce micropatterned/aligned forms. In this regard, we have also produced surface patterns with region-specific characteristics through plasma polymerization of appropriate monomers. Figure 13.6a shows the steps for plasma

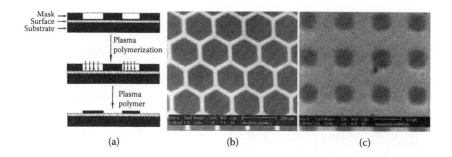

FIGURE 13.6

(a) Pattern formation by radio-frequency glow discharge. (b) A typical scanning electron micrograph (SEM) of the gold-coated mica sheets patterned by the MeOH-polymer with a transmission electron micrograph (TEM) grid consisting of hexagonal windows as the mask. (c) A typical SEM of gold-coated mica sheets patterned by the hexane polymer with a TEM grid consisting of square windows as the mask. (Adapted from Dai, L., Griesser, H. J., and Mau, A. W. H. Surface modification by plasma etching and plasma patterning. *Journal of Physical Chemistry B* 101, 9548–9554, 1997.)

patterning using a physical mask. As can be seen in Figures 13.6b and 13.6c, we prepared surface patterns by plasma polymerization of methanol (Figure 13.6b) and hexane (Figure 13.6c) on a gold-coated mica surface, respectively.[11] For MeOH polymerization, we choose a TEM grid consisting of hexagonal windows as the mask. Figure 13.6b shows a typical SEM micrograph of the patterns generated on a gold-coated mica surface by patterned polymerization of MeOH, with the dark areas representing MeOH polymer and bright regions being the uncovered gold surface. Similarly, we also performed hexane polymerization patterning with a TEM grid consisting of square windows as the mask. As shown in Figure 13.6c, the dark areas represent hexane polymer, and the bright regions are associated with the polymer-free gold surface.[11]

By extension, we developed a versatile method for obtaining patterned, conducting polymers by first depositing a thin patterned nonconducting (e.g., hexane) polymer layer onto a metal-sputtered electrode, and then performing electropolymerization of monomers (e.g., pyrrole, aniline) within the regions not covered by the patterned polymer layer.[11] Figure 13.7a represents a typical reflection light microscopic image of a polypyrrole pattern electrochemically deposited onto platinum-coated mica sheets, prepatterned with hexane polymer obtained by plasma deposition. It shows the same features as the pattern in Figure 13.6c but with inverse intensities. The bright regions characteristic of the uncovered metal surface in Figure 13.6c become dark in Figure 13.7a because of the presence of a dark layer of the newly electropolymerized polypyrrole film. The bright regions in Figure 13.7a represent a more reflective surface associated with the hexane polymer.

The cyclic voltammetric response of a polypyrrole pattern thus prepared (Figure 13.7a) is shown in Figure 13.7b, which clearly shows a quasi-reversible

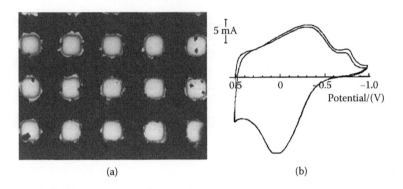

(a)   (b)

FIGURE 13.7

(a) Optical microscopy image of a polypyrrole pattern electrochemically polymerized onto a platinum-coated mica surface prepatterned by a hexane polymer deposited by a plasma. (b) Typical cyclic voltammogram of the polypyrrole patterns on platinum at 100 mV•s^{-1} in an aqueous solution containing 0.1 M sodium perchlorate. (Adapted from Dai, L., Griesser, H. J., and Mau, A. W. H. Surface modification by plasma etching and plasma patterning. *Journal of Physical Chemistry B* 101, 9548–9554, 1997.)

redox process with two reduction peaks for the polypyrrole film containing perchlorate. The first reduction peak seen in Figure 13.7b can be attributed to the presence of a cationic species (polarons) in the polypyrrole film, the second reduction peak arising from the coexistence of a dicationic species (bipolarons). As a control, the cyclic voltammetry measurement was carried out for a freshly prepared n-hexane polymer under the same conditions, which showed only small capacitive current with no peak attributable to the presence of any redox-active species. Therefore, the cyclic voltammogram shown in Figure 13.7b clearly indicates that the polypyrrole patterns prepared in this study are electrochemically active.

Subsequently, we produced conducting polymer microcontainers in a patterned fashion by prepatterning the working electrode surface with nonconducting polymers via plasma patterning.[12] Figure 13.8a schematically shows the steps for the patterning process. Prior to the patterning, the stainless steel electrode was cleaned by an aqueous solution of HNO_3 (15.8 M after being diluted to 1:1 v/v, H_2O/HNO_3) under ultrasonication for 30 min, followed by thoroughly washing with distilled water and acetone. For the plasma patterning, nonconducting polymers were produced on the stainless-steel electrode surface by plasma polymerization of an appropriate monomer (e.g., hexane at 150 kHz and 0.3 Torr for 2 min) with a mask (Figure 13.8a). The electrode with the prepatterned, nonconducting polymer film was then used for region-specific electrodeposition of polypyrrole microcontainers by the "soap bubble" template.[12] This involved the electrochemical polymerization of pyrrole around the wall of the gas bubbles generated within the polymer-free regions on the working electrode by the electrolysis of H_2O in the presence of surfactant (i.e., b-napthalenesulfonic acid, b-NSA). Figure 13.8b

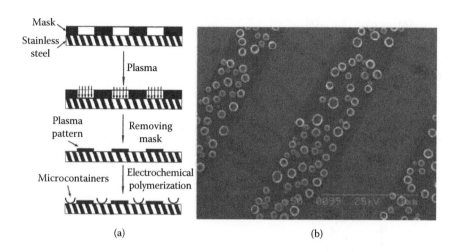

(a)   (b)

FIGURE 13.8
(a) The procedure for fabricating patterns of polypyrrole microcontainers by plasma pattern-ing. (b) Scanning electron micrograph (SEM) of polypyrrole microcontainer pattern obtained by plasma patterning. (Adapted from Bajpai, V., He, P., and Dai, L. Conducting-polymer micorcontainers: Controlled syntheses and potential applications. *Advanced Functional Materials* 14, 145–151, 2004, and references cited therein.)

shows a typical SEM image of the microcontainer patterns generated on a stainless-steel electrode by the plasma patterning method. As can be seen in Figure 13.8b, polypyrrole microcontainers can be region-specifically elec-trodeposited onto the electrode prepatterned with nonconducting polymer film. The patterned conducting polymer microcontainers thus prepared may find applications in sensing and optoelectronic devices.

Furthermore, some conjugated polymers have been directly deposited onto plasma-patterned surfaces via self-assembly. For example, we prepared light-emitting polymer patterns by first depositing hydrophilic (i.e., acetic acid) patterns onto hydrophobic substrates (e.g., perfluorinated ethylene-propylene copolymer films) with a plasma, and then performed selective adsorption from a solution of a 2,5-substituted poly(*p*-phenylene vinylene) derivative with methoxy-terminated oligo(ethylene oxide) side chains (i.e., EO_3-PPV, EO_3 = $O(CH_2CH_2O)_3CH3$, Figure 13.9).[13] The driving force for the pattern formation in this particular case was the polar–polar interaction between the EO_3 side chains and the micropatterned hydrophilic polymer. Under the fluorescence microscope, the EO_3-PPV-patterned regions gave flu-orescence emissions characteristic of the conjugated structure.[13]

13.3.2.2 Plasma Patterning of Carbon Nanotubes

In our investigation of carbon nanotube growth, we found that certain plasma polymerized patterns can be used not only for region-specific growth of the aligned carbon nanotubes but also for making patterns of nonaligned carbon

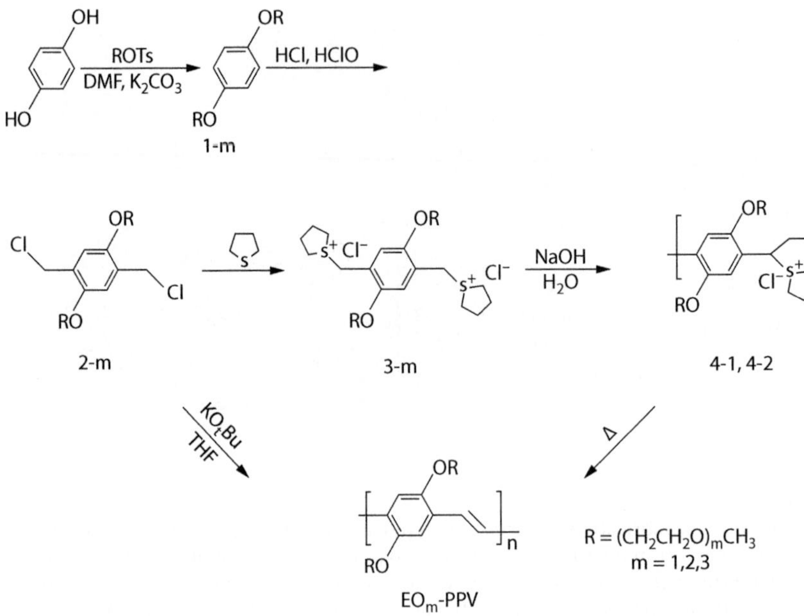

FIGURE 13.9

Synthesis of EO_3-PPV via a modified Gilch route [$R = (CH_2CH_2O)_3CH_3$]. (Adapted from Winkler, B., Dai, L., and Mau, A. W. -H. Novel poly(p-phenylene vinylene) derivatives with oligo(ethylene oxide) side chains: Synthesis and pattern formation. *Chemistry of Materials* 11, 704–711, 1999.)

nanotubes.[14] For instance, we generated surface patterns, either by nondepositing plasma treatment or by plasma polymerization, of $-NH_2$ groups onto a substrate (e.g., quartz glass plate, mica sheet, polymer film) and then performed region-specific adsorption of the COOH-containing carbon nanotubes from an aqueous medium through the polar–polar interaction between the COOH groups and the $-NH_2$ groups. The COOH-containing carbon nanotubes were prepared by acid treatment (HNO_3) of the FePc-generated nanotubes.[14]

Figures 13.10a and 13.10b show the steps used for plasma patterning of carbon nanotubes via the region-specific growth and adsorption process, respectively. The highly cross-linked structure of the polymer films ensured the integrity of the polymer layer, even without carbonization, at the high temperatures necessary for nanotube growth from FePc.[14] Therefore, the carbonization process involved in our previous work on photolithographic and soft-lithographic patterning of the aligned carbon nanotubes could be completely eliminated.[1,2] Owing to the generic nature of plasma polymerization, many other organic vapors could also be used to generate polymer patterns for patterned growth of aligned carbon nanotubes.

Figure 13.11a shows a SEM image of aligned nanotube micropatterns prepared on a n-hexane-polymer prepatterned quartz plate. Figure 13.11b

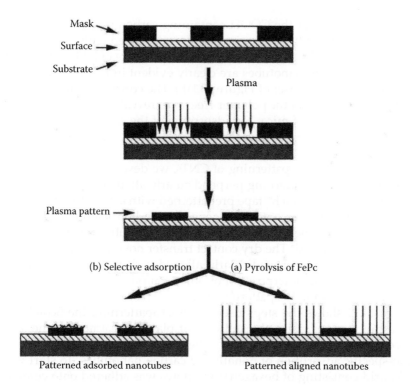

FIGURE 13.10
The procedure for fabricating patterns of carbon nanotubes by (a) plasma polymerization followed by aligned nanotube growth. (b) Plasma activation followed by region-specific adsorption. (Adapted from Chen, Q., and Dai, L. Plasma patterning of carbon nanotubes. *Applied Physics Letters* 76, 2719–2721, 2000).

FIGURE 13.11
Scanning electron micrographs (SEMs) of (a) aligned nanotube arrays growing from the polymer-free regions on an n-hexane-polymer patterned quartz plate, and (b) adsorbed COOH-containing carbon nanotubes (within the squared areas) on a heptylamine-patterned mica sheet. Inset gives a higher-magnification image of the covered areas, showing the individual adsorbed carbon nanotubes. (Adapted from Chen, Q., and Dai, L. Plasma patterning of carbon nanotubes. *Applied Physics Letters* 76, 2719–2721, 2000.)

shows a SEM image of the COOH-containing carbon nanotubes adsorbed (ca. 2.5 mg/10 mL H_2O) onto a mica sheet prepatterned with the heptylamine-polymer (200 kHz, 10 W, and a monomer pressure of 0.13 Torr for 30 s). The adsorbed carbon nanotubes are clearly evident in Figure 13.11b under higher magnification (inset of Figure 13.11b). The corresponding high-magnification SEM image for the polymer-free areas reveals a featureless smooth surface characteristic of mica. No adsorption of the carbon nanotubes was observed in a control experiment when a pure mica sheet was used as the substrate.

In addition to plasma patterning of CNTs, we developed a contact transfer method for micropatterning perpendicularly aligned carbon nanotubes by simply pressing a Scotch® tape prepatterned with a nonadhesive plasma-deposited polymer layer onto FePc-generated carbon nanotube films, followed by peeling of the nanotubes from the quartz substrate (designated as "dry contact transfer").[15] The dry contact transfer not only retains the structural integrity of the perpendicularly aligned carbon nanotubes but also allows region-specific interposition of other component(s) into the discrete areas interdispersed in the patterned nanotube structure.

Figure 13.12 shows the steps used for micropatterning the Scotch tape with a thin layer of silver, the subsequent plasma treatment of the silver surface, the contact transfer of the aligned carbon nanotubes, and the region-specific adsorption of nonaligned carbon nanotubes. To start with, a TEM grid consisting of hexagonal windows was adhered onto commercially available Scotch tape (3M, polypropylene-film-supported acrylic adhesive) as a mask, followed by sputter coating with silver through the mask (Figure 13.12a). Thereafter, the silver-patterned Scotch tape was subjected to plasma treatment to deposit a heptylamine film at 250 kHz, 30 W, and a monomer pressure of 0.18 Torr for 180 s (Figure 13.12a). After careful removal of the TEM grid (Figure 13.12b), a SEM image of the plasma-treated silver hexagons was recorded (Figure 13.13a). The corresponding energy-dispersive X-ray (EDX) line analyses of C KR and Ag KR given in Figure 13.13b clearly show the presence of a thin layer of silver-rich coating in the hexagonal region. This indicates the formation of hexagonal silver micropatterns on the adhesive layer of the Scotch tape. Aligned carbon nanotubes were then transferred onto the adhesive-covered area by pressing the Scotch tape on the as-synthesized aligned carbon nanotube film on a quartz plate, followed by peeling of the Scotch tape from the quartz substrate (Figures 13.12c and 13.12d). As expected, the nanotubes underneath the silver-free regions were selectively transferred onto the Scotch tape as a positive image of the TEM grid (Figure 13.13c), whereas those covered by the silver patterned areas remained on the quartz substrate as a negative pattern (Figure 13.13d). As can be seen in Figures 13.13c and 13.13d, the integrity of the carbon nanotubes (e.g., alignment, packing density) transferred onto the Scotch tape is almost the same as the as-grown nanotubes remaining on the quartz plate. However, the newly transferred carbon nanotubes

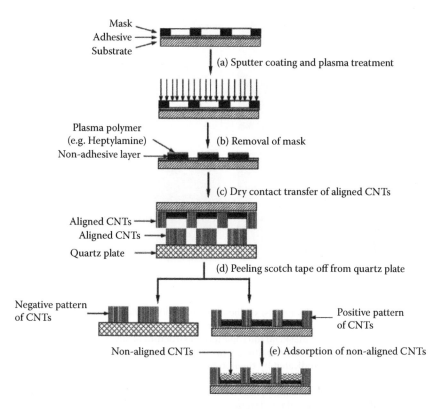

FIGURE 13.12

The procedures for fabricating multicomponent interposed carbon nanotube micropatterns by dry contact transfer, followed by region-specific adsorption. (Adapted from Yang, J., and Dai, L. Multicomponent interposed carbon nanotube micropatterns by region-specific contact transfer and self-assembling. *Journal of Physical Chemistry B* 107, 12387–12390, 2003.)

are supported by a flexible substrate. The crack edges seen within the hexagonal areas in Figure 13.13c were, most probably, caused by mechanical deformation of the silver layer when the overlaying flexible Scotch tape was pressed downward on the nanotube film during the transfer process. Although some care may be needed to prepare aligned carbon nanotube micropatterns with interposed crack-free silver patterns on flexible Scotch tape, the soft nature of the Scotch tape used for the dry contact transfer should allow us to develop multicomponent nanotube micropatterns for flexible device applications. This has been clearly demonstrated by our recent publications,[16,17] in which various multicomponent microarchitectures of vertically aligned carbon nanotubes have been developed through either the dry contact transfer, followed by region-specific adsorption,[16] or sequential multiple contact transfer.[17]

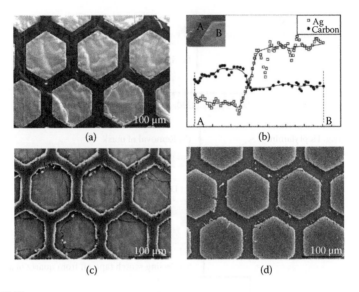

FIGURE 13.13

Scanning electron micrographs (SEMs) of the aligned carbon nanotube micropatterns produced by dry contact transfer with a transmission electron micrograph (TEM) grid consisting of hexagonal windows as the mask: (a) micropatterned structure of silver on Scotch® tape; (b) EDX profiles of Ag Kα (□) and C Kα (●) (the scanning path for the energy-dispersive spectroscopy (EDX) line analyses is indicated by the line from A to B in the inserted SEM picture); (c) aligned carbon nanotube patterns after being transferred onto the Scotch tape (i.e., positive pattern); (d) aligned carbon nanotube patterns left on the quartz plate after the dry contact transfer (i.e., negative pattern). (Adapted from Yang, J., and Dai, L. Multicomponent interposed carbon nanotube micropatterns by region-specific contact transfer and self-assembling. *Journal of Physical Chemistry B* 107, 12387–12390, 2003.)

13.3.3 Plasma Etching

We previously demonstrated that exposure of fluoropolymers, such as fluorinated ethylene propylene (FEP) and polytetrafluoroethylene (PTFE) films, to H_2O-plasmas under vapor pressures in the range of 0.3 to 0.5 Torr can lower the air/water contact angle.[1-3] However, topographical changes on the fluoropolymer surfaces were observed for polymer films treated with H_2O plasma under a "high" pressure (>0.5 Torr).[11] The driving force for the plasma-induced surface roughening arises, most probably, from enhanced atomic oxygen concentrations, which in turn leads to a high concentration of atomic oxygen in the "high"-pressure H_2O plasma that could make the water plasma very efficient for etching substrates. We treated freshly cleaved mica surfaces with a H_2O plasma under conditions similar to those used for the etching of the fluoropolymers. Surface topography of the mica sheets was greatly increased. Furthermore, we carried out the H_2O plasma treatment in a patterned fashion for the purpose of generating patterned surfaces. As shown in Figure 13.14, patterned etching was obtained with a close replication of the mask structure.[11]

In fact, plasmas can be used to effectively etch many substrates including mica sheets,[11] polymers, and carbon materials. For instance, we found that amorphous carbon could be selectively removed by H_2O plasma etching while carbon nanotubes remain largely unchanged.[18] As can be seen in Figure 13.15, the amorphous carbon layer was removed almost completely by H_2O plasma etching for 30 min. Prolonged etching, however, could partially remove graphitic sheets from the carbon nanotube structure. Because the nanotube tips are more reactive than tube walls, the end caps of perpendicularly aligned nanotubes exposed to a plasma for about 80 min were

FIGURE 13.14
Typical scanning electron micrograph (SEM) of freshly cleaved mica sheets patterned by H_2O plasma etching with a transmission electron micrograph (TEM) grid consisting of hexagonal windows as the mask. (Adapted from Dai, L., Griesser, H. J., and Mau, A. W. H. Surface modification by plasma etching and plasma patterning. *Journal of Physical Chemistry B* 101, 9548–9554, 1997.)

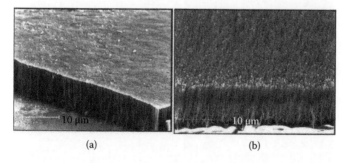

(a)   (b)

FIGURE 13.15
Scanning electron micrograph (SEM) images of an aligned nanotube film before and after H_2O plasma etching: (a) a large area aligned nanotube film covered by a thin amorphous carbon layer on a quartz glass plate before plasma treatment, (b) the same nanotube film as (a) after the H_2O plasma etching for 30 min at 250 kHz, 30 W, and 0.62 Torr. Note that the micrographs shown in (a) and (b) were not taken from the same spot due to technical difficulties. (Adapted from Huang, S., and Dai, L. Plasma etching for purification and controlled opening of aligned carbon nanotubes. *Journal of Physical Chemistry B* 106, 3543–3545, 2002.)

selectively opened (Figure 13.16). Plasma etching played an important role in opening the end tips of nanotubes and in introducing defects and oxygen functionalization to the nanotubes.[18] The etched carbon nanotubes showed a strong capacitive behavior in ionic liquid electrolytes.[19]

By using carbon nanotube arrays that are dominated by a straight body segment but with curly entangled top,[20] we also created gecko foot–mimetic dry adhesives that show macroscopic adhesive forces of ~100 Newtons per square centimeter, almost 10 times that of a gecko foot, and much stronger

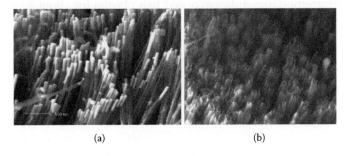

(a)   (b)

FIGURE 13.16
Scanning electron micrograph (SEM) images of the aligned nanotubes (a) before and (b) after plasma treatment for 80 min, followed by a gentle wash with HCl (37%) to remove the Fe catalyst residues, if any (see text). Note that the micrographs shown in (a) and (b) were not taken from the same spot due to technical difficulties. (Adapted from Huang, S., and Dai, L. Plasma etching for purification and controlled opening of aligned carbon nanotubes. *Journal of Physical Chemistry B* 106, 3543–3545, 2002.)

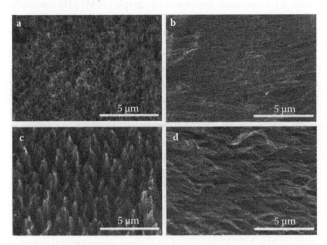

FIGURE 13.17
Effect of the plasma etching on adhesion forces. (a through d) Typical top view of vertically aligned multiwalled carbon nanotube (VA-MWNT) film before (a,c) and after (b,d) adhesion measurements, (a,b) without and (c,d) with oxygen etching. (Adapted from Qu, L., Dai, L., Stone, M., Xia, Z., and Wang, Z. Carbon nanotube arrays with strong shear binding-on and easy normal lifting-off. *Science* 322, 238–242, 2008.)

shear adhesion force than the normal adhesion force, which leads to strong binding along the shear direction and easy lifting in the normal direction. In that study,[20] we used oxygen plasma etching to physically remove the non-aligned nanotube segments and investigate their influence on the adhesion forces (Figure 13.17). The removal of the nonaligned nanotube segments from the top of a vertically aligned multiwalled carbon nanotube (VA-MWNT) array by plasma etching led to predominately point contacts, which largely eliminated the nanotube length dependence for both shear and normal adhesion forces within experimental error, and a concomitant decrease in adhesion forces. The plasma etching induced "bundle" formation (Figure 13.17c) together with surface functionalization that weakened the adhesion forces by reducing the number of effective contact points per unit area and the interaction energy per contact with the glass surface.

More recently, we developed a simple plasma etch process to effectively generate metal-free catalysts for efficient metal-free growth of undoped and nitrogen-doped single-walled carbon nanotubes.[21] As shown in Figures 13.18a

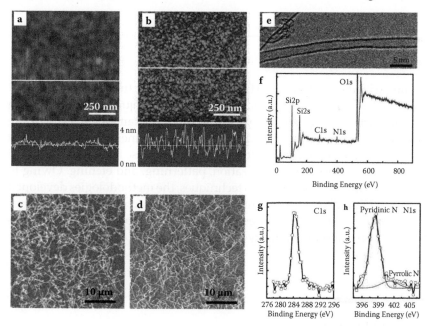

FIGURE 13.18
Atomic force microscopy (AFM) images of the SiO_2/Si surface (a) before and (b) after the H_2O plasma treatment. Scanning electron micrographs (SEMs) of (c) undoped and (d) N-doped carbon nanotubes (CNTs) grown on SiO_2/Si substrates. (e) high-resolution transmission electron microscopy (HRTEM) image, (f) x-ray photoelectron spectroscopy (XPS) survey, (g) XPS C1s, and (h) XPS N1s spectra of the N-doped CNTs. (Adapted from Yu, D., Zhang, Q., and Dai, L. Highly efficient metal-free growth of nitrogen-doped single-walled carbon nanotubes on plasma-etched substrates for oxygen reduction. *Journal of the American Chemical Society* 132, 15127–15129, 2010.)

and 13.18b, atomic force microscope (AFM) imaging of an etched SiO_2/Si substrate clearly shows the formation of homogenously distributed catalyst particles with an average size of <5 nm. The particle density was estimated to be ~150 particles/μm^2. Prior to nanotube growth, x-ray photoelectron spectroscopy (XPS) was used to confirm that the nanoparticles are free from metal. The metal-free SiO_2 nanoparticles were used to catalyze the growth of CNTs under a mixture flow of 100 sccm CH_4 and 100 sccm H_2 at 900°C for 20 min (Figure 13.18c). By introducing 50 sccm NH_3 during the metal-free CVD process, densely packed nitrogen-doped single-walled carbon nanotubes were produced on SiO_2/Si substrates (Figure 13.18d). In contrast, no nanotube deposition was seen for the pristine SiO_2/Si wafer under the same conditions. Compared with undoped CNTs, the newly produced metal-free nitrogen-containing CNTs were demonstrated to show relatively good electrocatalytic activity and long-term stability toward oxygen reduction reactions (ORRs) in an acidic medium.[21,22]

13.4 Concluding Remarks

We summarized our recent efforts to use plasmas for functionalization and modification of carbon nanomaterials, including conjugated polymer thin films and carbon nanotube arrays. Although the examples presented are focused on our own work, they demonstrate the numerous possibilities of functional structures and smart devices that could arise from the application of plasmas for surface functionalization, patterning, and etching. Owing to the highly generic nature of plasma techniques, the methodologies developed in these particular cases can be readily transferred to many other systems. Thus, there are vast opportunities for developing carbon-based multifunctional structures and materials by plasma techniques.

Acknowledgments

The authors thank our colleagues for their contributions to the work cited. The aim of this chapter is to summarize our recent work with no intention for a comprehensive literature survey of the subject. Therefore, we apologize to the authors of papers not cited here. We are also grateful for financial support from the NSF (CMMI-1047655, CMMI-1000768), Air Force Office of Scientific Research (AFOSR) (FA9550-10-1-0546, FA9550-09-1-0331, FA2386-10-1-4071), and Asian Office of Aerospace Research and Development AOARD-104055.

References

1. Dai, L., and Mau, A. W. H. Surface interface control of polymeric biomaterials, conjugated polymers, and carbon nanotubes. *Journal of Physical Chemistry B* 104, 1891–1915 (2000).
2. Dai, L. Radiation chemistry for microfabrication of conjugated polymers and carbon nanotubes. *Radiation Physics and Chemistry* 62, 55–68 (2001).
3. Chen, W., Dai, L., Jiang, H., and Bunning, T. J. Controlled surface engineering and device fabrication of optoelectronic polymers and carbon nanotubes by plasma processes. *Plasma Processes and Polymers* 2, 279–292 (2005).
4. Gong, X., Dai, L., Griesser, H. J., and Mau, A. W. H. Surface immobilization of poly(ethylene oxide): Structure and properties. *Journal of Polymer Science Part B: Polymer Physics* 38, 2323–2332 (2000).
5. Gong, X., Dai, L., Mau, A. W. H., and Griesser, H. J. Plasma-polymerized polyaniline films: Synthesis and characterization. *Journal of Polymer Science Part A: Polymer Chemistry* 36, 633–643 (1998).
6. Chen, Q., Dai, L., Gao, M., Huang, S., and Mau, A. Plasma activation of carbon nanotubes for chemical modification. *J. Phys. Chem. B* 105, 618–622 (2001).
7. Dai, L. et al. Biomedical coatings by the covalent immobilization of polysaccharides onto gas-plasma-activated polymer surfaces. *Surface and Interface Analysis* 29, 46–55 (2000).
8. Yang, Y., Huang, S., He, H., Mau, A. W. H., and Dai, L. Patterned growth of well-aligned carbon nanotubes: A photolithographic approach. *Journal of the American Chemical Society* 121, 10832–10833 (1999).
9. He, P., and Dai, L. Aligned carbon nanotube–DNA electrochemical sensors. *Chemical Communications*, 348–349 (2004).
10. Peng, Q., Qu, L., Da, L., Park, K., and Vaia, R. A. Asymmetrically charged carbon nanotubes by controlled functionalization. *ACS Nano* 2, 1833–1840 (2008).
11. Dai, L., Griesser, H. J., and Mau, A. W. H. Surface modification by plasma etching and plasma patterning. *Journal of Physical Chemistry B* 101, 9548–9554 (1997).
12. Bajpai, V., He, P., and Dai, L. Conducting-polymer micorcontainers: Controlled syntheses and potential applications. *Advanced Functional Materials* 14, 145–151 (2004), and references cited therein.
13. Winkler, B., Dai, L., and Mau, A. W. -H. Novel poly(p-phenylene vinylene) derivatives with oligo(ethylene oxide) side chains: Synthesis and pattern formation. *Chemistry of Materials* 11, 704–711 (1999).
14. Chen, Q., and Dai, L. Plasma patterning of carbon nanotubes. *Applied Physics Letters* 76, 2719–2721 (2000).
15. Yang, J., and Dai, L. Multicomponent interposed carbon nanotube micropatterns by region-specific contact transfer and self-assembling. *Journal of Physical Chemistry B* 107, 12387–12390 (2003).
16. Yang, J., Qu, L., Zhao, Y., Zhang, Q., Dai, L., Baur, J. W., Maruyama, B., Vaia, R. A., Shin, E., Murray, P. T., Luo, H., and Guo, Z. -X. Multicomponent and multidimensional carbon nanotube micropatterns by dry contact transfer. *Journal of Nanoscience and Nanotechnology* 7, 1573–1580 (2007).

17. Qu, L., Vaia, R. A., and Dai, L. Multilevel, multicomponent microarchitectures of vertically-aligned carbon nanotubes for diverse applications. *ACS Nano* (2010, DOI: 10.1021/nn102411s).
18. Huang, S., and Dai, L. Plasma etching for purification and controlled opening of aligned carbon nanotubes. *Journal of Physical Chemistry B* 106, 3543–3545 (2002).
19. Lu, W., Qu, L., Henrya, K., and Dai, L. High performance electrochemical capacitors from aligned carbon nanotube electrodes and ionic liquid electrolytes. *Journal of Power Sources* 189, 1270–1277 (2009).
20. Qu, L., Dai, L., Stone, M., Xia, Z., and Wang, Z. Carbon nanotube arrays with strong shear binding-on and easy normal lifting-off. *Science* 322, 238–242 (2008).
21. Yu, D., Zhang, Q., and Dai, L. Highly efficient metal-free growth of nitrogen-doped single-walled carbon nanotubes on plasma-etched substrates for oxygen reduction. *Journal of the American Chemical Society* 132, 15127–15129 (2010).
22. Gong, K., Du, F., Xia, Z., Dustock, M., and Dai, L. Nitrogen-doped carbon nanotube arrays with high electrocatalytic activities for oxygen reduction. *Science* 323, 760–764 (2009).

14

Plasma–Liquid Interactions for Fabrication of Nanobiomaterials

Toshiro Kaneko
Rikizo Hatakeyama

CONTENTS

14.1 Introduction

The interaction of plasma discharges with liquids [1,2] is a topic of great current interest in plasma science and technology. It has pioneered new research directions related to wastewater treatment, sterilization, microanalysis of liquids, and nanomaterial generation due to distinct properties such as ultrahigh density, high reactivity, high process rate, and so on. In particular, the boundary between plasmas and liquids, which activates physical processes and chemical reactions, has attracted much attention as a novel direction for nanobiomaterial synthesis. For example, nanoparticle synthesis using the plasma–liquid interface [3–7] is especially advantageous in that toxic stabilizers and reducing agents are unnecessary and the synthesis is continuous during the plasma irradiation. In these methods, although it has been reported that the metal salt is reduced by electrons or active hydrogen, the optimal plasma conditions in terms of the synthesis rate, size control, and simplicity remain unclear because the high-voltage discharge operated at atmospheric pressure and the consequential dynamic behavior of the gas–liquid interface prevent us from analyzing the precise properties of the plasma at the interfacial region.

359

We generated novel gas–liquid interfacial plasmas under a low gas pressure condition [8,9] by utilizing the unique properties of ionic liquids [10] that allow a fully ionized plasma state, extremely low vapor pressure, and high heat capacity. In this chapter, the dynamic behavior of charged particles and dissociation of the ionic liquid as a result of the gas–liquid interfacial plasma are discussed under conditions ranging from low gas pressures to atmospheric pressure. These studies are presented as a step toward nano-bio applications of plasma–liquid systems [11].

14.2 Gas–Liquid Interfacial Plasmas

Figure 14.1(a) shows a schematic diagram of the experimental setup that has a glass cell, 20 mm in diameter and 10 mm in depth, in a cylindrical glass chamber of 15 cm in diameter and 50 cm in length. A 15 mm diameter electrode plate made of platinum (Pt) is located inside the glass cell, and an ionic liquid (IL) (1-buthl-3-methyl-imidazolium tetrafluoroborate: $[C_8H_{15}N_2]^+[BF_4]^-$) in a liquid state at room temperature is introduced as a cathode electrode for the purpose of investigating the effects of the IL on the discharge. A grounded anode electrode made of a stainless steel (SUS) plate is placed in the gas phase (plasma) region at a distance of 60 mm from the cathode electrode. This discharge configuration, in which the cathode electrode is in the glass cell, is defined as "A-mode."

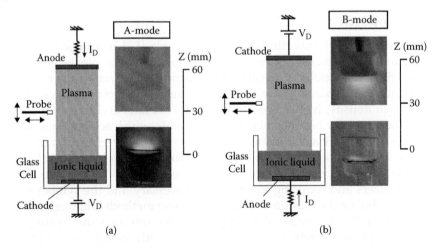

FIGURE 14.1
The experimental setup for direct current (DC) discharge plasmas, where the cathode is (a) an ionic liquid (IL) and (b) the gas plasma region. These configurations are defined as (a) A-mode and (b) B-mode, respectively.

In order to examine the effects of plasma irradiation on the IL, the cathode electrode is switched to the SUS plate located at the top of the gas plasma region, which is defined as "B-mode," and the anode electrode consisting of the IL in the glass cell is grounded as shown in Figure 14.1b. Removal of the water dissolved in the ionic liquid is performed under vacuum for 2 hours after introducing the IL into the glass chamber. A negative direct current (DC) voltage is supplied to the cathode electrode. The argon gas is adopted as a discharge medium, and the gas pressure P_{gas} is varied from 60 Pa to 40 kPa. A high-voltage probe is directly connected to the cathode electrode to measure its bias voltage. A Langmuir probe is inserted at the position of $z = 0 - 60$ mm to measure plasma parameters of the discharge in contact with the IL (the surface of the IL electrode in the glass cell is $z = 0$).

We achieved generation of IL incorporated plasmas at low gas pressures with high stability [9], similar to the normal glow discharge plasmas. Photos of a stable DC discharge plasma in A-mode in the regions below the anode electrode and above the cathode electrode are presented in Figure 14.1a. The cathode glow is clearly observed, with no emission observed in the region below the anode electrode. On the other hand, in B-mode (Figure 14.1b), a cathode glow and anode glow are observed. The volume of the anode glow becomes larger with an increase in the gas pressure (figure is not shown), most likely because recombination occurs more frequently at high pressures and the anode glow extends to the plasma region.

Discharge voltage, V_D, and current, I_D, characteristics are measured in both the A-mode and B-mode, which are presented in Figure 14.2, as a function of the electrode materials in the glass cell—that is, changing (a) the cathode electrode in A-mode and (b) the anode electrode in B-mode. When the cathode electrode material in A-mode is changed to SUS, nickel (Ni), and the IL (Figure 14.2a), the discharge voltage is found to be the smallest in the case of the IL, as compared to Ni and SUS at constant discharge current. On the

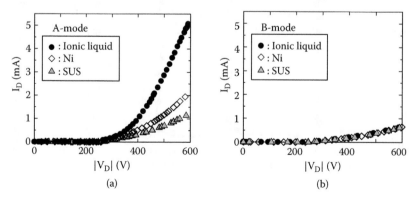

FIGURE 14.2
Dependence of the discharge voltage V_D minus current I_D characteristics on the following electrode materials: (a) the cathode in A-mode and (b) the anode in B-mode. $P_{gas} = 40$ Pa.

other hand, changing the anode electrode material in B-mode has no effect on the discharge characteristics as shown in Figure 14.2b. These results suggest that the IL works as an effective cathode electrode in A-mode and the secondary electrons are emitted from the IL more efficiently than SUS and Ni electrodes. For conventional DC glow discharges, the discharge voltage is known to depend on the amount of secondary electrons emitted from the cathode electrode. In B-mode, on the other hand, because the cathode electrode material remains the same, the discharge voltage and current characteristics do not change in spite of the various kinds of anode electrodes.

The decrease in the discharge voltage (and increase in the amount of secondary electrons) in the case of the IL cathode is attributed to the cathode sheath electric field distribution on the IL surface which is reported to have a string-shaped alkyl chain aligned toward the gas-phase region [12]. This electric field distribution enables efficient ion irradiation to the IL, resulting in the emission of a large amount of secondary electrons from the IL surface, even more than conventional metal cathodes such as SUS and Ni.

Figure 14.3 presents axial profiles of the plasma space potential Φ_s in (a) A-mode and (b) B-mode for $P_{gas} = 60$ Pa and $I_D = 1$ mA, where the cathode electrode materials are SUS, Ni, and IL in A-mode. In A-mode, the discharge voltages V_D (i.e., the potentials at the cathode electrode) are about -380, -470, and -540 V when the cathode materials are the IL, Ni, and SUS, respectively. The large potential difference between the plasma and the cathode electrode, namely, the large sheath electric field, leads to the acceleration of positive ions toward the cathode electrode [9].

In addition, it is found that in the case of the IL cathode, the space potential in the central region ($z = 10$ to 50 mm) is lower than in the case of SUS or Ni

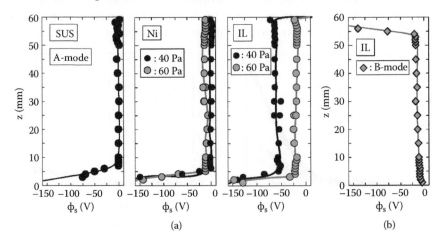

FIGURE 14.3
Axial profiles of the space potential Φ_s in the plasma region for the (a) A-mode with different cathode materials and (b) B-mode. $P_{gas} = 40$ and 60 Pa, $I_D = 1$ mA.

cathode. When the plasma ions irradiate the IL cathode, the anions (negative ion) of the IL are expected to be sputtered and accelerated toward the plasma region by the sheath electric field above the cathode electrode. The accelerated anions then collide with the anode electrode and cause emission of secondary electrons that form the anode sheath. This lowers the space potential in the central region. When the argon gas pressure is changed from 60 Pa to 40 Pa, the discharge voltage, namely, the irradiation energy of the plasma ions toward the cathode, becomes large as described later. This high-energy ion irradiation causes the extraction of large amounts of anions from the IL, and the space potential becomes lower as a result of the large anode sheath. This phenomenon is not observed in the case of the Ni cathode that cannot generate negative ions.

In B-mode, on the other hand, the discharge voltages are the same for the IL, Ni, and SUS electrodes as shown in Figure 14.2b, and the axial profiles of the space potential are almost the same. Figure 14.3b gives the axial profile of the space potential in the case of the IL anode electrode. Because the potential in the plasma region is about −10 V, the potential difference between the plasma and the IL is relatively small (~ 10 V) and the electric field direction is opposite to that in A-mode. Therefore, the electrons in the plasma are injected into the IL with small energy instead of the positive ion irradiation with high energy. It is also observed that the space potential in the case of the IL anode is slightly lower than that of the SUS anode. This result indicates that the secondary electrons are emitted from the anode electrode by the impact of injected electrons [13], and the secondary electron emission coefficient of the IL is larger than that of the SUS. Therefore, the electron-rich condition in the electron sheath region in front of the anode electrode is enhanced in the case of the IL anode, resulting in a lower space potential compared with the case of the SUS anode.

Figure 14.4a shows ultraviolet-visible (UV-vis) absorption spectra of the IL with the ion irradiation energy, E_i, as a parameter in A-mode, $I_D = 1$ mA, and the plasma irradiation time, $t = 2$ min. The ion irradiation energy E_i is determined by the potential difference between the plasma and the cathode potential, which is equivalent to the discharge voltage. E_i is varied by changing the argon gas pressure P_{gas} at constant discharge current I_D. The gas pressures are also indicated in the figures. The spectrum peak intensity at 297 nm gradually increases with an increase in E_i, which corresponds to the color change of the IL as it gradually changes to dark yellow. Because these phenomena are not observed in B-mode as shown in Figure 14.4b, where the electron irradiation energy to the IL anode is almost constant with gas pressure, we suggest that the increase in the absorption peak intensity in A-mode is caused by dissociation of the IL, which is enhanced by an increase in the ion irradiation energy. Therefore, the molecule structure of the IL can be modified by the ion irradiation energy in A-mode, in which the positive ions in the gas phase plasma are accelerated by the sheath electric field above the ionic liquid. We emphasize that this ion irradiation to the ionic liquid has the

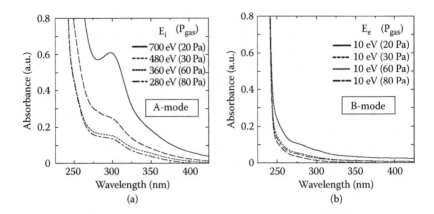

FIGURE 14.4
Ultraviolet-visible (UV-vis) absorption spectra of the ionic liquid (IL) with the ion irradiation energy E_i as a parameter for the (a) A-mode and (b) B-mode. $I_D = 1$ mA, $t = 2$ min. Here, E_i—that is, discharge voltage is controlled by changing argon gas pressure at constant discharge current.

potential to provide an effective reaction for material synthesis at the gas–liquid interface.

14.3 Synthesis of Nanoparticles Conjugated with Nanocarbons

The novel gas–liquid interfacial plasmas can be utilized for the synthesis of various kinds of nano-bio composite materials as shown in Figure 14.5a. We previously reported the formation of DNA encapsulated carbon nanotubes using a plasma ion irradiation method [14] and controlled the electrical properties of carbon nanotubes by changing the base sequence of the encapsulated DNA [15]. In the following sections, we discuss synthesis of metal nanoparticles using plasma irradiation of liquid electrodes [16,17] and conjugation of the metal nanoparticles with carbon nanotubes [18] to control particle size, interparticle distance, and optical and electrical properties of the hybrid material. Furthermore, we show that nanoparticles can be conjugated with DNA and inserted into the carbon nanotubes for applications in biosensors, drug delivery system, gene therapy, and so on.

Using ion irradiation, gold (Au) nanoparticles are synthesized in an ionic liquid by reducing Au chlorides such as $HAuCl_4$, as schematically shown in Figure 14.5b. Figure 14.6 shows transmission electron microscopy (TEM) images of the Au nanoparticles synthesized in (a) A-mode and (b) B-mode for $P_{gas} = 60$ Pa, $I_D = 1$ mA, and $t = 40$ min. In both cases, the Au nanoparticles can be formed; however, it is found that in A-mode the average diameter of

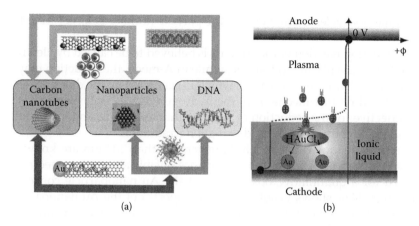

(a)   (b)

FIGURE 14.5
(a) Nano-bio conjugates composed of metal nanoparticles, carbon nanotubes, and DNA. (b) Synthesis approach for gold (Au) nanoparticles using a gas–liquid interfacial plasma.

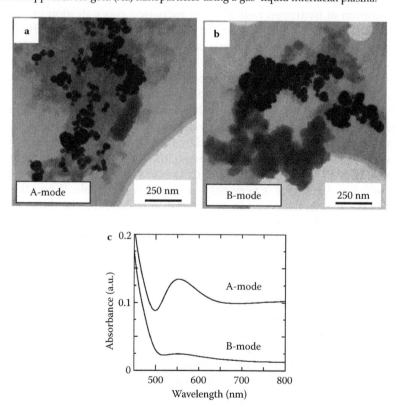

FIGURE 14.6
Transmission electron microscopy (TEM) images of Au nanoparticles synthesized in (a) A-mode and (b) B-mode. (c) Ultraviolet-visible (UV-vis) absorption spectra of Au nanoparticles. $P_{gas} = 60$ Pa, $I_D = 1$ mA, $t = 40$ min.

the Au nanoparticles is smaller and the particle number is larger than that in B-mode.

The reduction reaction of the Au ions is believed to be caused by electrons injected from the plasma in B-mode, while in A-mode, the reduction may be caused by the hydrogen radical H^*, which is generated by the dissociation of the IL. Based on this mechanism, the hydrogen radical is considered to be more effective for the reduction of Au ions than electrons, and efficient Au nanoparticle synthesis is realized using ion irradiation.

Because Au nanoparticles with diameter less than 100 nm are known to exhibit localized surface plasmon resonance, visible absorption spectra are obtained for a quantitative observation of the Au nanoparticle concentration. Figure 14.6c shows visible absorption spectra of the Au nanoparticles synthesized by an Ar plasma in A-mode and B-mode. The absorption peak appears around 550 nm, corresponding to the Au plasmon resonance, and the absorption-peak intensity in A-mode is obviously larger than that in B-mode. Ar ions with high energy can penetrate deep into the IL, promoting the generation of hydrogen radicals. The increased concentration of hydrogen radicals may reduce Au ions more effectively in A-mode than B-mode. The rate of Au nanoparticle synthesis could be controlled by the irradiation energy of inert gas ions such as Ar.

To synthesize well-dispersed nanoparticles (unagglomerated and separated), we attempted to make Au nanoparticles supported on functionalized single-walled carbon nanotubes (f-SWNTs). The SWNTs are dispersed in a new kind of ionic liquid (2-hydroxyethylammonium formate) that consists of carboxyl groups, and the IL is irradiated by the plasma. Using A-mode, high-energy ions dissociate the IL and the dissociated carboxyl groups bond to the surface of the SWNTs. This functionalization of the SWNTs using the gas–liquid interfacial plasmas is very easy, fast, and controllable. When the Au chloride ($HAuCl_4$) is dissolved in the IL with the f-SWNTs, the Au chloride is reduced by the IL and the Au nanoparticles are selectively synthesized on the carboxyl groups at the surface of the SWNTs. Because the density of carboxyl groups on the SWNTs can be controlled by discharge parameters, such as irradiation energy, flux, time, and so on, the density of Au nanoparticles can also be controlled.

Figure 14.7 presents transmission electron microscope (TEM) images of the Au nanoparticles synthesized on f-SWNTs, which have been treated in the IL by plasma irradiation for (b) $t = 1$ min and (c) $t = 10$ min. TEM image of the original f-SWNTs not treated by plasma irradiation are also presented as a reference in Figure 14.7a. It is found that monodispersed Au nanoparticles are synthesized on f-SWNTs when the SWNTs are treated by plasma irradiation, while only a few Au nanoparticles are observed on the f-SWNTs in the absence of plasma irradiation. In addition, the distance between the Au nanoparticles decreases with increasing irradiation time. This result indicates that the distance between the Au nanoparticles can be controlled by functionalization of the SWNTs using ion irradiation.

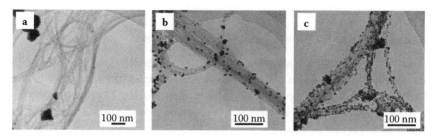

FIGURE 14.7
Transmission electron microscopy (TEM) images of Au nanoparticles synthesized on functionalized carbon nanotubes. (a) No plasma treatment, (b) plasma treatment for $t = 1$ min, (c) plasma treatment for $t = 10$ min. A-mode, $P_{gas} = 60$ Pa, $I_D = 1$ mA.

To synthesize size-controlled nanoparticles, we attempt to make Au nanoparticles inside carbon nanotubes [18]. Bundled carbon nanotubes are impregnated with Au chloride dissolved in the IL and exposed to the plasma. Here, we use B-mode because we need to prevent the carbon nanotubes from being damaged by high-energy ion irradiation. As a result, the Au chloride is reduced inside carbon nanotube bundles by electron irradiation at relatively low energy. The carbon nanotubes can inhibit agglomeration of the nanoparticles due to their small inner space allowing Au nanoparticles with uniform size distribution to be synthesized (Figure 14.8c). Figures 14.8a and 14.8b present TEM images of Au nanoparticles synthesized by irradiating carbon nanotubes for (a) $t = 10$ min and (b) $t = 60$ min. We find that monodispersed and small-sized (<2 nm) Au nanoparticles are synthesized by controlling the gas–liquid interfacial plasma. Furthermore, because the particle size for $t = 60$ min is almost the same as that for $t = 10$ min, the nanoparticles are believed to be synthesized inside the bundled carbon nanotubes.

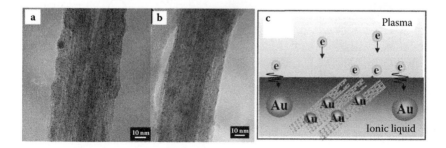

FIGURE 14.8
Transmission electron microscopy (TEM) images of Au nanoparticles synthesized inside bundled carbon nanotubes for irradiation times of (a) $t = 10$ min and (b) $t = 60$ min. $P_{gas} = 60$ Pa, $I_D = 1$ mA. (c) Au nanoparticles inside bundled carbon nanotubes.

14.4 Fabrication of Novel Nanobiomaterials

We synthesized water-soluble Au nanoparticles functionalized with DNA by using pure water and a pulsed power source instead of an ionic liquid and DC power source, respectively [19]. Single-stranded DNA with 30 guanine bases (denoted as G_{30}) is used as a stabilizing agent. The Ar gas pressure P_{gas} is 40 kPa. Figure 14.9 shows (a) TEM images for products synthesized at different G_{30} concentrations and (b) visible absorption spectra of the synthesized nanoparticles. The Au nanoparticles (~18 nm) synthesized without G_{30} precipitate quickly after synthesis. Interestingly, when G_{30} is used, the resultant water-soluble products consist of small-sized Au nanoparticles (~7 nm) (G_{30} concentrations less than 1 µM) or agglomerates (~31 nm) of the small-sized Au nanoparticles (~10 nm) (G_{30} concentrations of 1.75 µM), depending on the G_{30} concentration. The visible absorption spectra confirm that the absorbance peak shifts to longer wavelength when agglomeration of Au nanoparticles occurs.

It is possible to ascribe the solubility and assembly of synthesized Au nanoparticles to the presence of G_{30}. The interaction between DNA and Au nanoparticle surface has been well documented. Electrostatic binding occurs between the Au nanoparticle surface and phosphate groups of DNA as well as combinations of other DNA bases [20]. One G_{30} molecule can bind with multiple Au nanoparticles because it has many bases and phosphate groups. The conjugation of DNA on the Au nanoparticle surface makes the Au nanoparticles water soluble. Au nanoparticles may then bind to each other when the DNA molecules conjugating with Au nanoparticles attract one another. Therefore, a high G_{30} concentration in the liquid leads to assembly (i.e., agglomeration) of small-sized Au nanoparticles.

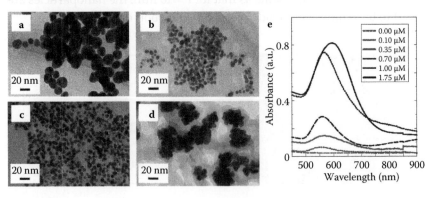

FIGURE 14.9
Transmission electron microscopy (TEM) images of Au nanoparticles synthesized with DNA at concentrations of (a) 0, (b) 0.35, (c) 0.7, (d) 1.75 µM, and (e) ultraviolet-visible (UV-vis) absorption spectra of DNA-nanoparticle conjugates.

The size of Au nanoparticles increases from 6.7 nm to 10.7 nm as the DNA concentration increases from 0.7 µM to 1.75 µM. This slight increase of Au nanoparticle size with increasing DNA concentration is explained as follows. When DNA is used in the synthesis process, the DNA molecules act as nucleation centers in the solution of $HAuCl_4$, and as a result, the number of initial Au nuclei reduced from Au ions is proportional to the number of DNA molecules in solution. Au nanoparticles then form from two parallel mechanisms (i.e., the continuous reduction of Au ions and the coalescence of initial Au nuclei). Therefore, higher DNA concentrations may lead to larger diameter Au nanoparticles.

We propose that DNA-Au nanoparticle conjugates could be encapsulated into or intercalated between carbon nanotubes by applying a DC electric field. Alternatively, because Au nanoparticles can be moved optically, the DNA-Au nanoparticle conjugates could be manipulated by light and inserted into carbon nanotubes. Potential applications of these conjugated materials composed of metal nanoparticles, carbon nanotubes, and biomolecules include highly efficient gas sensors, novel biocatalysts, solar cells, and drug delivery systems, to name a few.

References

1. Locke, B. R., Sato, M., Sunka, P., Hoffmann, M. R., and Chang J. -S. 2006. *Ind. Eng. Chem. Res.* 45: 882.
2. Bruggeman, P., and Leys, C. 2009. *J. Phys. D: Appl. Phys.* 42: 053001.
3. Hieda, J., Saito, N., and Takai, O. 2008. *J. Vac. Sci. Technol. A* 26: 854.
4. Torimoto, T., Okazaki, K., Kiyama, T., Hirahara, K., Tanaka, N., and Kuwabata, S. 2006. *Appl. Phys. Lett.* 89: 243117.
5. Meiss, S. A., Rohnke, M., Kienle, L., Zein El Abedin, S., Endres, F., and Janek, J. 2007. *Chem. Phys. Chem.* 8: 50.
6. Richmonds, C., and Sankaran, R. M. 2008. *Appl. Phys. Lett.* 93: 131501.
7. Xie, Y. B., and Liu, C. J. 2008. *Plasma Process. Polym.* 5: 239.
8. Baba, K., Kaneko, T., and Hatakeyama, R. 2007. *Appl. Phys. Lett.* 90: 201501.
9. Kaneko, T., Baba, K., and Hatakeyama, R. 2009. *J. Appl. Phys.* 105: 103306.
10. Rogers, R. D., and Seddon, K. R. 2003. *Science.* 302: 792.
11. Kaneko, T., and Hatakeyama, R. 2011. *Plasma Sources Sci. Technol.* 20:034014.
12. Sloutskin, E., Ocko, B. M., Tamam, L., Kuzmenko, I., Gog, T., and Deutsch, M. 2005. *J. Am. Chem. Soc.* 127: 7796.
13. Chapman, B. 1980. *Glow Discharge Processes—Sputtering and Plasma Etching.* New York: John Wiley and Sons, pp. 113–114.
14. Okada, T., Kaneko, T., and Hatakeyama, R. 2006. *Chem. Phys. Lett.* 417: 288.
15. Kaneko, T., and Hatakeyama, R. 2009. *Appl. Phys. Exp.* 2: 127001.
16. Baba, K., Kaneko, T., and Hatakeyama, R. 2009. *Appl. Phys. Express* 2: 035006.

17. Kaneko, T., Baba, K., Harada, T., and Hatakeyama, R. 2009. *Plasma Process. Polym.* 6: 713.
18. Baba, K., Kaneko, T., Hatakeyama, R., Motomiya, K., and Tohji, K. 2010. *Chem. Commun.* 46: 255.
19. Chen, Q., Kaneko, T., and Hatakeyama, R. 2010. *J. Appl. Phys.* 108: 103301.
20. Storhoff, J. J., Elghanian, R., Mirkin, C. A., Letsinger, R. L. 2002. *Langmuir.* 18: 6666.

15

Assembly and Self-Organization of Nanomaterials

Amanda Evelyn Rider
Kostya (Ken) Ostrikov

CONTENTS

15.1 Introduction

The increasing interest in nanoscience and nanotechnology has prompted intense investigations into appropriate fabrication techniques. Self-organized, bottom-up growth of nanomaterials using plasma nanofabrication techniques[1–10] has proven to be one of the most promising approaches for the construction of precisely tailored nanostructures (i.e., quantum dots,[11–13] nanotubes,[14–17] nanowires,[18–20] etc.) arrays. Thus the primary aim of this chapter is to show how plasmas may be used to achieve a high level of control during the self-organized growth of a range of nanomaterials, from zero-dimensional quantum dots (Section 15.2) to one- and two-dimensional nanomaterials (Section 15.3) to nanostructured films (Section 15.4).

This introductory section will begin with a brief discussion on the differences between self-assembly and self-organization (Section 15.1.1), will continue with an introduction to plasma nanoscience and complex systems (Section 15.1.2) and conclude with a description of how plasmas may be used to influence the self-organized growth of nanomaterials. An outline is provided for the rest of the chapter (Section 15.1.3).

15.1.1 What Is the Difference between Self-Assembly and Self-Organization?

Given the variety of fields that touch upon nanoscience, there is understandable confusion about what the difference is between *self-assembly* and *self-organization*. The terms *self-assembly* and *self-organization* are often used interchangeably in the scientific literature or take on different meanings depending on the background of the scientist using them.[21] For example, Whitesides (a chemist by background) and colleagues[22] avoided using the term *self organization* and instead defined the terms *static self-assembly* and *dynamic self-assembly* for systems that are at equilibrium and do not dissipate energy and systems that do dissipate energy and are not at equilibrium, respectively. On the other hand, Jones (a physicist by background)[23] designates *self-assembly* as a process occurring at equilibrium and *self-organization* as a process occurring away from equilibrium, where pattern formation is driven by energy input and dissipation.[23]

There are, of course, finer points to the distinctions, but to avoid straying too far from our focus on plasma-assisted growth of nanomaterials, we will instead refer the interested reader to a paper by Bensaude-Vincent[24] that discusses the various definitions and distinctions for self-organization and self-assembly at greater length. Throughout this chapter, the term *self-organization* will be used—the reasoning being that we consider the growth of nanomaterials via a low-temperature, thermally nonequilibrium, plasma-generated influx[25] to be a *nonequilibrium process*. In this case a stream of material at variable influx rates and ratios is directed at the deposition surface, and there is thus constant energy input and dissipation. The individual species deposited from the plasma will *self-organize* into a distinct pattern such as a nanostructure array. The characteristics of these arrays can be tailored by adjusting both plasma parameters and surface conditions.

15.1.2 Plasma Nanoscience and Complex Systems

Plasma nanoscience[26–30] is a relatively new research field that incorporates elements of plasma physics, nanoscience, materials science and engineering, physical chemistry and surface science, among others. It is focused on demonstrating the benefits of using an ionized gas during nanostructure growth and, subsequently, during nanodevice fabrication, with the ultimate aim being to show that plasmas have the potential to be used as an *all-in-one nanofabrication environment*. The simple reason why plasmas are useful during nanomaterial growth is that an ionized gas (plasma) environment influences the ways in which nearby species interact with each other—this influence can be manipulated to create an environment whereby relevant species will self-organize into tailored nanostructures with a reasonably high degree of *determinism*.

A deterministic fabrication process can be summed up as a process where it can be predicted with a great degree of certainty what type of nanostructure will be produced, from the type of material delivered to a surface and the manner in which it is deposited.[21] The framework we will use in this chapter is the notion of building and working units.[27] Building units (BUs) are species (atomic/ionic/molecular, etc.) that make up the nanostructured material, whereas working units (WUs) are species that prepare the surface for deposition (i.e., an Ar^+ WU activates and heats the deposition surface, atomic hydrogen WU will passivate dangling bonds, etc.[31]). Production rates of individual BUs and WUs in a complex plasma may be controlled by carefully modifying process parameters such as reactor pressure, input power, feed gas composition, and so forth. (This is discussed at greater length in Section 15.2.1 and Ostrikov, K., Yoon, H. J., Rider, A. E., and Ligatchev, V. 2007. Reactive species in $Ar+H_2$ plasma-aided nanofabrication: Two-dimensional discharge modeling. *Phys. Scr.* 76:187–195. Ostrikov, K., Yoon, H. J., Rider, A., and Vladimirov, S. V. 2007. Two-dimensional simulation of nanoassembly precursor species in $Ar+H_2+C_2H_2$ reactive plasmas. *Plasma Process. Polym.* 4:27–40. Ren, Y. P., Xu, S., Rider, A. E., and Ostrikov, K. 2011. Made-to-order nanocarbons through deterministic plasma nanotechnology. *Nanoscale* 3: 731–740. Complex systems science is concerned with understanding how the interaction of many seemingly disordered, disparate units/processes on a small scale can work together and bring about a degree of order on a larger scale.[21] This definition works equally well for the self-organized growth of nanomaterials.

There are two known directions[29,30] regarding self-organization in a plasma-based environment, namely,

1. The formation of functional BU and WUs in complex plasmas: Production of useful species (from atoms/ions to complex agglomerates) controlled by modification of the plasma process parameters
2. The plasma-surface interaction during self-organized nanomaterial growth: Individual building units organizing themselves into

useful surface-supported nanostructures mediated by the surface
conditions and the surrounding plasma environment

In each case, one can find several excellent examples of complex systems,
such as nonequilibrium plasma–solid systems. In this chapter, for the most
part, we will focus on the latter direction: surface-based self-organization of
nanostructures controlled by the plasma conditions.

15.1.3 Plasmas and Self-Organization: How Can We Control It?

Now that a definition of self-organization has been presented, an explanation
is needed of how such a myriad of complex interactions may be controlled
in a plasma environment to ensure that not only is a precisely controlled
nanostructure obtained, but also that the desired nanopatterned array and
eventually, the intended nanoarchitecture or element of a nanodevice may
be attained. Basically, one needs to know what control knobs in plasma processing are the most effective to manipulate in order to precisely tailor the
self-organized growth of nanomaterials.

Details about plasma-assisted self-organized growth of specific nanostructures (from 0D to nanostructured films) will be presented in Sections 15.2
through 15.4. In this section, we will focus on the existing approaches[27,34,35]
for nanoisland growth in low temperature, nonequilibrium plasmas.
These approaches can be relatively easily extended to account for other
nanostructures.

There are two ways one can tackle controlling self-organized nanomaterial growth by using a plasma: (1) adjust the influxes of building materials
and (2) adjust the surface conditions such as temperature or localized heat-
ing/cooling. These two approaches, however, are not mutually exclusive.
Adjusting the plasma influx will also alter the surface conditions (by influencing the surface temperature and the energy barrier for surface diffusion
as discussed later in this section), whereas adjusting the surface conditions
will influence the plasma influx (e.g., through "back-flux"[36]—during surface
reactions, some radical species may desorb and undergo recombination,
returning to the plasma bulk to undergo further gas-phase reactions[33,36]). To
achieve precisely controlled nanomaterials, one requires an approach that
makes these adjustments such that an appropriate balance between supply
and consumption of BUs is struck, as this balance will influence the way in
which nanostructure growth will proceed.

Levchenko et al.[35] examined the differences in the growth of Ge nanoislands in diffusion-controlled and supply controlled regimes. Assuming that
the main control knobs in nanoisland growth are surface temperature and
the rate of deposition, Levchenko et al. found that there were two different
growth routes that could occur, depending on the balance between the surface temperature and the BU influx rate. In this case, there would be either an
oversupply of BUs with respect to the surface diffusion rate (thus *diffusion-*

controlled growth) or the diffusion rate would be high and growth would be controlled by the supply of BUs (*supply controlled*).[35] They found that a diffusion-controlled mode (via lower temperatures and higher deposition rates, typical of low-temperature plasma processing) resulted in self-organized growth of dense patterns of small quantum dots (QDs) and enabled nonuniform QD growth as a result of avoiding the Strankski-Krastanov fragmentation.[35] This illustrated the idea that although plasmas were not typically considered the most conventional growth route for QDs, by properly understanding the interplay between the influx parameters and surface conditions, they could be effectively used to create dense, size-uniform QD patterns.

Consider the equation for surface diffusion speed, v_d, of an adatom:

$$v_d = \lambda v_o \exp[-\varepsilon_d / k_B T]$$

where λ is the lattice parameter, v_0 is the lattice oscillation frequency, and k_B is Boltzmann's constant. Clearly the parameters ε_d and T (surface diffusion activation energy and surface temperature, respectively) will have the most impact on v_d as they are in the exponent. If, as discussed in the case above, the self-organization of building units (such as adatoms) into a nanostructure is driven by surface diffusion (as mentioned, the diffusion-controlled regime is typical of low-temperature plasma processing[35]), then one of the most logical ways to influence self-organization is to adjust ε_d and T to ensure that the yield of desired nanostructures is enhanced. By precisely tailoring the plasma parameters such as the ionized influx and the surface bias, the surface temperature and the surface diffusion activation energy may be modified.

For example, for small nanostructures in a plasma environment, the energy barrier for surface diffusion, ε_d, is reduced by an amount,

$$\delta\varepsilon_d = \left|\partial E / \partial r\right| \lambda \tilde{p} \tag{15.1}$$

where $\tilde{p}$ is the adatom dipole moment, and E is the electric field.[34] For larger nanostructures, both the plasma sheath–surface electric field, E_λ, and the microscopic electric field, E_s (arising from the QDs becoming charged as a result of being in a plasma environment[34]) will influence ε_d. In this case, the change in ε_d may be expressed as in Ostrikov et al.[34]:

$$\delta\varepsilon_d = \frac{\lambda}{k_B T}\frac{\partial E}{\partial r}\left(\tilde{p} + \alpha\frac{\partial\phi}{\partial r}\right) = \frac{\lambda}{k_B T}\frac{\partial^2\phi}{\partial r^2}\left(\tilde{p} + \alpha\frac{\partial\phi}{\partial r}\right) \tag{15.2}$$

where ϕ is the potential and α is the polarizability of the adatom.[34]

In addition to influencing the energy barrier for surface diffusion, the presence of an ionized influx results in an elevation of surface temperature, T. Consider that ions possess kinetic energy—part of that kinetic energy results from the ion being accelerated toward the surface, through the plasma sheath, due to the application of a surface bias. The ion will hit the surface, giving up a portion of its energy to the substrate, and this energy is then dissipated throughout the top-most surface layers of the substrate as heat via lattice vibrations.[37] The resulting surface temperature may then be expressed as $T + \delta T$, with δT determined as[37]

$$\delta T = \frac{eU_s}{k_B} \tau^{3/4} \varphi^{3/4} \tag{15.3}$$

where e is the charge on an electron, U_s is the applied bias, φ is the ion influx rate, and $\tau = (1/\upsilon_0)\exp[\varepsilon_d/k_bT]$. Note that the surface temperature increase due to exposure to a plasma environment will vary depending on the type of plasma source and its configuration. More involved estimates (i.e., based on calculations of energy balance) of the increase in T due to a variety of plasma sources are available in the works of Kersten et al.[38,39] Therefore, by manipulating the plasma parameters including the ionized influx and applied bias, one can achieve larger $\delta \varepsilon_d$ and T which lead to increased υ_d, results in a higher adatom mobility, more frequent collisions, and eventually *faster self-organized QD growth* in a plasma environment as opposed to in a neutral gas.[40] This effect will be discussed in greater detail in Section 15.2.2.

So the questions we need to answer before proceeding with plasma-assisted self-organized growth of nanomaterials are as follows:

1. What type of material is being delivered, and how is it being delivered?
   a. What will be the deposition rate? Will a diffusion-controlled growth route or a supply controlled route dominate?
   b. What are the relative concentrations of BUs and WUs in the influx?
   c. How much of the material will be ionized, and what effect will this have on the diffusion rate?
2. What type of deposition surface is being considered?
   a. Is the surface stepped, are there any defects/dislocations?
   b. What sort of lattice mismatch does it have with the material that is being deposited (i.e., Ge and Si have a mismatch of 4%)? This mismatch will affect the growth mechanism.[41]
   c. What is the target temperature for the top-most layers of the surface? (How much heat will be supplied externally and how much will result from exposure to a plasma?)

Answering these questions (prior to fabrication) through a combination of numerical modeling of the plasma discharge and surface processes and pilot experiments, will help one to avoid expensive (both in terms of time and resources) trial-and-error-based methods and ultimately set up a procedural blueprint (discussed at greater length in Section 15.4.1) for choosing the most appropriate setup and process parameters for deterministic plasma-assisted self-organized nanomaterial growth and eventually nanodevice fabrication.

The rest of this chapter will proceed as follows: Section 15.2 will examine plasma-assisted self-organization of 0D nanomaterials, one- and two-dimensional nanomaterials will be briefly discussed in Section 15.3, Section 15.4 will cover nanostructured films and in particular present a case study on diamond-like nanocarbons showing how a framework for nanocarbon growth with control at the electronic level can be extended to form a procedural blueprint for plasma-assisted self-organized nanomaterial growth from materials design to device fabrication. Section 15.5 will first discuss the benefits of a plasma-assisted growth route in terms of cost, control, and commercial practicality and then consider the ways in which plasma-grown nanomaterials can be integrated into working devices, finally returning to the emphasis on the plasma as an all-in-one growth environment. This will be followed by a conclusion and outlook for future research on plasmas and self-organization in Section 15.6.

15.2 Zero-Dimensional Nanomaterials

Zero-dimensional (0D) nanostructures are nanomaterials where electron motion is confined in all three dimensions. In this section we will only very briefly cover nanoparticles and functional units generated in a complex plasma (Section 15.2.1), instead focusing on self-organized growth of surface-bound quantum dots and nanoislands.

15.2.1 Plasma-Generated Nanoparticles and Building Blocks

A complex plasma[42] contains a wealth of reactive species that can interact with each other to produce either functional nanoassemblies or undesired material. The key to obtaining useful functional species is to understand how varying plasma parameters such as operating pressure, partial pressures of feedstock gases, degree of ionization, input power, and so forth, will influence the yield of specific species in a discharge. The most apparent way to do this is through optical emission spectroscopy and quadrupole mass spectroscopy (for details on diagnostic methods in complex plasmas, please see Vladimirov et al.[42]). However, an alternative is to conduct fluid modeling beforehand, before even turning on the plasma discharge. By constructing fluid models that take into account the influence of the plasma parameters on the number densities and fluxes of species in the discharge, the yield of

BUs and WUs of interest for the growth of specific nanostructures may be optimized, and some of the expense associated with trial-and-error experiment only based methods may be negated.

This approach was taken in Ren et al.[33] (see also Section 15.4.1) for the growth of diamond-like nanocarbons where a combination of fluid modeling and experimental measurements enabled a link to be drawn between the plasma composition (specifically the ratio of sp^3- to sp^2-hybridized carbon-containing species) as a result of power and source gas influx variation and the characteristics of the deposited film. Similar previous work focused on describing how variations in plasma process parameters such as reactor pressure and partial pressure of the source gases, and so forth, influenced the yield of BUs and WUs involved in the self-organized growth of nanoparticles,[43] carbon nanotubes (CNTs),[31,32] and silane clusters.[44]

15.2.2 Surface-Bound Quantum Dots via Plasma-Assisted Self-Organization

It has been noted that plasmas are not typically considered to be suitable growth environments for the self-organized growth of delicate, ultra-small quantum dot nanopatterns due to the relatively high material influxes.[40] However, we showed in previous work[37,40–41] that by carefully adjusting the plasma influx and its influence on surface conditions, it is possible to grow highly tailored, self-organized quantum dot nanopatterns. In the following few examples, we will show that the size, elemental composition, and internal structure of a range of nanostructures can be fine-tuned by manipulating not only the influx rate, influx ratio, and substrate temperature, but also plasma-related effects such as the increased surface temperature and reduced surface diffusion activation energies (discussed in Section 15.1.3), through careful adjustment of the substrate bias and ionized influx. Here we will focus on the self-organized growth of Ge, Si, $Si_{1-x}C_x$, GaAs, and InSb QDs on Si(100) surfaces.

As noted in Section 15.1.3, stress-induced fragmentation of top-most surface layers (Stranski-Krastanov growth mode[35]) typically results in highly unpredictable QD nucleation sites and broad size-non-uniform QD patterns, which is a disadvantage when it comes to implementing the QD arrays in technological devices. We simulated the growth of ultra-small Ge/Si quantum dot (QD) nuclei (~1 nm) produced through deposition of atom-only influxes (representative of a neutral-gas growth route) and size-non-uniform cluster influxes (representative of a plasma-assisted growth route).[41] We found that the quantum dot nuclei produced via the plasma-based influx were more size-uniform than those produced via the atom-only influx route.[41]

This effort was extended in Seo et al.[45] to consider the growth of size-selected Si QDs on SiC(0001) via ion-assisted self-organization.[45] A combination of tailored nanocluster influxes, incoming fluxes of ionized species (low charge state Si^{1+} and high charge state Si^{3+}) with varying substrate temperature and applied bias was used. It was found that the highest percentage of the surface coverage by 1 and 2 nm Si QDs was achieved in a low temperature range (227 to 327°C) using Si^{1+} and the highest voltage considered (−20 V).[45] The changes in surface diffusivity with charge state and applied bias were explained using the ionization energy approximation theory.[46,47] Briefly, regarding charge state: Si^{1+} has a smaller ionization energy than Si^{3+}, and thus has a larger reduction in ε_d on the surface (see Seo et al.[45] and Arulsamy[47] for more rigorous details) and higher surface diffusivity. Regarding applied bias, due to directional polarization (a result of the electric field), there is a repulsive force between the ion and the substrate. The Si^{1+} experiences a stronger repulsive force than Si^{3+} (the electrons in Si^{1+} are more easily polarizable), which may be enhanced further through application of a larger bias. Ions are thus in a low charge state, with the higher applied bias exhibiting higher surface diffusivity, which will result in increased yields of self-organized, size-selected Si QDs on SiC(0001).[45] These results demonstrate that in addition to a reduction in the energy barrier for adatom surface diffusion and ion-induced surface heating, plasma-based self-organization can be tailored further by careful choice of the combination of ion charge state and surface bias. It is also notable that through modification of U_s and charge states, the yield of size-selected QDs was enhanced in relatively low temperature ranges.[45] The low-temperature growth of Si QDs in amorphous matrices[48,49] (e.g., see Figure 15.1) is of particular importance for

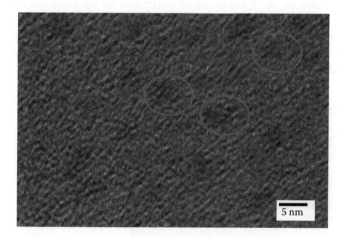

FIGURE 15.1
Transmission electron micrograph (TEM) of Si nanodots (three examples are circled) in an amorphous Si matrix grown via plasma-enhanced chemical vapor deposition (PECVD). (Figure courtesy of Q. J. Cheng, S. Xu, and K. Ostrikov.). (For more details about the growth of Si QDs via PECVD, please see Cheng, Q. J., Tam, E., Xu, S. Y., and Ostrikov, K. 2010. Si quantum dots embedded in an amorphous SiC matrix: Nanophase control by non-equilibrium plasma hydrogenation. *Nanoscale* 2:594; and Cheng, Q. J., Xu, S. Y., and Ostrikov, K. Single-step, rapid low-temperature synthesis of Si quantum dots embedded in an amorphous SiC matrix in high-density reactive plasmas. *Acta Mater.* 58:560.)

the next generation of Si-based photovoltaics.[50] This possibility is of particular interest in widening the range of substrates that may be processed through plasma-based techniques.[45]

Although QD size control is important, binary quantum dots, often in demand in applications ranging from biosensors to solar cells, present an additional complication in attempts to control their optoelectronic and mechanical properties as these properties will also be influenced by their elemental composition and internal structure.[37,40,51] In some cases, tuning elemental composition and internal structure is preferable to manipulating particle size. As indicated in Section 15.1.3, two of the main control knobs available are the substrate temperature and the influx. In the case of binary QDs, there is a balancing act between influx rate, influx ratio (i.e., Ψ_A/Ψ_B, where the QD may be expressed as $A_{1-x}B_x$), and substrate temperature. This balance is even more delicate, in particular, if one attempts to grow (see Figure 15.2)

- Compositionally graded QDs
- Core-multishell QDs
- Selected composition QDs

In a previous paper,[37] we numerically examined the plasma-assisted growth of tailored binary $Si_{1-x}C_x$ QDs on Si(100) surfaces by varying the influx rate, influx ratio, the percentage of ions in the influx, surface bias, and the substrate temperature. It was found that by using time-variable influx ratios and influx rates, one could achieve a good level of control over core-multishell structure. Moreover, by carefully modifying surface bias and ion influx it

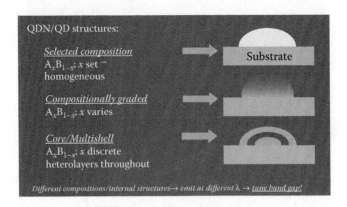

FIGURE 15.2
The range of quantum dot/quantum dot nuclei (QD/QDN) structures considered in Rider et al. (Rider, A. E., Ostrikov, K., and Levchenko, I. 2008. Tailoring the composition of self-assembled $Si_{1-x}C_x$ quantum dots: Using plasma/ion-related controls. *Nanotechnology*. 19: 355705), where "A" and "B" are the elements that make up the binary QD. Figure from Rider.[21]

was possible to achieve a selected composition earlier in the QD structure (i.e., a more homogeneous internal structure of the QD due to localized surface heating—increased T as discussed in Section 15.1.3) and to fine-tune the composition gradient. Additionally, by using ions in the influx, larger QDs could be obtained on average.[37] Thus manipulating ion influx can also be used as a way to tailor the average size of the QDs.

In the papers discussed above, the benefits of using ions during self-organized growth of nanomaterials were shown, but the effects of different ion sources on self-organized QD growth were not directly compared. This was done in a later paper,[40] where the self-organized growth of GaAs and InSb QDs on Si(100) in a neutral gas scenario was numerically compared with ionized gas scenarios. (A localized ionization source case and a background plasma case were considered.) It was found that not only was a stoichiometric QD composition (group III: group V = 1:1) achieved faster (and earlier on in the QD) in an ionized gas environment, compared to a neutral environment, but the stoichiometrization time was shorter for the background plasma case than for the localized ionization source case.[40] This result is demonstrated in Figure 15.3. This is important as the self-organization of nanomaterials is heavily influenced by the initial stages of growth, thus achieving a stoichiometric composition for III-V QDs at the very earliest opportunity is crucial in avoiding defects in more evolved QDs.[40]

Through these papers it has been shown that despite early misgivings about the suitability of plasmas as QD growth environments, plasmas are versatile tools for the self-organized growth of tailored QD nanopatterns.

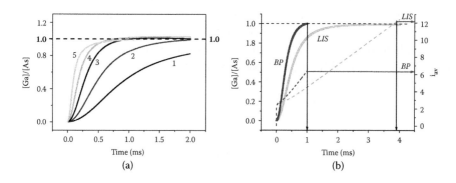

FIGURE 15.3
Results of numerical simulation of GaAs growth on Si(100) for (a) a neutral gas environment [Series 1 and 2] versus a background plasma [Series 3 through 5] environment and (b) localized ionization source [LIS] versus Background plasma [BP] source. Figure has been reprinted from Rider, A. E., and Ostrikov, K. 2009. The path to stoichiometric composition of III-V binary quantum dots through plasma/ion-assisted self-assembly. *Surface Science.* 603: 359–368. Copyright (2009), with permission from Elsevier.

15.3 One- and Two-Dimensional Nanomaterials

Much has already been said in previous chapters about the growth of one- and two-dimensional (1D and 2D, respectively) nanomaterials such as carbon nanotubes, nanowires, nanotips, and graphene; hence, in this section we will only very briefly touch on our recent work on carbon nanofibers (Section 15.3.1) and carbon nanowalls (Section 15.3.2), focusing on the benefits of a low-temperature plasma route in terms of what it means for self-organization.

15.3.1 Carbon Nanofiber Growth in Reactive Plasmas

In this section we will briefly discuss recent modeling efforts (incorporating plasma sheath, nanostructure growth, and thermal models) on CNF growth in $Ar+H_2+C_2H_2$ reactive plasmas[52,53] which examine the effect of plasma-based controls such as gas pressure, ion temperature, the C_2H_2:H_2 supply ratio, substrate potential, electron temperature and number density on the CNF growth rate, and heating of the catalyst particle, and how these parameters are of differing importance in low- and high-temperature regimes. We will then discuss what this means for plasmas and self-organized growth of carbon nanofibers.

The benefit of growing CNFs in plasmas as opposed to conventional CVD is that low-temperature growth is possible. This effect is due to the dissociation of the gas precursor molecules and plasma-based elementary processes in the gas phase and on the top-most surface of the catalyst particle. It is shown that plasma-related processes substantially increase the catalyst particle temperature, in comparison to the substrate and the substrate-holding platform temperatures.[53] This is important as higher catalyst particle temperature leads to increased surface diffusion rates. The increased surface diffusion rates in turn lead to higher CNF growth rates as follows from Equation (15.4) where the total CNF growth rate R_t is the sum of the surface diffusion rate R_s and bulk diffusion rate R_b[53]:

$$R_t = R_s + R_b = \frac{m_c}{(\rho A_p)}(J_s + J_v) \tag{15.4}$$

where m_c is the mass of a carbon atom, ρ is the CNF mass density, A_p is the surface area of the catalyst particle (which is taken as flat in this approximation, however extending the model to consider hemispherical catalyst particles is relatively simple and was done in a later paper on nanowires[54]),

$$J_s = -2\pi r_p D_s (d\tilde{n}_C / dr)\big|_{r=r_p}$$

is the carbon flux on the catalyst surface (with r_p the catalyst radius, $\tilde{n}_C$ the surface concentration of carbon, D_s the surface diffusion coefficient equal to $D_{s0}\exp(-E_{sd}/k_B T_C)$, where D_{s0} is a constant, E_{sd} is the energy barrier for surface diffusion of carbon atoms, T_C is the catalyst particle temperature[53]) and

$$J_v = \int_0^{r_p} \left(\pi \tilde{n}_C D_b / A_p \right) 2\pi r dr$$

is the carbon flux through the catalyst bulk, where D_b is the bulk diffusion coefficient for C diffusion through the catalyst nanoparticle bulk (equal to $(\upsilon A_p/\pi)\exp(-E_{bd}/k_B T_C)$, where υ is the thermal vibration frequency, E_{bd} is the energy barrier for C diffusion into the catalyst bulk[53]). Both D_s and D_b are functions with T_C, the catalyst temperature in the exponent. Thus, if one increases the catalyst particle temperature (by adjusting the plasma parameters), the CNF growth rate can be enhanced. By linking the bulk plasma parameters to the plasma sheath parameters and subsequently to the surface processes[53] one can find that gas pressure, ion temperature, and $C_2H_2{:}H_2$ supply ratio affect the CNF growth rate in all temperature ranges, whereas the substrate potential, electron temperature, and number density affect CNF growth at low catalyst temperatures.[53] This observation can be used to deterministically choose the plasma process parameters to enhance catalyst particle temperature and thus optimize the self-organized growth of CNFs during plasma-aided nanofabrication.

15.3.2 Carbon Nanowall Growth

Levchenko et al.[55] showed through numerical modeling of carbon nanowall (CNW) growth in plasma and neutral gas environments that more size uniform CNWs result from growth in a plasma environment, with uniformity better for the high-density plasma ($\sim 10^{18} \mathrm{m}^{-3}$ number density) case than for the lower density plasma ($\sim 10^{17} \mathrm{m}^{-3}$ number density) case.[55] This effect has been attributed to the fact that plasma-generated species can be delivered directly to the growing nanowalls, due to the irregular electric field around the CNWs which focuses the ion fluxes.[55] It was also shown that by manipulating the ratio of direct (to CNW) to diffusion (over the surface) fluxes, by changing the plasma density and degree of ionization (such flux tailoring is not possible in a neutral gas), it is possible to control the structure of CNWs even at low temperatures when surface diffusion may be negligible.[55] This observation is important for the self-organized growth of CNWs as it suggests that by deterministically choosing flux ratios, it is possible to grow regular, uniform CNW arrays at different temperature ranges. These calculations are consistent with recent experiments on CNWs conducted at the

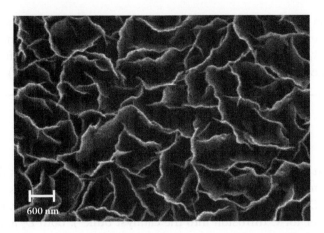

FIGURE 15.4
Micrograph of carbon nanowalls grown at Plasma Nanoscience Centre Australia (PNCA). Figure courtesy of S. Kumar, D. H. Seo and K. Ostrikov.

Plasma Nanoscience Centre Australia (PNCA), an example scanning electron micrographs of carbon nanowalls grown at PNCA is shown in Figure 15.4.

15.4 Nanostructured Films

Both surface conditions and modification of the influx of plasma-generated species within the plasma bulk through manipulation of power and gas inlet can directly affect the properties of nanostructured films. As noted in Section 15.1.2, it is a complex system. What needs to be done is to find a way to come up with a simple recipe-based approach for growing not only nano-structured films, but also exotic nanoassemblies and nanoarchitectures. In other words, we need to find a viable way to get from "controlled complexity to practical simplicity."[56] To fully understand what is happening during plasma-aided deposition of not only thin films, but also nanostructure arrays, one needs a procedural blueprint to take us from materials design through discharge and surface processes modeling to experimental realization.[21] An example of the first steps toward such a blueprint is presented as a case study in Section 15.4.1.

15.4.1 Case study: Diamond-Like Nanocarbons— Toward an Integrated Process Framework for Deterministic Plasma Nanodevice Fabrication

In diamond-like nanocarbons, the ratio of sp^2- to sp^3-hybridized carbons largely determines the optoelectronic and mechanical properties of the

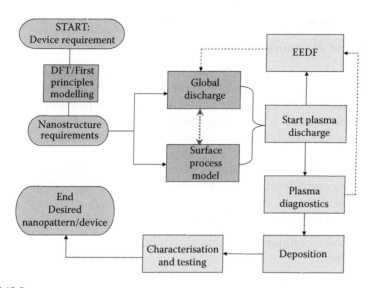

FIGURE 15.5
Procedural blueprint for nanodevice fabrication, extended and generalized from Ren, Y. P., Xu, S., Rider, A. E., and Ostrikov, K. 2011. Made-to-order nanocarbons through deterministic plasma nanotechnology. *Nanoscale* 3: 731–740.

material. Thus a high level of control even at the level of electrons is required during the synthesis of these films. Ren et al.[33] showed that such a level of control could be achieved during inductively coupled plasma CVD of diamond-like nanocarbon films. This control is made possible by modifying the plasma parameters to tailor the electron energy distribution function (EEDF), which will influence the production of ions and neutral radicals in the plasma bulk. These ions and neutral radicals will influence the film composition (i.e., sp^3/sp^2 ratio) and hence the film properties. By clearly elucidating the link between sp^3/sp^2 in the plasma and in the deposited film, it was possible to propose a combinatorial approach (including plasma species modeling and physical experiment) to plasma-aided fabrication of diamond-like nanocarbons.

This approach may be extended (as shown in Figure 15.5) to include materials modeling (i.e., density functional theory [DFT], or the ionization energy approximation theory[46] as a quick, coarse-grained alternative as suggested in our previous work[57]) and device testing that is essential if the nanomaterials are intended to be incorporated into a working device. Such an approach is applicable not only for carbon nanostructures and nanostructured films, but for a range of nanostructures of different materials, including binary and higher compounds.

15.5 Beyond Nanomaterials

We presented a lot of examples of the use of plasmas in the self-organized growth of nanostructures. However, it is important to clearly state why one should use plasmas for nanostructure growth, rather than say, a wet chemistry or a neutral gas route. This decision is something that must be based on the answers to the following questions that should be asked of fabrication techniques:

- Does the fabrication technique provide an acceptable level of control over nanostructure characteristics (i.e., size, shape, composition, internal structure, etc.)?
- Is the method low cost (in comparison to other fabrication techniques)?
- Does the technique have high throughput, efficiency, and deposition rates?
- Can the technique be scaled-up for high production volumes? Is it capable of processing large areas?
- Can the fabrication method be easily integrated into current manufacturing lines?
- What are the safety factors? Is the technique relatively safe to work with (i.e., no toxic reagents, etc.)? Is it "human health friendly"?
- What is the environmental impact of the process?

All of these considerations can be grouped under the categories of

1. Cost
2. Control
3. Commercial practicality
4. Process and nanosafety

In choosing to use a plasma-based nanofabrication technique over alternative methods, we have to balance the *control* with the *cost* with the *commercial practicality* and with the *process and nanosafety* of the technique, as visualized in the Venn diagram (Figure 15.6).

For example, consider atom-by-atom manipulation. Clearly, it has a high degree of control over nanostructure characteristics, but the process is intolerably slow, which makes it commercially impractical. Thermal chemical vapor deposition is another example, it is commercially practical enough. However, fragmentation-induced nucleation of nanoislands leads to unacceptably broad distributions of QD sizes. This is why this technique scores unacceptably low on the control front. Deterministic plasma-based self-organized growth, on the other hand, meets the requirements of comparatively lo- cost, high degree of control, commercial practicality, as well as process and nanosafety.

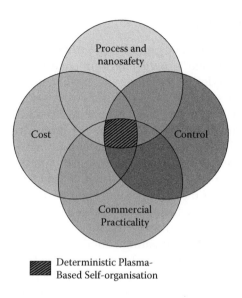

FIGURE 15.6
(See color insert.) Venn diagram for the four main considerations in nanofabrication, deterministic plasma-based self-organization satisfies all criteria.

Throughout this chapter we have shown that low-temperature plasmas offer a number of significant advantages over other self-organized bottom-up growth techniques (e.g., molecular beam epitaxy [MBE], thermal CVD, etc.) and prove more efficient and cost-effective than many contemporary top-down approaches. In particular, there is *a greater degree of controllability* over nanostructure size (both uniformity and selectivity), elemental composition, and internal structure than is possible in neutral-gas based processes. Plasmas are suited for large-area processing and high production volumes. Moreover, as the plasma-based technology is already in use in current semiconductor manufacturing processes, the approaches proposed in this chapter and related works[26,27,33] may be integrated into existing production lines. Additionally, understanding (and deterministically using) the control of energy and matter during plasma processing was highlighted in a recent review[58] as an important step toward matter- and energy-efficient nanofabrication. Plasma processes may be used to turn hazardous gases used during nanofabrication into useful by-products—an efficient handling of waste.[59] Moreover, plasma processing is a relatively benign technique in terms of human health, fabrication is carried out in a confined chamber—there is no significant leakage of radiation, gas, or by-products.[59] It has been noted[25,59] that the use of a vacuum during plasma nanofabrication restricts human exposure to the products of synthesis. Additionally, the surface-bound nanomaterials typically produced via plasma methods represent the lowest safety risk of contemporary nanomaterials.[25,59]

15.5.1 Nanoarchitectures

Before concluding, we will briefly revisit our earlier mention of the suitability of a plasma as an "all-in-one" growth environment (recall Section 15.1.2). This is apparent when considering the variety of exotic nanoassemblies that may be grown in plasma systems—from "nanoflowers" of multiwalled carbon nanotubes to silicon "nanotrees,"[60] but is most marked in the case of the formation of self-organized carbon connections between silver nanoislands [61].

Levchenko et al.[61] considered Ag nanoislands deposited on Si(100) via RF magnetron sputtering with carbon nanowires connections formed as a result of exposure to a Ar+CH$_4$ microplasma at atmospheric pressure. The self-organized growth of the carbon connections was driven mainly by the surface diffusion of carbon which was influenced by the nearby electric field, which could be adjusted by tailoring the microplasma.[61] This work can be viewed as the first step toward creating nanodevices incorporating circuitry, and so forth, using plasma sources. Thus we can see that plasmas have the potential to be an "all in one growth environment," acting as versatile tools[27] from generation of BUs and WUs in a plasma bulk, through to nanostructure nucleation, and further self-organized growth/post processing through to (eventually) device fabrication. It all comes down to knowing when to use which plasma, with which parameters—thus the need for a rigorous procedural blueprint as noted in Section 15.4.1.

15.6 Conclusion/Outlook to Future Research

In this chapter we discussed how low-temperature, nonequilibrium plasmas may be used for the self-organized growth of tailored nanopatterned arrays. We elucidated the benefits of using plasmas during the self-organized growth of 0D, 1D, and 2D nanomaterials as well as nanostructured films. Moreover, the first steps toward establishing a procedural blueprint for a self-organized nanodevice growth from materials design to discharge and surface processes modeling to experimental realization were presented. Developing such approaches is crucial if there is to be an evolution from plasma nanoscience to a fully realized plasma-based nanotechnology.

It was explained why plasmas are promising growth environments in terms of cost, control, commercial practicality and process, and nanosafety considerations. The branching out of plasma nanofabrication from nanomaterials through to exotic nanoassemblies and nanodevices has been briefly discussed and linked to the suitability of plasmas as all-in-one growth environments— or rather versatile tools for all stages of the nanofabrication process.

In terms of future research, the most important direction is what we briefly mentioned in Section 15.4.1—finding a way to get from *controlled complexity to practical simplicity.* In order to do that, a thorough understanding of every single stage of the plasma nanofabrication process is needed so the effect of changing the plasma discharge parameters on the properties of self-organized nanomaterials and eventually, on the response from nanodevices can be predicted through the use of a simple, procedural blueprint for plasma-assisted self-organized growth.

References

1. Melechko, A. V., Merkulov, V. I., McKnight, T. E. et al. 2005. Vertically aligned carbon nanofibers and related structures: Controlled synthesis and directed assembly. *J. Appl. Phys.* 97:041301.
2. Zheng, J., Yang, R., Xie, L., Qu, J. L., Liu, Y., and Li, X. G. 2010. Plasma-assisted approaches in inorganic nanostructure fabrication. *Adv. Mater.* 22:1451.
3. Gresback, R., Holman, Z., and Kortshagen, U. 2007. Nonthermal plasma synthesis of size-controlled, monodisperse, freestanding germanium nanocrystals. *Appl. Phys. Lett.* 91:093119.
4. Hash, D. B., Bell, M. S., Teo, K. B. K., Cruden, B. A. , Milne, W. I., and Meyyappan, M. 2005. An investigation of plasma chemistry for dc plasma enhanced chemical vapour deposition of carbon nanotubes and nanofibres. *Nanotechnology* 16:925.
5. Huang, H., Tan, O. K., Lee, Y. C., Tran, T. D., Tse, M. S., and Yao, X. 2005. Semiconductor gas sensor based on tin oxide nanorods prepared by plasma-enhanced chemical vapor deposition with postplasma treatment. *Appl. Phys. Lett.* 16:163123.
6. Meyyappan, M. 2009. A review of plasma enhanced chemical vapour deposition of carbon nanotubes. *J. Phys. D: Appl. Phys.* 42:213001.
7. Teo, K. B. K., Hash, D. B., Lacerda, R. G. et al. 2004. The significance of plasma heating in carbon nanotube and nanofiber growth. *Nano. Lett.* 4:921.
8. Mariotti, D., and Sankaran, R. M. 2010. Microplasmas for nanomaterials synthesis. *J. Phys. D: Appl. Phys.* 43:323001.
9. Dumpala, S., Safir, A., Mudd, D., Cohn, R. W., Sunkara, M. K., and Sumanasekera, G U. 2009. Controlled synthesis and enhanced field emission characteristics of conical carbon nanotubular arrays. *Diam. Relat. Mater.* 18:1262.
10. Kato, T., Jeong, G. H., Hirata, T., Hatakeyama, R., Tohji, K., and Motomiya, K. 2003. Single-walled carbon nanotubes produced by plasma-enhanced chemical vapor deposition. *Chem. Phys. Lett.* 381:422.
11. Das, D., and Samanta, A. 2011. Photoluminescent silicon quantum dots in core/shell configuration: Synthesis by low temperature and spontaneous plasma processing. *Nanotechnology* 22:055601.
12. Hori, Y., Oda, O., Bellet-Amalric, E., and Daudin, B. 2007. GaN quantum dots grown on $Al_xGa_{1-x}N$ layer by plasma-assisted molecular beam epitaxy. *J. Appl. Phys.* 102:024311.
13. Huang, S. Y., Xu, S. Y., Long, J. D. et al. 2006. Separated Al_xIn_{1-x} quantum dots grown by plasma-reactive co-sputtering. *Physica E.* 31:200.

14. Kato, T., Kuroda, S., and Hatakeyama, R. 2011. Diameter tuning of single-walled carbon nanotubes by diffusion plasma CVD. *J. Nanomater.* 2011:490529.

15. Kaneko, T., and Hatakeyama, R. 2009. Control of carbon nanotube semiconducting properties by DNA encapsulation using electrolyte plasmas. *Appl. Phys. Express* 2:127001.

16. Robertson, J., Zhong, G., Telg, H. et al. 2008. Growth and characterization of high-density mats of single-walled carbon nanotubes for interconnects. *Appl. Phys. Lett.* 93:163111.

17. Shashurin, A., and Keidar, M. 2008. Factors affecting the size and deposition rate of the cathode deposit in an anodic arc used to produce carbon nanotubes. *Carbon.* 46:1826.

18. Yu, L. W., O'Donnell, B., Alet, P. J., and Cabarrocas, P. R. I. 2010. All-in-situ fabrication and characterization of silicon nanowires on TCO/glass substrates for photovoltaic application. *Sol. Energ. Mat. Sol. C.* 94:1855.

19. Zardo, I., Yu, L., Conesa-Boj, S. et al. 2009. Gallium assisted plasma enhanced chemical vapor deposition of silicon nanowires. *Nanotechnology.* 20:155602.

20. Cvelbar, U., Chen, Z. Q., Sunkara, M. K., and Mozetic, M. 2008. Spontaneous growth of superstructure alpha-Fe_2O_3 nanowire and nanobelt arrays in reactive oxygen plasma. *Small.* 10:1610.

21. Rider, A. E. 2011. *Tailored self-organised nanopatterns: A plasma nanoscience approach.* PhD thesis: University of Sydney, Australia.

22. Whitesides, G. M., and Grzybowski, B. 2002. Self-assembly at all scales. *Science.* 295:2418.

23. Jones, R. *Self-assembly vs self-organisation—can you tell the difference?* Retrieved January 10, 2011, from http://www.softmachines.org/wordpress/.

24. Bensaude-Vincent, B. 2009. Self-assembly, self-organization: Nanotechnology and vitalism. *Nanoethics.* 3:31.

25. Ostrikov, K. 2011. Control of energy and matter at nanoscales: Challenges and opportunities for plasma nanoscience in a sustainability age. *J. Phys. D: Appl. Phys.* 44:174003.

26. Ostrikov, K., and Murphy, A. B. 2007. Plasma-aided nanofabrication: Where is the cutting edge? *J. Phys. D: Appl. Phys.* 40:2223–2241.

27. Ostrikov, K. 2005. Colloquium: Reactive plasma as a versatile nanofabrication tool. *Rev. Mod. Phys.* 77:489–511.

28. Ostrikov, K. 2007. Plasma nanoscience: From nature's mastery to deterministic plasma-aided nanofabrication. *IEEE Trans. Plasma Sci.* 35:127–136.

29. Ostrikov, K., and Xu, S. 2007. *Plasma-aided nanofabrication: From plasma sources to nanoassembly.* Weinheim, Germany: Wiley-VCH.

30. Ostrikov, K. 2008. *Plasma nanoscience: Basic concepts and applications of deterministic nanofabrication.* Weinheim, Germany: Wiley-VCH.

31. Ostrikov, K., Yoon, H. J., Rider, A. E., and Ligatchev, V. 2007. Reactive species in $Ar+H_2$ plasma-aided nanofabrication: Two-dimensional discharge modeling. *Phys. Scr.* 76:187–195.

32. Ostrikov, K., Yoon, H. J., Rider, A., and Vladimirov, S. V. 2007. Two-dimensional simulation of nanoassembly precursor species in $Ar+H_2+C_2H_2$ reactive plasmas. *Plasma Process. Polym.* 4:27–40.

33. Ren, Y. P., Xu, S., Rider, A. E., and Ostrikov, K. 2011. Made-to-order nanocarbons through deterministic plasma nanotechnology. *Nanoscale* 3:731–740.

34. Ostrikov, K., Xu, S., and Levchenko, I. 2008. Self-organized nanoarrays: Plasma-related controls. *Pure Appl. Chem.* 80:1909–1918.

35. Levchenko, I., Ostrikov, K., and Murphy, A. B. 2008. Plasma-deposited Ge nano-island films on Si: Is Stranski-Krastanow fragmentation unavoidable? *J. Phys. D: Appl. Phys.* 41:092001.

36. Möller, W. 1993. Plasma and surface modelling of the deposition of hydroge-nated carbon films from low-pressure methane plasmas. *Appl. Phys. A.* 56:527.

37. Rider, A. E., Ostrikov, K., and Levchenko, I. 2008. Tailoring the composition of self-assembled $Si_{1-x}C_x$ quantum dots: Using plasma/ion-related controls. *Nanotechnology.* 19: 355705.

38. Kersten, H., Deutsch, H., Steffen, H., Kroesen, G. M. W., and Hippler, R. 2001. The energy balance at substrate surfaces during plasma processing. *Vacuum.* 63:385.

39. Rohde, D., Pecher, P., Kersten, H., Jacob, W., and Hippler, R. 2002. The energy influx during plasma deposition of amorphous hydrogenated carbon films. *Surf. Coat. Technol.* 149:206.

40. Rider, A. E., and Ostrikov, K. 2009. The path to stoichiometric composition of III-V binary quantum dots through plasma/ion-assisted self-assembly. *Surface Science.* 603: 359–368.

41. Rider, A. E., Levchenko, I., Ostrikov, K., and Keidar, M. 2007. Ge/Si quantum dot formation from non-uniform cluster fluxes. *Plasma Process. Polym.* 4:638–647.

42. Vladimirov, S. V., Ostrikov, K., Samarian, A. A. 2005. *Physics and Applications of Complex Plasmas.* London: Imperial College Press.

43. De Bleecker, K., Bogaerts, A., and Goedheer, W. 2004. Modeling of the formation and transport of nanoparticles in silane plasmas. *Phys. Rev. E.* 70:056407.

44. De Bleecker, K., Bogaerts, A., Goedheer, W., and Gijbels, R. 2004. Investigation of growth mechanisms of clusters in a silane discharge with the use of a fluid model. *IEEE Trans. Plasma Sci.* 32:691.

45. Seo, D. H., Rider, A. E., Arulsamy, A. D., Levchenko, I., and Ostrikov, K. 2010. Increased size selectivity of Si quantum dots on SiC at low substrate tempera-tures: An ion-assisted self-organization approach. *J. Appl. Phys* 107:024313.

46. Arulsamy, A. D., and Ostrikov, K. 2009. Diffusivity of adatoms on plasma-exposed surfaces determined from the ionization energy approximation and ionic polarizability. *Phys. Lett. A.* 373:2267–2272.

47. Arulsamy, A. D. 2010. *Many-body Hamiltonian based on the ionization energy con-cept: A renormalized theory to study strongly correlated matter and nanostructures.* PhD thesis: University of Sydney, Australia.

48. Cheng, Q. J., Tam, E., Xu, S. Y., and Ostrikov, K. 2010. Si quantum dots embedded in an amorphous SiC matrix: Nanophase control by non-equilibrium plasma hydrogenation. *Nanoscale.* 2:594.

49. Cheng, Q. J., Xu, S. Y., and Ostrikov, K. Single-step, rapid low-temperature syn-thesis of Si quantum dots embedded in an amorphous SiC matrix in high-den-sity reactive plasmas. *Acta Mater.* 58:560.

50. Song, D., Cho, E. C., Conibeer, G., Flynn, C., Huang, Y., and Green, M. A. 2008. Structural, electrical and photovoltaic characterization of Si nanocrys-tals embedded SiC matrix and Si nanocrystals/c-Si heterojunction devices. *Sol. Energy Mater. Sol. Cells.* 92:474.

51. Bailey, R. E., and Nie, S. 2003. Alloyed semiconductor quantum dots: Tuning the optical properties without changing the particle size. *J. Am. Chem. Soc.* 125:7100.
52. Mehdipour, H., Ostrikov, K., and Rider, A. E. 2010. Low- and high-temperature controls in carbon nanofiber growth in reactive plasmas. *Nanotechnology.* 21:455605.
53. Mehdipour, H., Ostrikov, K., Rider, A. E., and Han, Z. J. 2011. Heating and plasma sheath effects in low-temperature, plasma-assisted growth of carbon nanofibers. *Plasma Process. Polym.* 8:386–400.
54. Ostrikov, K., and Mehdipour, H. 2011. Energy and matter-efficient size-selective growth of thin quantum wires in a plasma. *Appl. Phys. Lett.* 98:033104.
55. Levchenko, I., Ostrikov, K., Rider, A. E., Tam, E., Vladimirov, S. V., and Xu, S. 2007. Growth kinetics of carbon nanowall-like structures in low-temperature plasmas. *Phys. Plasmas.* 14:063502.
56. Ostrikov, K. 2010. *Plasma nanoscience: Nanoscale control of energy and matter for a sustainable future*, the Walter Boas Medal award talk. The bi-annual congress of the Australian Institute of Physics, Melbourne, Australia, 5–9 December 2010.
57. Arulsamy, A. D., Rider, A. E., Cheng, Q. J., Xu, S., and Ostrikov, K. 2009. Effect of elemental composition and size on electron confinement in self-assembled SiC quantum dots: A combinatorial approach. *J. Appl. Phys.* 105:094314.
58. Ostrikov, K. 2011. Nanoscale transfer of energy and matter in plasma–surface interactions. *IEEE Trans. Plasma. Sci.* 39:963–970.
59. Han, Z. J., Levchenko, I., Kumar, S. et al. 2011. Plasma nanofabrication and nanomaterials safety. *J. Phys. D: Appl. Phys.* 44:174019.
60. Ostrikov, K., Kumar, S., Cheng, Q. J. et al. 2011. Different nanostructures from different plasmas: Nanoflowers and nanotrees on silicon. *IEEE Trans. Plasma. Sci.* doi:10.1109/TPS.2011.2159022.
61. Levchenko, I., Ostrikov, K., Mariotti, D., and Švrček, V. 2009. Self-organized carbon connections between catalyst particles on a silicon surface exposed to atmospheric-pressure Ar+CH$_4$ microplasmas. *Carbon* 47:2379.

Index